C0-AJY-759

Practical Microwave Electron Devices

Practical Microwave Electron Devices

T. Koryu Ishii

Department of Electrical, Computer, and Biomedical Engineering
Marquette University
Milwaukee, Wisconsin

Academic Press, Inc.

Harcourt Brace Jovanovich, Publishers

San Diego New York Boston
London Sydney Tokyo Toronto

This book is printed on acid-free paper. ∞

Academic Press, Inc.
San Diego, California 92101

United Kingdom Edition published by
Academic Press Limited
24–28 Oval Road, London NW1 7DX

Library of Congress Cataloging-in-Publication Data

Ishii, T. Koryu (Thomas Koryu).
Practical microwave electron devices / T. Koryu Ishii.
p. cm.
Includes bibliographical references.
ISBN 0-12-374700-7 (alk. paper)
1. Microwave devices--Design and construction. I. Title.
TK7876.I74 1990
621.381'3--dc20 89-17903
CIP

Printed in the United States of America
90 91 92 93 9 8 7 6 5 4 3 2 1

ΕΙΣ ΤΗΝ ΑΓΑΠΗ ΣΟΥ

Contents

Appendices

Preface

Electronics engineers with a knowledge of microwave technology have always been in demand by industry, and this is likely to continue. In addition to electronics engineers, physicists, chemists, and even mathematicians often need knowledge and "know-how" of practical microwave electron devices and their applications. In many cases, such knowledge must be acquired quickly by independent study. Therefore, there is a need for a practical, self-teaching book with immediacy and utility to provide such people with information on microwaves at a sufficient level and depth.

To the author's knowledge, such a book has not been available in the past. Most books have been either highly theoretical at the post-graduate level or overly simplified technician-level works. Highly sophisticated books are certainly needed in microwave electronics, but they are not practical in meeting the requirements of immediacy and utility. Books that are too practical may provide basic explanations of the materials but do not give the reader the creative design capability which is one of the essential qualifications of a good engineer.

This book has been written so that practicing engineers, scientists, and technicians may acquire practical know-how of microwave electron devices and their applications at an adequate level through self-study. This book is also suitable for use as a college textbook in upper-division courses for seniors and first-year graduate students in electrical engineering.

All areas of microwave electron devices are covered in this book. In the area of microwave solid-state devices, popular microwave transistors and both passive and active diodes are covered. Other areas presented include quantum electron devices, thermionic devices (including relativistic thermionic devices), and ferrimagnetic electron devices. The design of each of these devices is covered as well as their applications, including oscillation, amplification, switching, modulation, demodulation, and parametric interactions. Numerous design examples and case studies are presented throughout the book.

One of the best ways to learn how to design is by looking at current designs and investigating case studies. When each microwave electron device is covered, typical design examples or case studies are presented first and followed by qualitative or quantitative explanations. The fundamental theory of each device is summarized along with the underlying principles of the design. Each summary is presented so that the design techniques can be applied to other specific cases, designs, and applications. Review questions are included with each chapter to stimulate creative thinking and enhance the acquisition of knowledge and design skills.

The author thanks Sabine Teich, Dolores Marrari, and M. Michael Ishii for their assistance in manuscript preparation. The author also thanks George Zdasiuk, Christine Oshanick, and Brian L. Jones, who supplied valuable illustrations and data to enhance the effectiveness of this book.

T. Koryu Ishii

1

Introduction

1.1 Microwaves and Electronics

Microwaves are electromagnetic waves with free space wavelengths in a range of 0.1 mm to 100 cm [1,16,17]. Ranging in frequency from 300 MHz to 3 THz, microwaves are employed in microwave telecommunications, radar navigational aides, industrial and domestic heating and material processing, telemetering, and sensing. In addition, they provide a vehicle for scientific and biological studies. Microwave electronics techniques provide generation, modulation, control, amplification, demodulation, and multiplexing of microwaves. It is essential to study the practical aspects of microwave electronics to acquire knowledge of such techniques. Before getting into their details, some microwave electron devices involved in this book are briefly and qualitatively reviewed in this chapter to obtain some idea of typical microwave electron devices and their applications.

1.2 Generation of Microwave Power

Microwaves are generated either by solid-state devices (Chapters 2–10, 12) or thermionic devices (Chapters 14–16), depending on

Fig. 1.1 Photograph of a sample microwave transistor [7].

the frequency and power level of microwaves required. Today, high frequency and high power, especially in ranges of super high power levels, are difficult to generate with microwave solid-state devices. In such high power and high frequency applications, thermionic devices are employed. Microwave solid-state devices include microwave transistors, tunnel diodes, Gunn diodes, and IMPATT diodes. Microwave thermionic devices include klystrons, magnetrons, backward wave tubes, and gyrotrons.

An example of a microwave transistor is shown in Fig. 1.1 [9, 18]. This transistor is designed to operate at 6 GHz, with the output of 10 mW. The specification is listed in Table 1.1. The transistor is designed to be mounted on a microstripline resonator. Though the appearance is quite different from one of its low frequency counterparts, it operates on the same basic principle of the metal–semiconductor field effect transistor (MESFET; Chapter 2) [2]. When a dc power supply of 5 V is applied, the mechanism of positive feedback, together with voltage amplification produced by the transistor action, causes the transistor to oscillate. Thus, microwaves are generated by the transistor. In the package shown in Fig. 1.1, the slant cut lead strip is the gate. The lead strip directly against the gate is the drain. The two crossing lead strips are the source. If this is a bipolar junction transistor (BJT), the slant cut lead strip is the base. Directly against it is the collector and the crossing lead strips are the emitter. The dot seen in Fig. 1.1 on the transistor package signifies either the drain or

Table 1.1 Specifications of the AT–8050 and AT–8051 Microwave GaAs FETs[a]

Drain-to-source voltage	5 V
Gate-to-source voltage	−8 V
Drain current	100 mA
Thermal resistance	200 °C/W
Channel temperature	125 °C
Continuous dissipation	400 mW
Storage temperature	−65 to +125 °C
Operating frequency	4 to 6 GHz
Output power	10 mW
Gain	9 to 11 dB
Noise figure	1.8 to 2.2 dB

[a] From Avantek [7].

the collector, depending on whether the transistor is either an FET or BJT.

An example of a microwave tunnel diode is shown in Fig. 1.2, and its specifications are listed in Table 1.2. The tunnel diode is a heavily doped degenerate p–n junction diode [3]. When forward biased properly it produces a negative resistance due to the tunnel effect (Chapter 3). On the one hand, positive resistance consumes microwave power. The negative resistance, on the other hand, "negatively consumes," or increases, microwave power when a

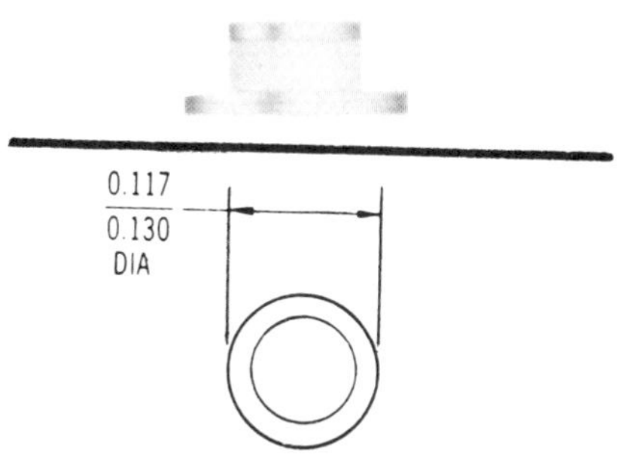

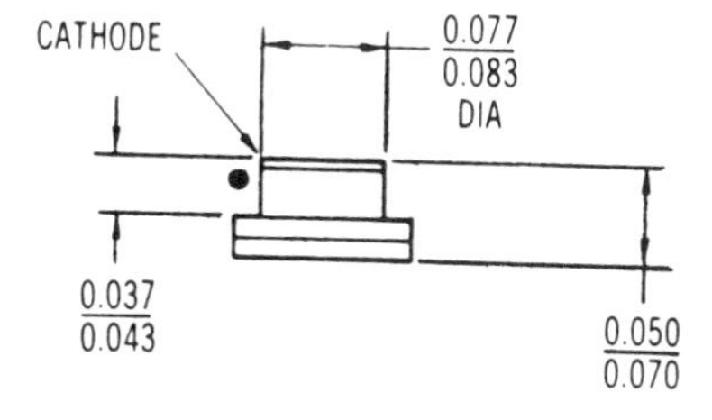

Fig. 1.2 Photograph (*top*) and dimensions (*bottom*) of a microwave tunnel diode [8].

Table 1.2 Specifications of the D4966 Tunnel Diode

Peak dc diode current	100 mA
Peak-to-valley current ratio	9.1
Junction capacitance	71.44 pF
Series resistance	0.13 Ω
Parallel resistance	1.2 Ω
Oscillation frequency	5.06 GHz

resonator and the impedance of the attached circuit are properly matched with the negative resistance. Thus, with the proper resonator and circuit attached, and the power supply adjusted to supply a proper bias, the tunnel diode generates microwaves.

A photograph of a Gunn diode and its waveguide mount is shown in Fig. 1.3, and the specifications for this diode are listed in Table 1.3. A Gunn diode is basically a properly doped n–type GaAs semiconductor [4,20,21]. When this is biased properly, there will be bunched electrons in the semiconductor due to the momentum difference of electrons in a nonuniform electric field distribution. The bunched electrons, drifting through the bulk semiconductor, produce a pulse current in the connected circuit when it is discharged at the anode. The process repeats at microwave frequencies. Thus, microwaves are generated when a proper amount of dc bias is applied to a Gunn diode (Chapter 4).

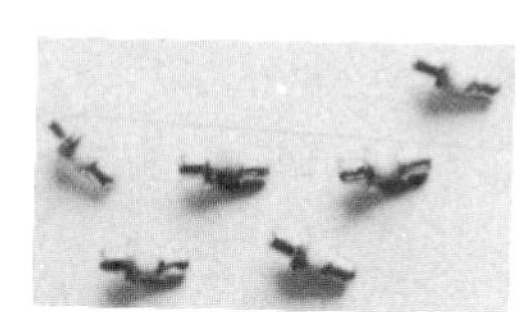

Fig. 1.3 Photograph of Gunn diodes (*left*) and a waveguide mount (*right*) [8,9].

Table 1.3 Specifications of the DGB3–6204 Gunn Effect Diode[a]

CW power output	200 mW
Frequency range	8 to 12.5 GHz
Efficiency	2%
Threshold voltage	4 V
Operating voltage	10 V
Operating current	1 A
Maximum dc input power	12 W

[a] From Alpha Industries [8].

The bunched carriers can also be formed by the avalanche effect. Charge carriers are injected into a drift region of a reverse biased p–n junction diode and the diode is biased at the verge of avalanche breakdown. When a resonator is connected in series with the diode, the avalanche breakdown occurs at every positive half cycle to inject bunched carriers into the drift region. When the resonator and the bias voltage are adjusted properly, the reverse biased p–n junction diode produces microwave power. Such a diode is called an impact avalanche transit time (IMPATT) diode (Chapters 5 and 6) [20,21]. A photograph of a typical IMPATT diode and the waveguide mount is shown in Fig. 1.4 and its specifications are given in Table 1.4. As seen from a comparison of Tables 1.4 and 1.3, IMPATT diodes produce more power than Gunn diodes. Tunnel diodes have a lower power capacity, but they have the advantage of a low operating power supply voltage and are usually low in noise.

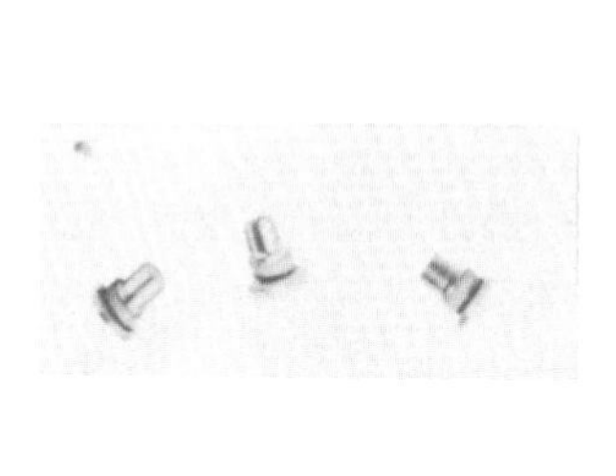

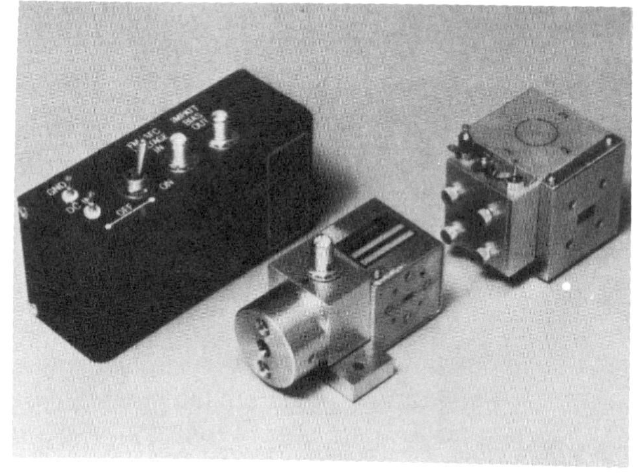

Fig. 1.4 Photograph of IMPATT diodes (*left*) and waveguide mounts (*right*) [10,11].

Table 1.4 Specifications of the 5082–0611 IMPATT Diode[a]

Operating frequency	10 to 14 GHz
Minimum midband output	2.5 W
Typical efficiency	10%

[a] From Hewlett Packard [10].

The active region of microwave solid-state devices is small, as narrow as the cross section of a human hair and as thin as onion skin. Due to its size, if high power operations are desired, problems with heat dissipation occur. Therefore, microwave solid-state devices have some limitations for high power operations at high frequencies.

Microwave generation can be achieved using these discrete solid-state devices. But these devices can be monolithically integrated with associated microstrip circuitry in the form of a hybrid microwave integrated circuit (HMIC) or a monolithic microwave integrated circuit (MMIC) [29–31]. These MMICs are often packaged in a fashion similar to that seen in Fig. 1.1 or housed in a standard TO-type transistor housing.

Thermionic devices, on the other hand, have no such high power restrictions. For example, a domestic microwave oven uses a magnetron to produce microwave power. A photograph of a cooking magnetron is shown in Fig. 1.5, and its specifications are tabulated in Table 1.5. It will be a long time before the magnetron is replaced by a solid-state device at a reasonable cost.

Electrically, a magnetron is a vacuum diode [1,19]. It consists of a hot cathode and an anode with a number of cavity resonators. When an axial magnetic field is applied, electrons rotate around the cathode by the Lorentz magnetodynamic force. By velocity modulation due to cavity resonators around the anode, the rotating electrons are bunched and form electron poles. The rotating electron poles induce a microwave field in the cavities which are directly attached to the anode by electrostatic induction. Thus, microwaves are generated in the anode, and one of the anode cavities is coupled to a waveguide for the external use of generated microwave power (Chapter 12).

The output of most electron devices can be controlled by modulating the power supply voltage. When the power supply voltage is modulated, the modulation occurs in both the frequency and the oscillator output power. To obtain pure FM or pure AM from these electron devices, a separate modulator in the micro-

Fig. 1.5 Photograph of a Litton L–5261A magnetron for a domestic microwave oven. By courtesy of Litton Industries.

wave circuit or transmission line is needed rather than direct modulation to the power supply for the electron devices.

1.3 Amplification of Microwaves

In microwave telecommunications, microwave amplification is important to increase the sensitivity of a microwave receiver. The transmitted power is also boosted by microwave amplification to extend the distance of communication or to improve communication reliability and quality. If positive feedback is properly con-

Table 1.5 Specifications of the L–5261A CW Magnetron[a]

Operating frequency	2455 MHz
Output power	625 W
Anode voltage	3.5 kV
Anode current	360 mA
Heater voltage	4.6 V
Heater current	16 A
Permanent magnet	

[a]By courtesy of Litton Industries.

trolled or eliminated, MESFET and tunnel diodes are readily adaptable for amplification. Other solid-state diodes, such as Gunn diodes and IMPATT diodes, may be used for amplification, but they are primarily oscillators which depend on electron transit time effects. It is therefore not easy to use them for amplification. Packaged transistors in the form of HMICs or MMICs are commercially available [29–31].

Varactor diodes are used for microwave amplification. A varactor diode is a variable capacitance diode. The junction capacitance of a p–n junction is a function of voltage across the diode. A photograph of a microwave amplifier which employs a variable capacitance diode is shown in Fig. 1.6. This type of amplifier is called a parametric amplifier [1,22] (Chapter 9). When the varactor is excited by a local oscillator, called the pump oscillator, the energy of the pump oscillator is transferred to the signal frequency channel through the varactor. Thus amplification of input signals

Fig. 1.6 Photograph of a parametric amplifier (*left*) and varactors (*right*) [8,12].

Table 1.6 Specifications of the D 5146 Varactor Diode

Power dissipation at 25 °C	300 mW
Operating temperature	+175 °C (max)
Breakdown voltage	5.5 V
Cutoff frequency	90 GHz
Maximum junction capacitance	0.3 to 8 pF
Rated max/min capacitance ratio	8 min

Fig. 1.7 Photograph of a Varian VA–953 UHF-TV klystron. By courtesy of Varian Associates.

Table 1.7 Specifications of the VA–953 UHF–TV Klystron[a]

Operating frequency	470 to 566 MHz
Output power	55 kW
Gain	48 dB
Tuning range	96 MHz
1 dB bandwidth	8 MHz
Number of cavities	5
Beam voltage, dc	24 kV
Beam current	7.5 A
Heater voltage	7.3 V
Heater current	18 A
Electromagnet focused	
Air cooled	

[a]By courtesy of Varian Associates [38].

(Chapter 9) is accomplished. An example of specification of a varactor diode is presented in Table 1.6.

Microwave amplification at high power levels is handled by klystrons and crossed field amplifiers [1,9,21,22]. An example of a klystron is shown in Fig. 1.7, and the specifications are tabulated in Table 1.7. When an input microwave signal is fed to a klystron, the signals cause velocity modulation in the electron beam. The velocity modulated electron beam produces bunched electrons, and bunched electrons induce large output signals in the output circuit by electrostatic induction. In this way, microwave input signals are amplified (Chapter 11).

A photograph of a crossed field amplifier is shown in Fig. 1.8. Its specifications are given in Table 1.8. In the crossed field amplifier in the magnetron, the rotating electron poles are formed by radial electric fields, axial magnetic fields, and input microwave signals in the anode structure. The anode is structured to accommodate traveling waves which travel along the anode structure. The anode structure surrounds the cathode of the crossed field amplifier.

If the speed of the traveling waves on the anode structure and the rotating speed of the electron poles are adjusted to be nearly equal, then there will be an interaction between the traveling microwave signals in the anode structure and the rotating electron poles. When mutual speed and phase are properly adjusted, the kinetic energy of the rotating electron poles transfers to the microwave signals by electrostatic induction and the microwave signals are amplified [1,19,21,22] (Chapter 16). Microwaves are also amplified quantum mechanically, as exemplified by masers

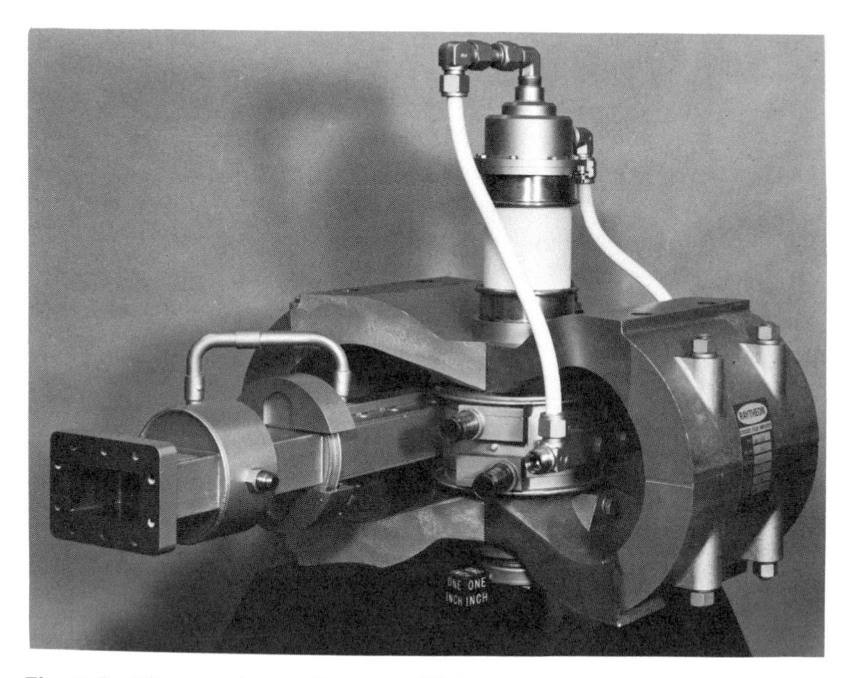

Fig. 1.8 Photograph of a Raytheon QKS1606 S-band crossed field amplifier. By courtesy of Raytheon Company.

Table 1.8 Specifications of the QKS 1606 Crossed Field Amplifier[a]

Operating frequency	3100 to 3500 MHz
Average output power	75 kW
Peak output power	0.8 MW
Peak anode current	70 A
Typical anode peak voltage	32 to 36 kV
Pulse width	100 μs
Peak microwave drive	75 kW
Heater	None
Weight	200 lb

[a] By courtesy of the Raytheon Company [39].

(Chapter 12) or quasi-quantum mechanically, as in the case of gyrotron amplifiers (Chapter 16).

1.4 Demodulation of Microwaves

The Schottky barrier diode is made from a metal–semiconductor contact. It is used for microwave rectification and mixing (Chapter 7) because of the low capacitance at the contact which is called

Fig. 1.9 Photograph of a Schottky barrier diode (*left*) and waveguide mount (*right*) [10,13].

the Schottky barrier. A photograph of various Schottky barrier diodes and a waveguide mount is shown in Fig. 1.9. This particular photograph is for the mixer. A detector has a similar appearance. The microwave input will be fed through the waveguide and detected. Video or beat frequency output is taken from the coaxial output. The specifications of the HP5082–2200 detector diode are shown in Table 1.9, and the specifications of a D5818A mixer diode are presented in Table 1.10.

Another method of demodulating by detection or mixing is based on the nonlinearity fo a semiconductor diode (Chapter 8). A special type of PIN diode called the step recovery diode, a junction diode with a heavily doped p–n junction called the back diode, and tunnel diodes [23,24] are highly nonlinear in their current–voltage characteristics. These diodes are also employed for detection and mixing (Chapters 7 and 8).

These microwave detectors and mixers for AM, FM, PCM, FSK, PSK, and OOK can be waveguide type, coaxial line type, or microstripline type. As with other assorted microwave compo-

Table 1.9 Specifications of the HP 5082–2200 Schottky Barrier Diode[a]

Noise figure	6 dB
Tangential sensitivity	−55 dBm at 10 GHz, 2 MHz bandwidth
Voltage sensitivity	5 mV/μW
Operating frequency	X-band

[a]From Hewlett Packard [10].

Table 1.10 Specifications of the D5818A Beam Lead Schottky Barrier Mixer Diode[a]

Noise figure	6.5 dB (max)
Zero bias junction	
Capacitance	0.1 to 0.35 pF
Series resistance	12 Ω
Operating frequency	X-band
Local oscillator power	1 mW

[a]From Alpha Industries [8].

nents, integrated units in the form of HMICs or MMICs are commercially available [32–34].

Other microwave electron devices which are remotely related to microwave demodulation are thermistors and barretter bolometers. These devices are termed bolometric detectors [6]. Their operation depends upon resistance changes due to microwave heating. Therefore, their response is usually slow, with low sensitivity, and the devices are extremely fragile electrically. This is the prime reason that thermistors and barretters are not used for detection or mixing in communications. They are primarily used for precision microwave power measurement. These devices are used in a Wheatstone bridge to sense the resistance change caused by microwave heating of the element. Thermistors are made of a semiconductor material, and the temperature coefficient of the resistance is negative. Barretters are made of a fine tungsten wire, and the temperature coefficient of the resistance is positive. A

Fig. 1.10 Photograph of microwave thermistor waveguide mounts (486 series) [14].

Table 1.11 Specifications of the 486 Series Microwave Thermistor Mounts[a]

Frequency range	Maximum VSWR	Operating resistance
3.95 to 18 GHz	1.5	100
18 to 40 GHz	2	200

[a]From Hewlett Packard [14].

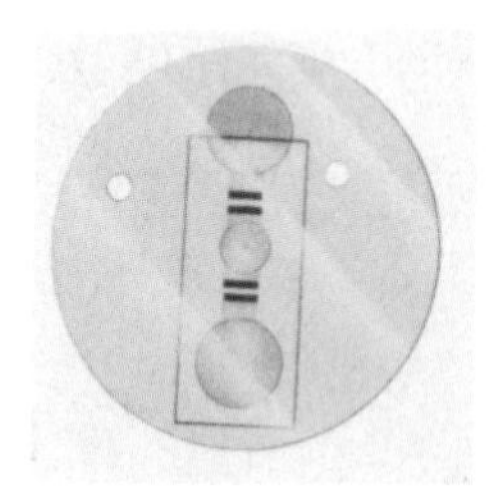

Fig. 1.11 Photograph of a microwave barretter (*left*) and the mount (*right*) [15].

Table 1.12 Specifications of the P603–4.5 Microwave Barretter[a]

Type	Mica disc
Frequency range	500 MHz to 10 GHz
dc bias current	4.5 ± 0.5 mA
dc resistance at the bias point	200 Ω
Maximum power handling capacity	17 mW
Pulse sensitivity	0.1667 Ω/mW-μs
Time constant	350 μs

[a]From Narda Microwave Corp. [15].

photograph of typical thermistor waveguide mounts is shown in Fig. 1.10 and specifications are listed in Table 1.11. A photograph of a typical barretter is shown in Fig. 1.11, with specifications listed in Table 1.12.

1.5 Microwave Multiplexing

Microwave power can be switched from one circuit to another or turned on and off by a reverse biased PIN diode [25,26] (Chapter 11). A photograph of a microwave PIN diode and associated circuitry is shown in Fig. 1.12, and specifications of the diode are given in Table 1.13.

The PIN diode is a diode with a thin layer of intrinsic semiconductor (i) between p- and n-type semiconductors. This is then reverse biased at the threshold of avalanche breakdown. When microwaves are fed to the junction, it breaks down and the avalanche breakdown current flows. When the microwave power is removed, the avalanche breakdown ceases and no current flows. The microwave PIN diode is also called a microwave switching diode (Chapter 11) [27]. Microwaves can also be turned on or off by controlling the dc bias voltage. These solid-state microwave

Fig. 1.12 Photograph of PIN diodes (*left*) and associated circuitry (*right*) [10,11].

switches are available in the HMIC or MMIC form as well as waveguide, coaxial line, and microstripline forms [35–37].

Microwave multiplexing, or switching, can also be done by microwave ferrite circulators (Chapter 13). A photograph of a microwave circulator is shown in Fig. 1.13, and the specifications are given in Table 1.14. When a piece of ferrite is magnetized, it produces asymmetrical tensor permeability [1,28]. This makes microwave propagation nonreciprocal in the ferrite media. Microwave switching can then be accomplished by utilizing the nonreciprocal propagation. These circulators can also take a form of waveguide type, coaxial line type, microstripline type, HMIC type, or MMIC type.

1.6 Features of Microwave Electronics

Due to an extremely high operating frequency and a very short operating wavelength, the size of an electron device is usually

Table 1.13 Specifications of the 5082–3306 PIN Diode

Series resistance	0.5 Ω[a]
Switching time	5 ms[a]
Switching current	20 mA[a]
Isolation	15 dB (min)[b]
Insertion loss	1.5 dB[b]

[a] From Hewlett Packard [10].
[b] From Hughes Aircraft [11].

Fig. 1.13 Microwave ferrite circulator and the associated circuitry. By courtesy of Raytheon Company.

extremely small in comparison with its counterpart in lower frequency systems which function at equal operating power levels. The electron transit time at interelectrodes is significant and not at all negligible in microwave electronics. Electron devices are designed to minimize the junction capacitances, and the lead inductances are made as small as possible. The lead inductance tends to impede the entrance or exit of microwave energy and also to act as an antenna. This is one source of energy loss and electromagnetic interference. The junction capacitance also tends to lower the input or output impedance of the device. The low input or

Table 1.14 Specifications of the CK_uH7 Ferrite Circulator

Frequency range	15.55 to 15.85 GHz
Average operating power	700 W CW
Minimum isolation	20 dB
Maximum insertion loss	0.4 dB
Maximum input VSWR	1.15
Approximate weight	$3\frac{1}{4}$ lb
Length	9 in
Cooling	Air cooled

output impedance usually makes the coupling between the electron device and the circuit weak.

Velocity modulation is the basic concept used in microwave thermionic devices. High power *and* high frequency operations are performed satisfactorily by thermionic devices. The challenge of solid-state devices to thermionic devices in this area remains to be met in the future.

Problems

1.1 Define microwaves.

1.2 Describe the relationship between wavelength, frequency, and the propagation phase velocity of microwaves. Explain why it is so.

1.3 List the applications of microwaves.

1.4 Define microwave electronics.

1.5 Categorize the available microwave electron devices.

1.6 State the reason why thermionic devices still occupy the area of high power *and* high frequency technology.

1.7 List different features of microwave FET's from low to high frequency FETs.

1.8 Itemize the features of the microwave tunnel diode which are distinct from the low frequency tunnel diode.

1.9 Describe the mechanism of microwave oscillation for a tunnel diode.

1.10 Describe the mechanism of microwave oscillation for a Gunn diode.

1.11 Explain the principle of IMPATT oscillation.

1.12 Explain the principle of magnetron oscillation.

1.13 List the reasons a magnetron is suited for high microwave power oscillation.

1.14 Describe the schemes of modulating microwave electron devices.

1.15 Propose methods of modulating microwave electron devices in order to produce pure FM or pure AM.

1.16 Name microwave electron devices readily fit for microwave amplification.

1.17 Describe the microwave varactor principle.

1.18 Describe the principles of microwave parametric amplifiers.

1.19 Describe the principles of a klystron.

1.20 Describe the principles of a crossed field amplifier.

1.21 How do microwave Schottky diodes differ from their low frequency counterparts, and why?

1.22 Describe the principles of microwave mixing by a Schottky barrier diode. Explain why a p–n junction diode does not fit well into microwave demodulation.
1.23 Using examples, explain bolometric detectors.
1.24 List microwave electron devices used for microwave diplexing.
1.25 Using the on–off characteristics of PIN diodes, design schematically a microwave diplexer using a PIN diode.
1.26 Using the nonreciprocal propagation characteristics of a microwave ferrite material under dc magnetization, schematically design a microwave diplexing system.
1.27 List the features of microwave electron devices.
1.28 Explain why the junction capacitance and the lead inductance must be minimized in microwave electron devices.
1.29 In certain packaged microwave transistors or integrated circuits, lead strips are identifiable by observation. Describe the identification.
1.30 Calculate the efficiency of a microwave AT–8050 GaAs FET.
1.31 Calculate the dc input of the AT–8050 and compare it with continuous dissipation and microwave output power. Can the AT–8050 be driven continuously with 100 mA drain current?
1.32 Calculate the input power level of the AT–8050 when the output power is 10 mW.
1.33 Calculate the output signal-to-noise ratio of the AT–8050 knowing that the noise figure is equal to the signal-to-noise ratio at the input to signal-to-noise ratio at the output. The input noise level is calculated by multiplying the Boltzmann constant, the frequency bandwidth, and the absolute temperature.
1.34 Calculate the capacitive reactance due to the junction capacitance of the D4966 tunnel diode. Compare it with the parallel resistance and series resistance.
1.35 Calculate the junction resistance of the D4966 to oscillate with a 1 Ω load.
1.36 By inspecting the specification sheet for the DGB–6204 Gunn Effect Diode, examine for discrepancies and contradictions as to the efficiency and the input–output relationship.
1.37 Find the necessary dc input power to operate the 5082 IMPATT diode.
1.38 Calculate the efficiency of the L–5261A CW magnetron.
1.39 Assuming 50% efficiency from an ac line to a dc power supply for the L–5261A magnetron, calculate the ac input

power required to operate this magnetron in order to produce 625 W microwave output power.

1.40 Calculate possible operating microwave voltage and current of the D5146 varactor diode.
1.41 Calculate the driving input power of the VA–953 klystron.
1.42 Calculate the efficiency of the VA–953 klystron.
1.43 Calculate the electron velocity in the electron beam of the VA–953 klystron.
1.44 Calculate the detected output voltage of the HP 5082–2200 Schottky barrier diode with an input of 10 GHz at −55 dBm.
1.45 Calculate the impedance across the junction capacitance of the beam lead for the D5818 Schottky barrier mixer diode and compare it with the series resistance.
1.46 Calculate the voltage reflection coefficient, the voltage transmission coefficient, the power reflection coefficient, and the power transmission coefficient of 486-series thermistor mounts.
1.47 At 10 GHz, if 10 mW is applied to the P603–4.5 barretter for the period of the time constant, how many cycles of microwaves are fed to this barretter during this period? Calculate the resistance change during this period.
1.48 Calculate the number of cycles of microwaves and the microwave energy that passes during the switching time of the 5082–3306 PIN diode operating at 6 GHz.
1.49 If the 5082–3306 PIN diode is operated as a microwave on–off switch of 10 W, calculate the microwave power loss during the on period. Calculate the amount of power leak during the off period.
1.50 Calculate the average power leak and power loss if the CK_uH7 ferrite circulator is operated at 700 W CW. Calculate the amount of power reflected back at the input. Determine an adequate cooling method.

References

1 T. K. Ishii, "Microwave Engineering." Ronald Press, New York, 1966.
2 A. Bar-Lev, "Semiconductors and Electronic Devices." Prentice-Hall, London, 1979.
3 L. Esaki, New phenomenon in narrow PN junctions. *Phys. Rev.* **109**, 602–603 (June 1958).
4 J. B. Gunn, Instabilities of current and of potential distribution in GaAs and InP. *In* J. Bok, ed., "Plasma Effects in Solids." Dunod, London, 1965.
5 W. T. Read, A proposed high frequency negative resistance Diode. *Bell Syst. Tech. J.* **37**, 401–446 (March 1958).

6 "IEEE Standard Dictionary of Electrical and Electronics Terms." Wiley (Interscience), New York, 1972.
7 Avantek, "Transistor Data Book." Avantek, Santa Clara, California.
8 Alpha Industries, "Alpha Microwave Diodes," D-72. Alpha Industries, Woburn, Massachusetts.
9 Microwave Associates, "MA 86541 Pulsed Doppler Transceiver," Bull. 7622B. Microwave Associates, Burlington, Massachusetts.
10 Hewlett Packard, "Diodes and Transistors," short form catalog. Hewlett Packard, Palo Alto, California, October 1976.
11 Hughes Aircraft, Millimeter wave transmitter products. *Microwaves* **17**(4), 33–52 (1978).
12 Scientific Communications, The signal source. *Microwave Syst. News* **6**(5), 135 (October/November 1976).
13 Norsal Industries, Waveguide balanced mixers. *Microwaves* **17**(4), 118 (1978).
14 Hewlett Packard, "Electronic Instruments and Systems," p. 428. Hewlett Packard, Palo Alto, California, 1980.
15 Narda Microwave Corp., "Catalog 20." Narda Microwave Corp., Plainview, New York, 1976.
16 S. Y. Liao, "Microwave Devices and Circuits," 2nd ed. Prentice-Hall, Englewood Cliffs, New Jersey, 1985.
17 P. A. Rizzi, "Microwave Engineering: Passive Circuits." Prentice-Hall, Englewood Cliffs, New Jersey, 1988.
18 Avantek, *Microwave Syst. News Commun. Tech.* **18**(6), 12–13 (1988).
19 S. Y. Liao, "Microwave Electron-Tube Devices." Prentice-Hall, Englewood Cliffs, New Jersey, 1988.
20 S. Y. Liao, "Microwave Solid-State Devices." Prentice-Hall, Englewood Cliffs, New Jersey, 1985.
21 J. T. Coleman, "Microwave Devices." Reston Publ., Reston, Virginia, 1982.
22 D. Roddy, "Microwave Technology." Prentice-Hall, Englewood Cliffs, New Jersey, 1986.
23 FEI Microwaves, *Microwave J.*, **31**(6), 38 (1988).
24 J. M. Gering, T. J. Rudnick, and P. D. Coleman, Microwave detection using resonant tunneling diode. *IEEE Trans. Microwave Theory Tech.* MTT-**36**(7), 1145–1150 (1988).
25 M/A Com, *Microwave J.*, **31**(6), 86–87 (1988).
26 Frequency Sources, *Microwave J.*, **31**(6), 231 (1988).
27 Daico Industries, *Microwave J.*, **31**(6), 108 (1988).
28 R. F. Soohoo, "Microwave Magnetics." Harper & Row, New York, 1985.
29 Avantek, *Microwaves RF* **27**(6), 107 (1988).
30 Watkins-Johnson Co., "RF and Microwave Component Designers' Handbook—1988/89." Watkins-Johnson Co., Palo Alto, California, 1989.
31 Mini-Circuits, *Microwave Syst. News Commun. Tech.* **18**(2), 50 (1988).
32 Mini-Circuits, *Microwaves RF* **27**(6), 2 (1988).
33 MA/Com, *Microwave Syst. News Commun. Tech.* **18**(2), 2 (1988).
34 Lorch Electronics, *Microwave J.*, **31**(6), 60 (1988).
35 Mini-Circuits, *Microwave J.* **31**(6), 7 (1988).
36 Adams Russell Electronics, *Microwave Syst. News Commun. Tech.* **18**(2), 18 (1988).
37 Joslyn Defense Systems, *Microwaves RF* **27**(6), 78 (1988).
38 Varian Associates, Specification sheet of VA–953. Palo Alto, California.
39 Raytheon Company, Specification sheet of QKS 1606. Waltham, Massachusetts.

2

Microwave Transistors

2.1 Transistors for Microwave Applications

Microwaves are amplified or generated by the use of microwave transistors. In low microwave frequencies, bipolar junction transistors (BJT) and junction field effect transistors (JFET) are functional. They are the microwave versions of modified bipolar transistors and junction field effect transistors. For microwave operations, the size of the active device is smaller. The lead wire and the package configurations are carefully designed to reduce stray inductance, stray capacitance, and interelectrode capacitance. In low ranges of microwave frequencies, the structure of microwave transistors is made small enough to eliminate the effect of interelectrode transit time [8]. The transistors can be fabricated using silicon, germanium, or gallium arsenide. Though the outward appearance of packaged microwave transistors is quite different from their low frequency counterparts, the operating principle of these microwave bipolar transistors and junction field effect transistors is the same [1–3]. In high microwave frequencies, however, simply reducing stray inductance, capacitance, and interelectrode capacitance is not enough. The interelectrode transit time must be sufficiently small in comparison with the period of operating microwave frequency. The mobility of electrons must be kept sufficiently high. At the same time, high thermal conductivity of the material is required so that the heat generated by the device may be conducted away as quickly as possible [8]. Silicon and germanium, when compared with gallium arsenide (GaAs), are less favorable materials with respect to thermal conductivity and electron mobility. For high electron mobility in high frequency applications, high electron mobility transistors (HEMT) are used [9]. An HEMT is a heterojunction FET with a high electron mobility channel [9].

To minimize the interelectrode capacitance, the configuration of the field effect transistor with an insulated gate is applied. In a MESFET, the insulated gate is created by a depletion layer due to metal–semiconductor contact. There will be a Schottky carrier depletion layer around the gate. This layer acts as the insulator for the gate. Such metal–semiconductor field effect transistors are called MESFET. In this chapter, structural design and applications of microwave BJTs, JFETs, MESFETs, and HEMTs are reviewed.

2.2 Structure of Microwave Transistors

2.2.1 Microwave BJT

An example of a cross-sectional view of the active part of a microwave BJT is shown in Fig. 2.1 [10]. This BJT is developed on a wafer of p-type Si with crystallographic orientation [100] and a

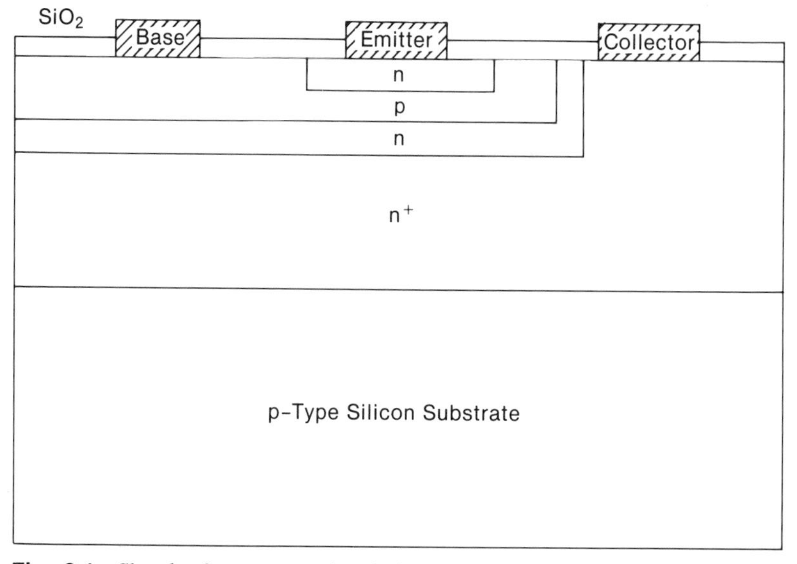

Fig. 2.1 Sketch of a cross-sectional view of a microwave BJT.

resistivity of 10 Ω-cm. This is an n–p–n transistor. The sketch is not proportional to actual size. In a BJT, all metal–semiconductor contacts are ohmic contacts [11]. The n-type emitter is 1 μm wide, 6 μm long, and 0.11 μm thick. This is doped with $1 \times 10^{20}/\text{cm}^3$ arsenic ion implantation. The active layer of p-type base is 0.1 μm thick. This area is doped with $3 \times 10^{17}/\text{cm}^3$ boron ion implantation. The active base area is again 1 μm wide and 6 μm long, just below the emitter layer. The rest of the p-layer is considered to be a lead to the metallized base electrode. Collector n- and n^+- layers are formed by antimony (Sb) diffusion 0.4 to 1.8 μm thick. The collector n-layer is doped with 4×10^{14} to $1.2 \times 10^{16}/\text{cm}^3$ Sb ions. The n^+-layer is doped with $2 \times 10^{19}/\text{cm}^3$ Sb ions. Thickness of the n-layer is 0.03 to 0.6 μm. The n^+-layer must have a good contact with the metallized collector electrode. The entire top surface is protected by a 30-nm thick SiO_2 film.

The working collector current density for this type of BJT is about 1 $\text{mA}/\mu\text{m}^2$. The electron transit time, τ, from the emitter to collector in the active region is approximately 4×10^{-12} s. If the active region is 2 μm thick, then the average drift velocity of electrons in this transistor is $u = 2\ \mu\text{m}/4 \times 10^{-12}\ \text{s} = 0.5 \times 10^6$ m/s which is approximately 10^{-3} times the velocity of light in a vacuum.

If the operating microwave frequency is f and the period is T, it is desirable that T is greater than 10 times τ, the electron transit time between the emitter and the collector in the active region. For $\tau = 4 \times 10^{-12}$ s, $T > 4 \times 10^{-11}$ s. Then $f = 1/T < 25$ GHz.

Since the active area of this transistor is about 1 $\mu\text{m} \times 6$ $\mu\text{m} = 6\ \mu\text{m}^2$, the collector current density is 1 $\text{mA}/\mu\text{m}^2$. This means that the collector current of this transistor is $I_c = 6$ mA. If the total number of carriers is n, the electronic charge is e, and the drift velocity is u, then $I_c = nue$ or the total number of carriers involved is $n = I_c/ue = 6 \times 10^{-3}\ \text{A}/(0.5 \times 10^6\ \text{m/s})(1.602 \times 10^{-19}\ \text{C}) = 7.5 \times 10^{10}$.

When higher power is needed, multiples of the basic unit shown in Fig. 2.1 are developed on a wafer and connected in parallel, either monolithically or with gold wire bonds.

The metallization for the emitter, base, and collector are usually wire bonded to appropriate lead strips of a standard packaging of the transistor. If this transistor is a part of MMIC, then the electrode metallization is monolithically connected to the appropriate microstriplines from the outside.

In microwave frequency operation, it is desirable that the interelectrode capacitance is small. Knowing that the active region has an area of $A = 6\ \mu m^2$, the interelectrode capacitance between the emitter and base with thickness $t_E = 0.11\ \mu m$ and $\varepsilon_r = 11.8$ for silicon [11] is at least $C_{EB} = \varepsilon_o \varepsilon_r\ A/t_E = (8.85\ pF/m)\ (11.8)(6\ \mu m^2)/(0.11\ \mu m) = 0.006$ pF. This is considered to be small even at microwave frequencies. Capacitive susceptance of C_{EB} at $f = 25$ GHz is $B_{EB} = 2\pi f C_{EB} = 2\pi(25 \times 10^9)(0.006 \times 10^{-12}) = 0.942 \times 10^{-3}$ S. The shunt reactance is then $X_{EB} = 1/B_{EB} = 1.06 \times 10^3\ \Omega$. If the transistor is in a 50 Ω system, 1.06 kΩ is significantly large. But in the real world there will be additional interlead capacitance and additional capacitance due to fringing fields, junction formation, and high frequency effects such as dispersion, radiation, dissipation, and polarization delay time. Therefore, by rule of thumb, the actual capacitance between the emitter lead to base lead $C_{EB}^A = 10C_{EB} = 0.06$ pF. This brings X_{EB} down to 106 Ω. This is indeed comparable to a 50 Ω impedance system.

The capacitance between the base and the collector in the active region is $C_{BC} = (8.85\ pF/m)(11.8 \times 6\ mm^2)/1.9\ \mu m = 0.0003$ pF. $X_{BC}^A = 1/(2\pi f C_{BC}) = [1/(2\pi \cdot 25 \times 10^9 \cdot 0.0003 \times 10^{-12})] \cdot 10 = 2.12$ kΩ.

The capacitance between the emitter and the collector is then $C_{EC} = (8.85\ pF/m)(11.8 \times 6\ \mu m^2)/(2\ \mu m) \approx C_{BC}$. Therefore $X_{EC}^A \approx 2.12$ kΩ.

The transistors may be used in the common emitter configuration or common base configuration. These interelectrode capacitances are important parameters to be considered for microwave amplifier, oscillator, and switching design.

The metallization of the electrode, bondwire, and lead microstrip can be considered part of a microwave transmission line to the outside world. For example, at $f = 25$ GHz, the free space wavelength is $\lambda_o = c/f = (3 \times 10^8\ m/s)/25 \times 10^9\ Hz = 12$ mm. The effective wavelength in silicon is $\lambda = \lambda_o/\sqrt{\varepsilon_r} = 12\ mm/\sqrt{11.8} = 3.5$ mm. The wave number for the metallized electrode, which is only $l = 6\ \mu m$ long, is $l/\lambda = (6\ \mu m)/(3500\ \mu m) = 1.7 \times 10^{-3}$ wavelengths. Considering that wave number 0.1 is the upper limit of the lumped parameter in "rule of thumb" microwave engineering, the metallization and bond wire of this order of magnitude can be treated as lumped parameters. The current carrying conductors of this order of magnitude present some inductance. The inductance of a conductor of length $l = 6\ \mu m$ can be calculated by $L = 2.3(\mu_o/2\pi)l = 2.3[(4\pi \times 10^{-7}\ H/m)/2\pi)] \cdot 6 \times 10^{-6}\ m = 8.6$ pH. Therefore, in most cases, the inductance may be omitted. However, the inductance of this order of magnitude presents a reactive impedance of $X_L = 2\pi(25 \times 10^9)(8.6 \times 10^{-12}) = 1.35\ \Omega$. This is not so small. For a 50 Ω system, however, this may be omitted. For lower system impedance, the lead impedance 1.35 Ω may be significant.

For aluminum metallization with conductivity $\sigma = 3.82 \times 10^7$ S/m of thickness $t = 60$ nm, width $w = 1$ μm, and length $l =$ 6 μm, the resistance $R = l/(\sigma wt) = 6 \times 10^6/(3.82 \times 10^7 \times 10^{-6} \times 60 \times 10^{-9}) = 260$ Ω. This is a significant amount of resistance. When the metallization is used as a lumped electrode, the current configuration is different from the case in which the metallization is used as a lead strip. As a lumped electrode, the resistance of 260 Ω will not enter directly into the design of the device function. However, if the metallization or bond wire of this magnitude is used as a lead transmission line, a significant amount of resistance will be encountered with the design and device function. Lower resistance is obtainable by wider and thicker metallization.

As a user or designer of a BJT of this size, it is interesting to know the resistance between the emitter and the base (R_{EB}), the resistance between the base and the collector (R_{BC}), and the resistance between the emitter and the collector (R_{EC}). The values of these resistances vary greatly by the doping concentration and the magnitude and direction of bias voltages. For reference regarding the intrinsic silicon, the resistivity is $\rho = 2.5 \times 10^5$ Ω-cm [11].

In the electrode configuration shown in Fig. 2.1, if the emitter–base center-to-center distance $l_{\mathrm{EB}} = 4$ μm, the thickness of n–p layer is $t_{\mathrm{EB}} = 1$ μm, and the length of the metallization $l =$ 6 μm, then the intrinsic resistance R_{EB} is at least $R_{\mathrm{EB}} = \rho l_{\mathrm{EB}}/l t_{\mathrm{EB}}$ $= (2.5 \times 10^5$ Ω-μm)(4 μm)/(6 μm)(1 μm) = 167 kΩ. A similar order of magnitude of resistance can be expected between the emitter and the collector as in intrinsic silicon.

A resistance as much as twice that of R_{EB} can be expected between the base and the collector. These resistance values are so high in comparison with the resistance of the associated component (such as the biasing resistors) that the effect of the intrinsic resistance can be omitted. It can be treated as an insulator as far as the transistor performance and the circuit designs are concerned. However, these silicon layers are doped and biased when operated. When it is in operation, the emitter–base junction is usually forward biased in such a way that the emitter is negative and the base is positive. The base–collector junction is reverse biased in such a way that the base is negative and the collection is positive. Because of the doping with impurity and junction formation, the forward resistance between the base and emitter is as low as 0.5 kΩ and the reverse resistance between the collector and the base is as high as 3.5 kΩ [10].

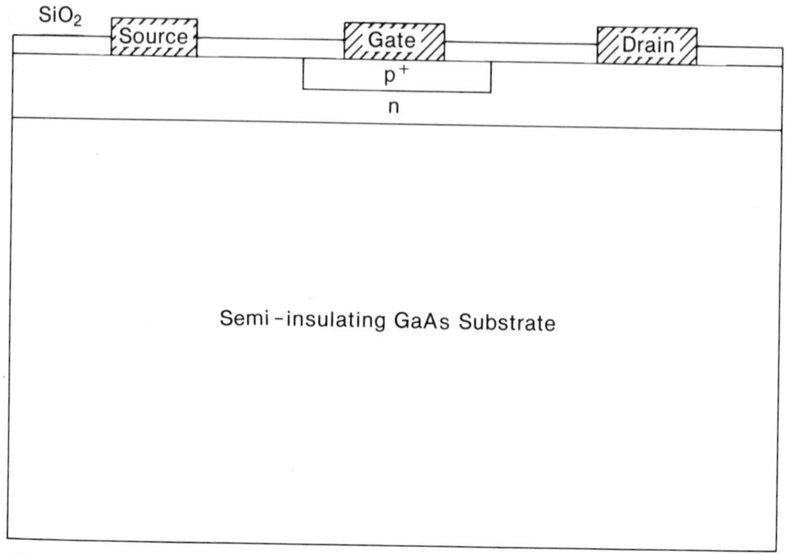

Fig. 2.2 Sketch of a cross-sectional view of a microwave JFET.

2.2.2 Microwave JFET

A cross-sectional view of a microwave junction field effect transistor (JFET) is shown in Fig. 2.2. All metal–semiconductor contacts are ohmic contacts, as in the case of the BJT [11]. Most microwave JFETs are n-channel FETs. The source injects electrons which are then drained from the drain after the electrons have passed through a narrow n-channel. The electron flow is controlled by the negative potential applied to the gate. The gate is reverse biased. The carrier depletion region at the p–n junction is controlled by the reverse bias voltage across the gate. The n-channel thickness is controlled by the reverse bias gate voltage. Therefore the drain current is controlled by the gate voltage. So, as for any other transistor, if the proper amount of current-sensing impedance is connected in series with the drain, the output voltage appears across that impedance.

For a microwave JFET, the p^+-layer can be as thin as $t_p = 0.1$ μm. The thickness of the n-layer can be as little as $t_n = 0.2$ μm. The channel length, which is the shortest distance between the source and drain, is as short as 2 μm for low microwave frequency JFETs and 0.5 μm for high microwave frequency JFETs [6]. A popular substrate is semi-insulating GaAs. The donor concentration in the n-layer is on the order of 10^{17} donors/cm^3. The electron saturation velocity in this sort of n-type GaAs is known to be about 100 km/s. The electron transit time across the channel length for 0.5 μm is 0.5 μm/100 km/s = 5 ps and for 2 μm is 2 μm/100 km/s = 20 ps. Using the "rule of thumb" factor 10, the working microwave frequency should be in the range of 1/50 ps = 20 GHz to 1/200 ps = 5 GHz for the upper working frequency limit with the given channel length. Electron mobility in GaAs is 8500 cm^2/V-s which is larger than electron mobility in Si, 1350 cm^2/V-s [11]. In spite of that, it is of interest to note that the specific resistivity of GaAs 4×10^8 Ω-cm is larger than the specific resistivity of Si, which is 2.5×10^5 Ω-cm. The relative permittivity of GaAs is 13.2 and is comparable to the relative permittivity of Si, which is 11.8 [11]. Therefore interelectrode capacitances of a Si FET and a GaAs FET of comparable size would have comparable values of capacitance. For example, in a microwave GaAs JFET, if the metallization strip is $w_m = 1$ μm wide and $l_m = 6$ μm long and these are separated by $d_e = 2$ μm, center-to-center, then the minimum interelectrode capacitance between the source and the gate C_{SG}, and the minimum interelectrode capacitance between the gate and the drain C_{GD} can be estimated. For the n-layer thickness $h_n = 0.2$ μm as $C_{SG} = C_{GD} = \varepsilon_o \varepsilon_r h_n l_m / d_e$ = (8.85 pF/m)(3.8)(0.2 μm)(6 μm)/2 μm = 73 μpF. The capacitive reactance $X_{SG} = X_{GD} = 1/(2\pi f C_{SG}) = 1/(2\pi f C_{GD})$. At $f = 20$ GHz, $X_{SG} = X_{GD} = 1/(2\pi \times 20 \times 10^9 \times 73 \times 10^{-6} \times 10^{-12}) = 109$ kΩ. This amount of reactance is considered too high for usual 50 Ω systems. Therefore that much interelectrode capacitance will not enter into consideration of the transistor performance. The interelectrode capacitance between the source and the drain C_{SD} can be estimated to be half of C_{SG} or C_{GD}. So it can also be omitted for the design and performance analysis of the device.

A microwave JFET of this size can operate with a drain current $I_D = 0.9$ mA [13]. The cross section of the n-channel is $l_m h_n$ = (6 μm)(0.2 μm) = 1.2 μm^2, and the current density in the channel J_D is approximately $J_D = I_D / l_m h_n$ = 0.9 mA/1.2 μm^2 = 0.75 mA/μm^2. It is interesting to note that the working current density in GaAs JFET is on the same order of magnitude as the working current density in Si BJT presented in Sec. 2.2.1.

The effective wavelength λ_{eff} in GaAs at an operating frequency of $f = 20$ GHz is $\lambda_{eff} = c/(\sqrt{\epsilon_r} f) = 3 \times 10^{-8}$ m/s/$(\sqrt{13.2} \times 20 \times 10^9)$ = 4.12 mm. One tenth of λ_{eff} is 412 μm and thus the length of metallized electrodes of the source, gate, and drain can be as long as $l_m = 412$ μm, without considering the standing wave effect on the electrodes. For higher power, a longer electrode is needed [13].

In the $l_m = 6\ \mu m$ and $w_m = 1\ \mu m$ device, the drain current $I_D = 0.9$ mA is reached when the drain–source voltage $V_{DS} = 5$ V with the channel wide open. At this voltage, the equivalent drain–source resistance $R_{DS} = V_{DS}/I_D = 5\ V/0.9\ mA = 5.6\ k\Omega$. This means that the resistivity of the channel is $\rho_{DS} = R_{DS} l_m (h_n - h_p)/2d_e = (5.6\ k\Omega)(6\mu m)(0.2 - 0.1\ \mu m)/(2 \times 2\ \mu m) = 0.84\ k\Omega\text{-}\mu m$. This compares with intrinsic GaAs resistivity 400 $M\Omega\text{-}\mu m$ [11]. If the channel is narrowed by biasing the gate negatively, the effective resistivity would be higher.

For higher power operation, many single units as shown in Fig. 2.2 can be connected in parallel either monolithically or by wire bonding. On the other hand, the length of the electrodes l_m can be increased. If l_m is increased to 60 μm instead of 6 μm, R_{DS} would be 560 Ω instead of 5.6 $k\Omega$. The current capacity would increase 10 times and the power handling capability under the same $V_{DS} = 5$ V dc would increase 10 times if a proper heat sink is provided. The dc power loss of $l_m = 6\ \mu m$ device is $P_{DC} = V_{DS} I_D = (5\ V)(0.9\ mA) = 4.5$ mW. At most, $\eta = 30\%$ efficiency can be expected. Therefore the upper limit of $l_m = 6\ \mu m$ JFET of this type would be $P_{RF} = \eta P_{DC} = 0.3 \times 4.5\ mW = 1.35\ mW = 10 \log_{10}(1.35\ mW/1\ mW) = 1.3$ dBm. This means that if $l_m = 60\ \mu m$, the output power can be as high as 13.5 mW or 11.3 dBm.

2.2.3 MESFET

A simplified cross-sectional view of the active portion of a GaAs MESFET is schematically shown in Fig. 2.3. The diagram may give the reader a false impression with respect to size, but the actual size outlined in Fig. 2.3 is about one tenth the thickness of a human hair. Therefore, it is difficult to see the active portion with the naked eye. This size transistor is designed for operating frequencies in a range of 10–20 GHz. The period of 20 GHz is 50 ps. To avoid electron transit time effect, it is desirable that electrons travel from the source to the drain in less than one tenth the period of operating frequency. In this case, that is only 5 ps or less. The drift velocity of electrons in the channel between the source and the drain is approximately 100 km/s under working conditions. Thus, the distance between the source and the drain must be at most 0.5 μm. This is the reason that an almost invisibly small size is needed. Though the active portion of the GaAs MESFET is "invisibly" small, the packaged transistor is of a somewhat more comfortable size to handle. An example of the top view pattern of bonding pads by which leads are attached to the

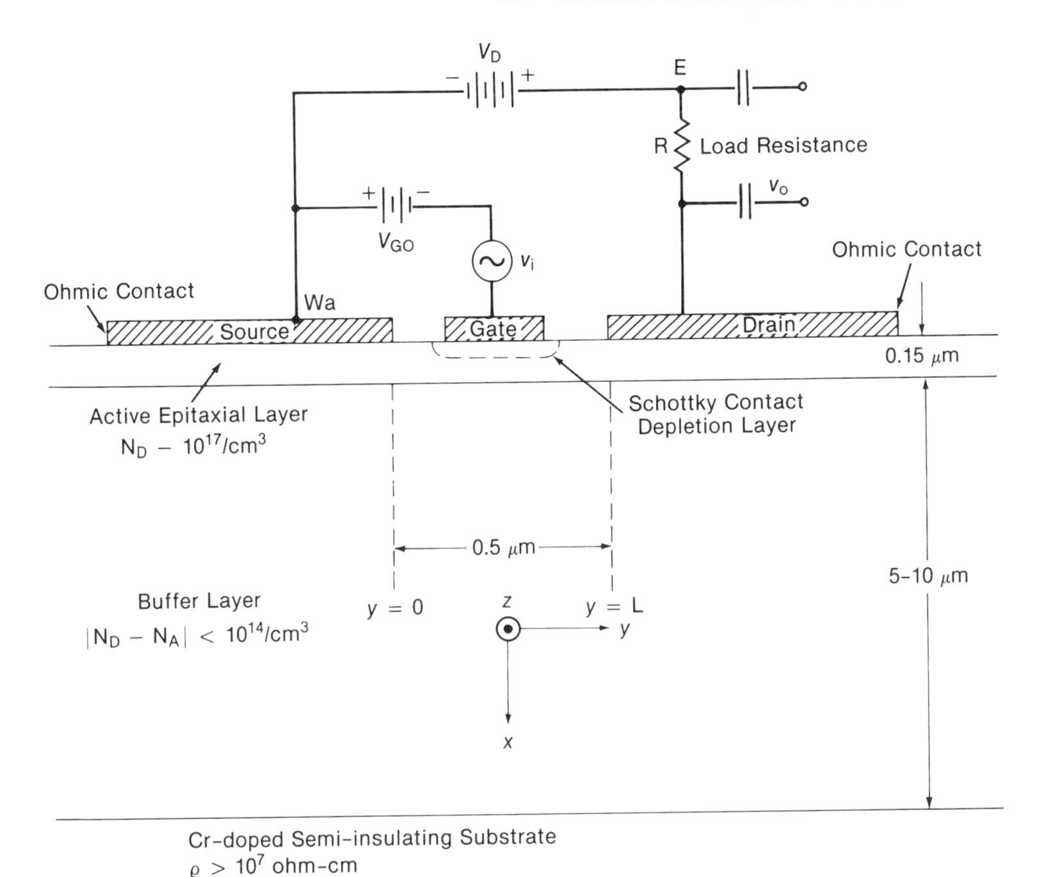

Fig. 2.3 Example of the active portion of a microwave MESFET [6].

active portion is sketched in Fig. 2.4. Using a properly shaped bonding pad and bonding wires, the transistor chip is packaged and connected to the external circuit.

It should be noted that both the drain and the source are ohmic contacts to the active n-type epitaxial layer as seen in Fig. 2.3, but the gate is a Schottky contact. Therefore, when negatively biased with respect to the source as shown in Fig. 2.3, a controllable carrier depletion region will develop. Under this condition, the gate is practically insulated and the action of the transistor resembles the action of a metal oxide semiconductor field effect transistor (MOSFET), which is one of the insulated gate field effect transistors (IGFET) [11,12]. Unlike the MOSFET, the insulation region around the gate of a MESFET is controllable, and it squeezes and pinches the electron flow in the channel. In Fig. 2.3, a general biasing scheme and microwave signal input voltage v_i and output voltage v_o are indicated respectively. Many of the

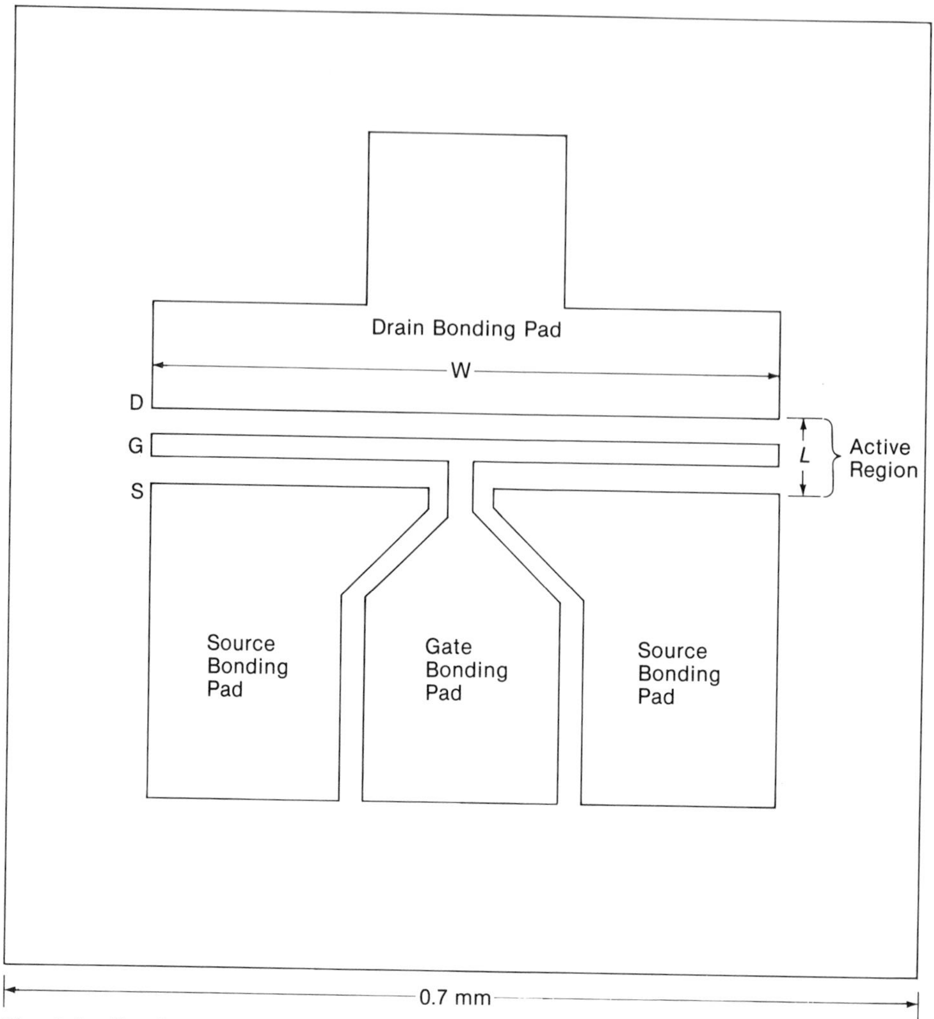

Fig. 2.4 Sketch of the top view of a 20 GHz GaAs MESFET chip [6,7].

example analyses of the device given in Section 2.2.2 on JFETs apply to MESFETs also [14].

2.2.4 HEMT

There are many various structures of high electron mobility transistors (HEMT). Most HEMTs are heterojunction FETs. It is impossible to review all existing designs, so one example of HEMT design is detailed in Fig. 2.5. The main feature of an HEMT is its high frequency adaptability [15]. The source metallization and the drain metallization are extended by ohmic contact

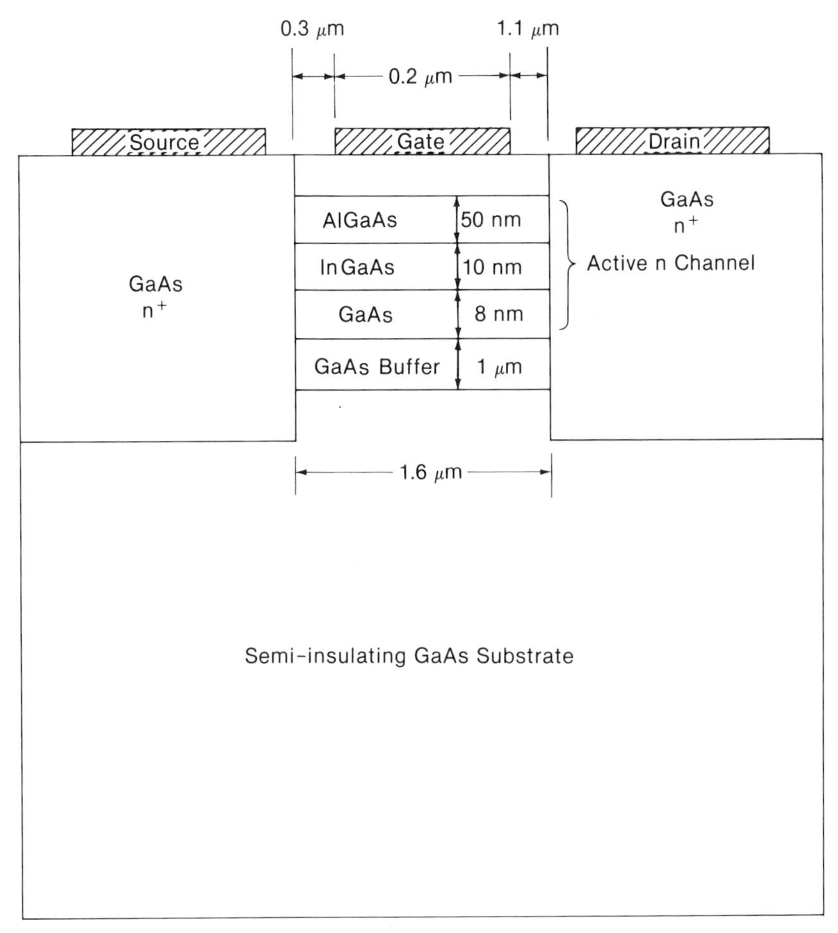

Fig. 2.5 Cross-sectional view of a HEMT. The AlGaAs (50 nm), InGaAs (10 nm), and GaAs (8 nm) layers are doped n-type.

to n^+-type GaAs. Thus a heterojunction is clearly defined, as shown in Fig. 2.5. A heterojunction is a junction formed by different materials. In this design, the channel length from the source n^+ GaAs to the drain n^+ GaAs is $l_{SD} = 1.6$ μm. This channel is made of layers of $2 \times 10^{18}/\text{cm}^3$ Cr doped heterojunctions of AlGaAs, InGaAs, and GaAs, as shown in Fig. 2.5. Between the top layer and undoped GaAs layer is an insulating layer for the gate. The fifth layer of 1 μm thick undoped GaAs is a buffer layer. Note that the active layers are a total of 68 nm thick, or the channel thickness $t_c = 0.068$ μm. High mobility electrons are provided between the layers of AlGaAs and InGaAs, InGaAs and doped GaAs, and within the InGaAs layer itself [15]. The gate metallization is only $l_g = 0.2$ μm wide and $l_m = 50$ μm long. These are fabricated by electron beam lithography [15,16]. The channel is an n-channel with length $l_c = 1.6$ μm [15]. The gate is located closer to the source. This greater proximity provides better gate control over electron injection from the source. It is located only 0.3 μm from the source and 1.1 μm from the drain. The maximum drain current, 800 mA per mm gate length, is obtained at the gate voltage $V_{SG} = 0.6$ V with the drain voltage $V_D = 4.3$ V [15]. This means that for the actual $l_m = 50$ μm, the drain current I_D is $I_D = (800 \text{ mA})(50\ \mu\text{m})/1000\ \mu\text{m} = 40$ mA. In the same HEMT, $I_D = 24$ mA is reported when $V_{SG} = 0$ V [15].

The cross-sectional area of the active n-channel is $A_c = t_c l_m = (0.068\ \mu\text{m})(50\ \mu\text{m}) = 3.4\ \mu\text{m}^2$. When $V_{SG} = 0$ V, the channel resistance $R_{SD} \approx V_D/I_D = 4.3 \text{ V}/24 \text{ mA} = 179\ \Omega$. The resistivity of this channel is then $\rho = R_{SD}A_c/l_c = (179\ \Omega)(3.4\ \mu\text{m}^2)/1.6\ \mu\text{m} = 380\ \Omega\text{-}\mu\text{m}$. This is extremely low in comparison with ρ of intrinsic GaAs which is $\rho = 4 \times 10^8\ \Omega\text{-}\mu\text{m}$ [11].

If the electron charge is q (C), the average drift velocity in the channel is u (m/s), the donor density is N_D (m^{-3}), and the cross-sectional area of the channel is A (m^2), the drain current I_D (A) is $I_D = N_D quA$. Therefore the average drift velocity of electrons is $u = I_D/N_D qA$, using $V_G = 0$ V, $V_D = 4.3$ V, $u = 24 \text{ mA}/(2 \times 10^{18} \text{ cm}^{-3})(1.6 \times 10^{-19} \text{ C})(3.4\ \mu\text{m}^2) = 2.2 \times 10^7$ m/s. This is very fast when compared with the saturation electron velocity in degenerate n-type GaAs, which is about 10^5 m/s. Electrons move more than 100 times faster than n-channel GaAs FET.

The average electric field strength E in the channel is $E \approx V_D/l_c = 4.3 \text{ V}/1.6\ \mu\text{m} = 2.7 \text{ V}/\mu\text{m} = 2.7 \times 10^4$ V/cm. Electron mobility μ_m (cm^2/V-s) $= u/E = (2.2 \times 10^7 \text{ m/s})(10^2 \text{ cm})/(2.7 \times 10^4 \text{ V/cm}) = 8.15 \times 10^4$ (cm^2/V-s). This compares with 1.35×10^3 cm^2/V-s in Si, 3.9×10^3 cm^2/V-s in Ge, 8.5×10^3 cm^2/V-s in GaAs, and 2.26×10^4 cm^2/V-s in InAs [11]. The electron mobility in this HEMT is approximately ten times the electron mobility in an ordinary GaAs MESFET.

The operating frequency is reported to be 55 GHz [11]. With the electron velocity $u = 2.2 \times 10^7$ m/s, the electron transit time across the channel of thickness $l_c = 1.6$ μm is $\tau = l_c/u = 1.6\ \mu\text{m}/(2.2 \times 10^7 \text{ m/s}) \times 10^6\ \mu\text{m/m}) = 0.073 \times 10^{-12}$ s. The electron transit angle for this τ is then $\theta = \omega\tau = 2\pi(55 \times 10^9)(0.073 \times 10^{-12}) = 2.5 \times 10^{-2}$ radians or $\theta = 360° \times 2.5 \times 10^{-2}/2\pi = 5.73$ degrees. This is well within the "rule of thumb" value of 36 degrees. Therefore it is possible to operate at 55 GHz.

The output of this HEMT is reported to be 0.85 W per mm gate length. Since the actual gate length is $l_m = 50\ \mu$m, the actual 55 GHz output power is $P = (0.85\ \text{W/mm})(10^{-3}\ \text{mW/W})(50\ \mu\text{m}) =$ 42.5 mW. Since this HEMT is considered to be an insulated gate FET (IGFET), the resistance between the source and the gate and the gate and the drain is extremely high in a 50 Ω system. However, as noted before, $R_{SD} = 179\ \Omega$. This is a significant value to be considered in a 50 Ω system design. The interelectrode capacitances and inductances arising because of metallization of the source, gate, and drain of this HEMT are, as other types of microwave transistors—even at 55 GHz—not significant for design consideration. This is because of the small size of the structure.

2.3 Transistor Current Control

In a microwave BJT, the injected base current I_b, controls collector current I_c. In a microwave JFET, MESFET, or HEMT, the voltage applied to the gate controls the drain current. A common configuration to all transistors is that a current-sensing impedance is connected in series with the collector circuit or drain circuit most of the time, and the voltage across the current-sensing impedance (often called the load impedance of the transistor) is considered to be the output voltage. The current-sensing impedance or the load impedance is sometimes connected in series with the emitter or the source. Such a circuit is called the emitter follower or the source follower. For a microwave BJT, the current gain $\beta = I_C/I_B$ is a matter of interest. For the FET and HEMT, the transconductance $g = I_D/V_G$ is a matter of concern. Large values of β or g mean high gain of the transistor for a given load impedance.

2.3.1 Microwave BJT

In a microwave n-p-n BJT as shown in Fig. 2.1, the emitter–base junction is forward biased; the emitter is negative and the base is positive. The collector–base junction is reverse biased; the collector is positive and the base is negative. The collector current is defined as I_C, the base current is I_B, and the emitter current is I_E. Kirchhoff's current law is applied to the entire transistor as a current junction with the bias directions taken into consideration,

$$I_C + I_B - I_E = 0 \tag{2.3.1}$$

where the current flowing into the device is considered to be positive. Then the current gain β is

$$\beta \equiv \frac{I_C}{I_B} = \frac{I_E}{I_B} - 1 = \frac{I_E}{I_E - I_C} - 1 = \frac{I_C}{I_E - I_C} \gg 1 \tag{2.3.2}$$

In most transistors, $I_E \approx I_C$. A part of I_E is lost in the base current by recombination with holes injected into the base. The rest of the I_E carrier electrons reach the collector and appear as the collector current I_C. For example, if I_E lost 1% of carrier in the base region due to the injected I_B, then $I_C = 0.99I_E$. Then $\beta = 0.99I_E/(I_E - 0.99I_E) = 0.99/0.01 = 99$. For a high beta transistor, β on the order of 100 is common. Now

$$I_C = \beta I_B \tag{2.3.3}$$

so the collector current I_c is controlled by the base current I_B.

In a linear region where β is constant, $\partial I_c = \beta_d \partial I_B$ and

$$\beta_d = \frac{\partial I_c}{\partial I_B} \tag{2.3.4}$$

This knowledge is useful since in most cases the base must be forward dc biased and signal current is superimposed on the dc bias current. For example, if the transistor is properly dc biased and β_d is a constant, and the injected signal current in the base is

$$\partial I_B = I \sin \omega t \tag{2.3.5}$$

then the collector signal current is

$$\partial I_c = \beta_d I \sin(\omega t + \phi) \tag{2.3.6}$$

where ϕ is the phase delay due to finite hole–electron recombination time in the base region and the effective electron transit time in the base–collector region, if it is significant. Thus the collector current is controlled by the base injection current.

2.3.2 JFET, MESFET, and HEMT

For JFETs and MESFETs the drain current I_D in the n-channel is controlled by a negatively biased gate voltage. The control mechanism is two-fold. The control is accomplished both by Coulomb forces acting on electrons in the channel and by the effective channel thickness, which depends on the negatively biased depletion layer of the gate p–n junction for JFET and the Schottky depletion layer at the metal–semiconductor contact for the MESFET. For the HEMT, the current is primarily controlled by the Coulomb force. The gate is insulated and located close to the source. The gate voltage can swing to positive in order to pull more electrons out of the source. The n-channel of the HEMT is well insulated and the electrons are therefore pulled out from the source by the gate and the drain voltage and are forced to go to the drain with high velocity. The velocity of the electrons can be controlled by the Coulomb force of the gate voltage.

The Coulomb force on an electron in the n-channel of a FET is

$$m\frac{\mathrm{d}u}{\mathrm{d}t} = qE = q(E_\mathrm{G} + E_\mathrm{D}) \tag{2.3.7}$$

where m is the electron mass, u is the electron velocity along the channel, q is the electron charge, and E is the electric field strength in the channel along the direction of the channel which consists of the sum of the electric field strengths due to the gate E_G and the drain E_D. Then the steady state electron velocity is

$$u = \mu_\mathrm{n}E = \mu_\mathrm{n}(E_\mathrm{G} + E_\mathrm{D}) \tag{2.3.8}$$

where μ_n is the electron mobility in the channel. The drain current due to the Coulomb force is

$$I_\mathrm{D} = l_\mathrm{m}t_\mathrm{c}N_\mathrm{D}qu = l_\mathrm{m}t_\mathrm{c}N_\mathrm{D}q\mu_\mathrm{n}(E_\mathrm{G} + E_\mathrm{D}) \tag{2.3.9}$$

where l_m is the channel width, t_c is the channel thickness and N_D is the carrier concentration. If the channel length is l_c, the distance between the source and the gate is l_SG, the gate voltage is V_G, and the drain voltage is V_D, then

$$I_\mathrm{D} \approx \mathrm{k}l_\mathrm{m}t_\mathrm{c}N_\mathrm{D}q\mu_\mathrm{n}\left(\frac{V_\mathrm{G}}{l_\mathrm{SG}} + \frac{V_\mathrm{D}}{l_\mathrm{c}}\right) \tag{2.3.10}$$

where k is a correction factor due to E_G not being exactly equal to $V_\mathrm{G}/l_\mathrm{SG}$ and E_D not being exactly equal to $V_\mathrm{D}/l_\mathrm{c}$. This equation shows clearly that the drain current is controlled by the gate voltage. When V_G is positive I_D increases, and when V_G is negative I_D decreases. The static transconductance is then

$$G_\mathrm{tr} \equiv \frac{I_\mathrm{D}}{V_\mathrm{G}} = \mathrm{k}l_\mathrm{m}t_\mathrm{c}N_\mathrm{D}q\mu_\mathrm{n}\left(\frac{1}{l_\mathrm{SG}} + \frac{1}{l_\mathrm{c}} \cdot \frac{V_\mathrm{D}}{V_\mathrm{G}}\right) \tag{2.3.11}$$

For example, in the HEMT presented in Section 2.2, $l_m = 50\ \mu$m, $t_c = 0.068\ \mu$m, $N_D = 2 \times 10^{18}/\text{cm}^3$, $q = 1.6 \times 10^{-19}$ C, $\mu_n = 8.15 \times 10^4\ \text{cm}^2/\text{V-s}$, $l_{SG} = 0.3\ \mu$m, $l_c = 1.6\ \mu$m, $V_D = 4.3$ V, $V_G = 0.6$ V, and $I_D = 40$ mA [15]. In this case

$$\begin{aligned} G_{tr} = {} & \text{k}(50 \times 10^{-6}\ \text{m})(0.068 \times 10^{-6}\ \text{m})(2 \times 10^{18}\ \text{cm}^{-3}) \\ & \times (10^{-6}\ \text{m}^{-3}/\text{cm}^{-3})(1.6 \times 10^{-19}\ \text{C}) \\ & \times (8.15 \times 10^4 \times 10^{-4}\ \text{m}^2/\text{V-s}) \\ & \times \left[\frac{1}{0.3} \times 10^6\ \text{m}^{-1} + \left(\frac{1}{1.6} \times 10^6\ \text{m}^{-1}\right)\frac{4.3\ \text{V}}{0.6\ \text{V}}\right] = 69.2\text{k} \end{aligned}$$

Also,

$$G_{tr} = \frac{I_D}{V_G} = \frac{40 \times 10^{-3}\ \text{A}}{0.6\ \text{V}}\ \text{Siemens}.$$

Therefore $\text{k} = 0.964 \times 10^{-3}$, hence $G_{tr} = 66.6 \times 10^{-3}$ S.

The dynamic transconductance is, from Eq. 2.3.10,

$$g_{tr} \equiv \left.\frac{\partial I_D}{\partial V_G}\right|_{V_D = \text{constant}} = \text{k} l_m t_c N_D q \mu_n / l_{SG} \tag{2.3.12}$$

It is of interest to note that a larger g_{tr} is obtained by making l_{SG} smaller.

For the HEMT in question, g_{tr} at $V_D = 4.3$ V is $g_{tr} = 28.4 \times 10^{-3}$ Siemens [15]. This means that if the gate voltage changes 1 V, the drain current changes by 28.4 mA.

For JFETs and MESFETs, t_c is a function of V_G. It is well known that at V_G the thickness of the carrier depletion region w at the reverse biased junction or Schottky contact area is proportional to $\sqrt{V_o + V_G}$ where V_o is the contact potential at the gate p–n junction of the JFET or at the Schottky contact of MESFET [11]. If k_c is a proportionality constant, then

$$w = \text{k}_c \sqrt{V_o + V_G} \tag{2.3.13}$$

If the channel thickness when $V_G = 0$ is t_{co}, then under reverse biased conditions,

$$t_c = t_{co} - w = t_{co} - \text{k}_c \sqrt{V_o + V_G} \tag{2.3.14}$$

Combining Eq. 2.3.14 with Eq. 2.3.11, the static transconductance is (with the understanding that the gate voltage is negative and is used only to control the channel thickness)

$$G_{tr} = kl_m\left(t_{co} - k_c\sqrt{V_o + V_G}\right)N_D q\mu_n\left(\frac{1}{l_c}\frac{V_D}{V_G}\right) \quad (2.3.15)$$

To obtain the dynamic transconductance from Eq. 2.3.10 with the understanding that the gate voltage is negative,

$$I_D = kl_m\left(t_{co} - k_c\sqrt{V_o + V_G}\right)N_D q\mu_n\frac{V_D}{l_c} \quad (2.3.16)$$

Here again I_D is controlled by V_G. Then,

$$g_{tr} = \left|\frac{\partial I_D}{\partial V_G}\right| = kk_c l_m N_D q\mu_n\frac{V_o}{2l_c}(V_o + V_G)^{-1/2} \quad (2.3.17)$$

Both G_{tr} and g_{tr} are nonlinear with respect to V_G.

For example, for the case of a JFET presented in Section 2.2.2, $I_D = 8.7$ mA, $l_m = 6\ \mu$m, $N_D = 10^{17}\ \text{cm}^{-3}$, $t_{co} = 0.2\ \mu$m, $q = 1.6 \times 10^{-19}$ C, $l_{SG} = 2\ \mu$m, $l_c = 4\ \mu$m, $V_G = 0$ V, $V_D = 5$ V, $V_o = 0.5$ V, and $\mu_n = 5.8 \times 10^3\ \text{cm}^2/\text{V-s}$ [7,13]. It is also known from the V_D–I_D curve of this JFET, it cuts off at $V_G = 1.5$ V and $I_D \approx 0$ [13].

In Eq. 2.3.16 $I_D = 0$ when

$$t_{co} - k_c\sqrt{V_o + V_G} = 0 \quad (2.3.18)$$

so

$$k_c = \left(\frac{t_{co}}{\sqrt{V_o + V_G}}\right) \quad (2.3.19)$$

In this example

$$k_c = \left(\frac{0.2\ \mu\text{m}}{\sqrt{0.5\ \text{V} + 1.5\ \text{V}}}\right) = 0.1414\ \mu\text{m}/\text{V}^{1/2}$$

For $I_D = 8.7$ mA with $V_G = 0$, from Eq. 2.3.16,

$$I_D = (8.7 \times 10^{-3}\ \text{A}) = k \cdot (6 \times 10^{-6}\ \text{m}) \cdot (10^{17} \times 10^6\ \text{m}^{-3})$$
$$(1.6 \times 10^{-19}\ \text{C})(5.8 \times 10^3 \times 10^{-4}\text{m}^2/\text{V-s})$$
$$\left\{0.2 \times 10^{-6}\ \text{m} - \left[0.1414 \times 10^{-6}\ \text{m}/\text{V}^{1/2}\right)\sqrt{0.5\ \text{V}}\right]\right\}$$
$$\cdot\frac{5\text{V}}{4 \times 10^{-6}\ \text{m}} \quad \text{or,} \quad (8.7 \times 10^{-3}\ \text{A}) = (6.96 \times 10^{-3}\ \text{A})k$$

so k = 1.25.

Knowing both values of k_c and k, the value of static transconductance G_{tr} and dynamic transconductance g_{tr} are calculated using Eqs. 2.3.15 and 2.3.17, by choosing the quiescent point (Q-point) at $V_D = 5$ V and $-V_G = -1.0$ V.

$$G_{tr} = 1.25 \cdot (6 \times 10^{-6}\ \text{m}) \cdot (10^{17} \times 10^{6}\ \text{m}^{-3})$$

$$\cdot \Bigg(1.6 \times 10^{-19}\ \text{C} \cdot (5.8 \times 10^{3} \times 10^{-4}\ \text{m}^2/\text{V-s})$$

$$\left[0.2 \times 10^{-6}\ \text{m} - (0.1414 \times 10^{-6}\ \text{m/V}^{1/2})\sqrt{0.5\ \text{V} + 1.0\ \text{V}}\right]$$

$$\cdot \frac{1}{4 \times 10^{-6}\ \text{m}} \frac{5\ \text{V}}{1\ \text{V}}\Bigg)$$

$$= 2.33 \times 10^{-3}\ \text{Siemens}$$

and

$$g_{tr} = 1.25 \cdot (0.1414 \times 10^{-6}\ \text{m/V}^{1/2}) \cdot (6 \times 10^{-6}\ \text{m})$$

$$\cdot (10^{17} \times 10^{6}\ \text{m}^{-3}) \cdot (1.6 \times 10^{-19}\ \text{C})$$

$$\cdot (5.8 \times 10^{3} \times 10^{-4}\ \text{m}^2/\text{V-s})$$

$$\cdot \frac{5\ \text{V}}{2 \times 4 \times 10^{-6}\ \text{m}\sqrt{0.5\ \text{V} + 1\ \text{V}}}$$

$$= 5.02 \times 10^{-3}\ \text{Siemens}$$

An important mechanism is that the drain current in an HEMT is mainly controlled by the Coulomb force from the gate, and the drain current in a JFET or MESFET is mainly controlled by the thickness of the depletion layer as a result of the negative voltage applied to the gate.

2.4 Voltage Gain

In a bipolar junction transistor, the collector current is controlled by the current injected into the base. Usually a current-sensing impedance or load impedance is connected to the collector. The output signal voltage is developed across the load impedance $\dot{Z}_c$. The base current is a combination of the dc base current I_b and microwave signal current $\dot{i}_b$. The collector current is a combination of the dc collector current I_c and microwave signal current $\dot{i}_c$. The microwave collector current $\dot{i}_c$ and microwave base current $\dot{i}_b$

are related by the dynamic current gain $\dot{\beta}_d$

$$\dot{i}_c = \dot{\beta}_d \dot{i}_b \tag{2.4.1}$$

If the output microwave voltage across the load impedance $\dot{Z}_L$ is $\dot{v}_o$, then

$$\dot{i}_c = \frac{\dot{v}_o}{\dot{Z}_L} \tag{2.4.2}$$

If the input impedance to the base is $\dot{Z}_{BE}$ and the input voltage across the base–emitter terminal of the transistor is $\dot{v}_i$, then

$$\dot{i}_b = \frac{\dot{v}_i}{\dot{Z}_{BE}} \tag{2.4.3}$$

Substituting both Eqs. 2.4.2 and 2.4.3 into Eq. 2.4.1,

$$\frac{\dot{v}_o}{\dot{Z}_L} = \dot{\beta}_d \frac{\dot{v}_i}{\dot{Z}_{BE}} \tag{2.4.4}$$

The voltage gain, or the voltage amplification A, is then

$$A \equiv \frac{\dot{v}_o}{\dot{v}_i} = \dot{\beta}_d \frac{\dot{Z}_L}{\dot{Z}_{BE}} \tag{2.4.5}$$

For example, if a high β transistor of $\beta = 100$ is chosen and $\dot{Z}_L$ is chosen to be equal to $\dot{Z}_{BE}$, then $\dot{A} = 20 \log_{10}(100)\,(\dot{Z}_{BE}/\dot{Z}_{BE}) = 40$ dB. This is quite a high gain for a single stage amplifier.

For JFETs, MESFETs, or HEMTs, the microwave drain current $\dot{i}_d$ is related to the microwave gate voltage or microwave input voltage $\dot{v}_i$ at the gate by the dynamic transconductance g_{tr} by

$$g_{tr} = \frac{\dot{i}_d}{\dot{v}_i} \tag{2.4.6}$$

If the current-sensing impedance or load impedance $\dot{Z}_L$ is connected in series with the drain, the output voltage $\dot{v}_o$ across the load impedance is then

$$\dot{v}_o = \dot{Z}_L \dot{i}_d \tag{2.4.7}$$

Combining Eq. 2.4.6 with Eq. 2.4.7,

$$g_{tr} = \frac{\dot{v}_o}{\dot{Z}_L \dot{v}_i} \tag{2.4.8}$$

The voltage gain, or the voltage amplification $\dot{A}$, for the JFET, MESFET, or HEMT is then

$$\dot{A} \equiv \frac{\dot{v}_o}{\dot{v}_i} = g_{tr}\dot{Z}_L \tag{2.4.9}$$

For example, if a microwave FET with $g_{tr} = 5 \times 10^{-3}$(S) is used with $Z_L = 1000\ \Omega$, then $\dot{A}$(dB) $= 20\log_{10}\dot{A} = 20\log_{10}(5 \times 10^{-3} \times 1000) = 14$ dB.

In microwave transistor engineering, in addition to $\dot{\beta}_d$ and g_{tr}, the performance of a transistor, including the voltage gain, is expressed by the S-parameters [4,12]. *S*-parameters are basically the voltage reflection coefficients and the voltage transmission coefficients viewed from the input and output of the transistor. A transistor is viewed as a two-port network. The input is Port 1 and the output is Port 2. When the transistor is in action, there are voltage waves converging toward the transistor and voltage waves diverging from the transistor. For example, the converging voltage wave toward Port 1, $\dot{V}_1^c$, is the incident microwave voltage wave. The diverging voltage wave from Port 1, $\dot{V}_1^d$, consists of the reflected waves at Port 1, $\dot{V}_1^r$, caused by $\dot{V}_1^c$ and transferred wave $\dot{V}_1^{tr}$ from Port 2. At Port 2, the diverging wave $\dot{V}_2^d$ consists of the transferred or amplified voltage wave $\dot{V}_2^{tr}$ from Port 1 and the reflected wave $\dot{V}_2^r$ at Port 2 due to the converging voltage wave $\dot{V}_2^c$, which in most cases is the reflection from the load impedance back into the transistor.

Hence,

$$\dot{V}_1^d = \dot{V}_1^r + \dot{V}_1^{tr} \tag{2.4.10}$$

$$\dot{V}_2^d = \dot{V}_2^{tr} + \dot{V}_2^r \tag{2.4.11}$$

If the voltage reflection coefficient at Port 1 is $\dot{\rho}_1$, then by definition,

$$\dot{V}_1^r = \dot{\rho}_1\dot{V}_1^c \tag{2.4.12}$$

If the voltage transfer coefficient from Port 2 to Port 1 is $\dot{\tau}_{12}$, then by definition

$$\dot{V}_1^{tr} = \dot{\tau}_{12}\dot{V}_2^c \tag{2.4.13}$$

If the voltage reflection coefficient at Port 2 is $\dot{\rho}_2$, then by definition,

$$\dot{V}_2^r = \dot{\rho}_2\dot{V}_2^c \tag{2.4.14}$$

If the voltage transfer coefficient from Port 1 to Port 2 is $\dot{\tau}_{21}$, then, by definition,

$$\dot{V}_2^{\mathrm{tr}} = \dot{\tau}_{21}\dot{V}_1^{\mathrm{c}} \tag{2.4.15}$$

If Eqs. 2.4.12 and 2.4.13 are substituted in Eq. 2.4.10,

$$\dot{V}_1^{\mathrm{d}} = \dot{\rho}_1\dot{V}_1^{\mathrm{c}} + \dot{\tau}_{12}\dot{V}_2^{\mathrm{c}} \tag{2.4.16}$$

If Eqs. 2.4.14 and 2.4.15 are substituted in Eq. 2.4.11,

$$\dot{V}_2^{\mathrm{d}} = \dot{\tau}_{21}\dot{V}_1^{\mathrm{c}} + \dot{\rho}_2\dot{V}_2^{\mathrm{c}} \tag{2.4.17}$$

Eqs. 2.4.16 and 2.4.17 can be combined and expressed in matrix form such as [4]

$$\begin{pmatrix} \dot{V}_1^{\mathrm{d}} \\ \dot{V}_2^{\mathrm{d}} \end{pmatrix} = \begin{pmatrix} \dot{\rho}_1 & \dot{\tau}_{12} \\ \dot{\tau}_{21} & \dot{\rho}_2 \end{pmatrix} \begin{pmatrix} \dot{V}_1^{\mathrm{c}} \\ \dot{V}_2^{\mathrm{c}} \end{pmatrix} \tag{2.4.18}$$

The S-parameter of any two-port network is expressed in this form [4,12]. By definition,

$$\begin{pmatrix} \dot{V}_1^{\mathrm{d}} \\ \dot{V}_2^{\mathrm{d}} \end{pmatrix} = \begin{pmatrix} \dot{S}_{11} & \dot{S}_{12} \\ \dot{S}_{21} & \dot{S}_{22} \end{pmatrix} \begin{pmatrix} \dot{V}_1^{\mathrm{c}} \\ \dot{V}_2^{\mathrm{c}} \end{pmatrix} \tag{2.4.19}$$

So, by comparing Eqs. 2.4.18 and 2.4.19,

$$\dot{S}_{11} = \dot{\rho}_1 \tag{2.4.20}$$

$$\dot{S}_{22} = \dot{\rho}_2 \tag{2.4.21}$$

$$\dot{S}_{12} = \dot{\tau}_{12} \tag{2.4.22}$$

$$\dot{S}_{21} = \dot{\tau}_{21} \tag{2.4.23}$$

In the case of a microwave transistor, $\dot{V}_1^{\mathrm{c}}$ is the microwave incident voltage applied across the input terminal, which can be the base–emitter terminal or the gate–source terminal. $\dot{V}_2^{\mathrm{c}}$ is the reflection from the load at the collector or drain. $\dot{V}_1^{\mathrm{d}}$ is the voltage wave going back toward the signal source at the input terminal. $\dot{V}_2^{\mathrm{d}}$ is the voltage wave going toward the load at the collector or the drain. If $\dot{V}_1^{\mathrm{c}}$ is the transistor input voltage, then the voltage gain discussed here is

$$\dot{A} \equiv \frac{\dot{V}_{\mathrm{o}}}{\dot{V}_{\mathrm{i}}} \equiv \frac{\dot{V}_2^{\mathrm{d}} - \dot{S}_{22}\dot{V}_2^{\mathrm{c}}}{\dot{V}_1^{\mathrm{c}}} = \frac{\dot{S}_{21}\dot{V}_1^{\mathrm{c}}}{\dot{V}_1^{\mathrm{c}}} = \dot{S}_{21} \tag{2.4.24}$$

Usually the S-parameters of a microwave transistor are measured and listed in a specification sheet provided by the manufacturer.

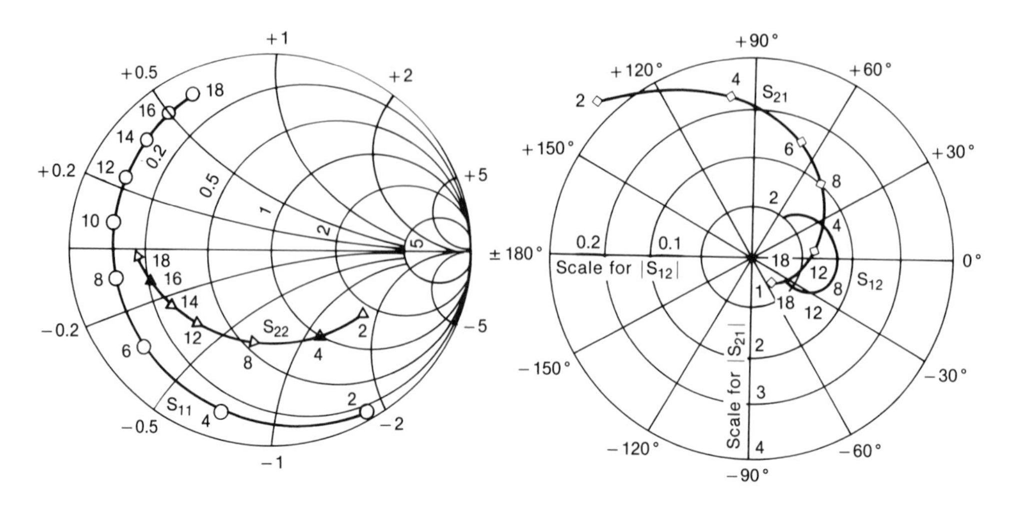

Frequency (GHz)	S_{11}		S_{12}		S_{21}		S_{22}	
	Magnitude	Phase Angle	Magnitude	Phase Angle	Magnitude	Phase Angle	Magnitude	Phase Angle
2	0.94	−59	0.053	53	4.37	135	0.57	−34
3	0.89	−85	0.070	38	3.86	115	0.53	−49
4	0.85	−107	0.080	25	3.36	98	0.50	−61
5	0.82	−126	0.086	14	2.93	83	0.48	−72
6	0.80	−142	0.088	5	2.56	69	0.47	−82
7	0.79	−156	0.088	−3	2.25	57	0.47	−92
8	0.79	−169	0.086	−10	1.98	45	0.47	−101
9	0.79	180	0.083	−17	1.76	34	0.48	−110
10	0.80	170	0.079	−22	1.57	23	0.49	−118
11	0.81	161	0.074	−27	1.40	13	0.51	−126
12	0.81	153	0.069	−31	1.25	4	0.52	−135
13	0.82	145	0.065	−34	1.12	−6	0.55	−142
14	0.84	138	0.060	−37	1.01	−15	0.57	−150
15	0.85	132	0.055	−38	0.90	−23	0.59	−157
16	0.86	126	0.051	−39	0.81	−32	0.62	−165
17	0.87	121	0.047	−38	0.73	−39	0.64	−172
18	0.88	116	0.044	−37	0.65	−47	0.67	−178

Fig. 2.6 Specifications of the JS8851-AS microwave FET. $V_{DS} = 10$ V, $I_{DS} = 100$ mA. Courtesy of Matcom [17].

An example is shown in Fig. 2.6 [17]. In this figure, all *S*-parameters on microwave power for the S8851 JS8851–AS GaAs FET are plotted and the magnitudes and the phase angles are tabulated against the operating frequency. This is a GaAs power FET. The magnitude of $\dot{S}_{21}$ is not so large, but the output can be as high as 24 dBm. At 2 GHz, $|\dot{S}_{21}| = 4.37$. This means that the voltage gain is $20 \log |\dot{S}_{21}| = 20 \log 4.37 \approx 13$ db. At 8 GHz, $20 \log 1.98 \approx 6$ dB and at 11 GHz, $20 \log 1.40 \approx 3$ dB.

As seen from Eqs. 2.4.5 and 2.4.9, to calculate the voltage gain or to design a microwave amplifier, it is necessary to know the equivalent circuit of the transistor and specific values of the circuit parameters. These are available from a manufacturer's specification sheet. For example, the equivalent circuit of the JS8851–AS transistor and the circuit parameters are shown in Fig. 2.7 [17]. In

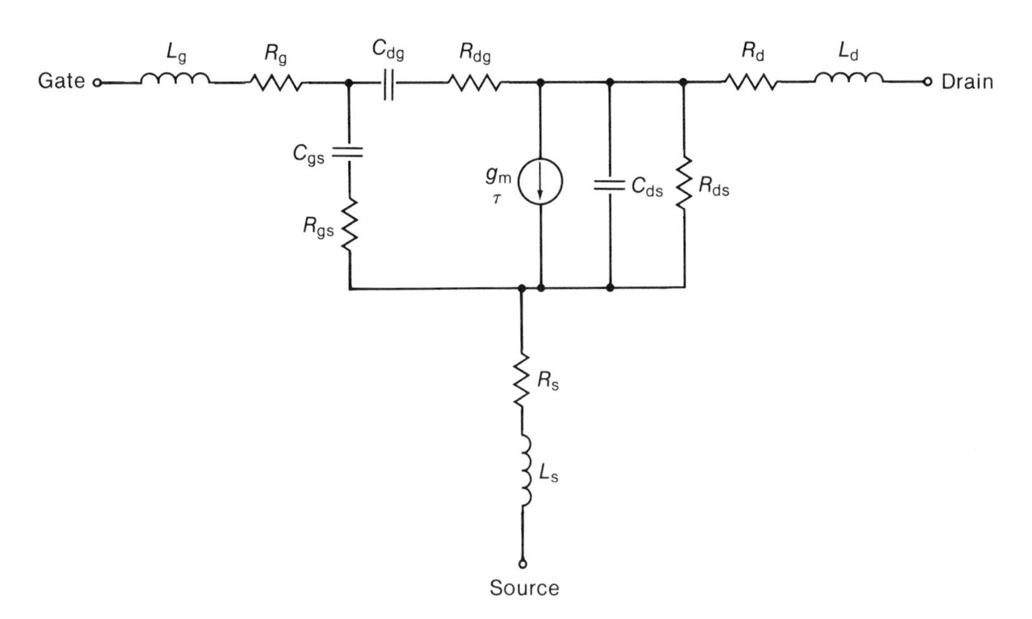

Bonding Wire Condition

	Number (pcs)	Length (mm)	Diameter (μm)
Source	2	0.3	625
Drain	1	0.4	625
Gate	1	0.4	625

Fig. 2.7 Equivalent circuit, including normal bonding wires, for the JS8851-AS microwave FET. $V_{DS} = 10$ V, $I_{DS} = 100$ mA. The components have the following values: $g_m = 65$ mS; $\tau = 5.3$ ps; $R_g = 1.39\ \Omega$; $R_s = 0.76\ \Omega$; $R_d = 1.30\ \Omega$; $R_{gs} = 1.42\ \Omega$; $R_{dg} = 0.01\ \Omega$; $R_{ds} = 197\ \Omega$; $C_{gs} = 0.69$ pF; $C_{dg} = 0.06$ pF; $C_{ds} = 0.26$ pF; $L_g = 0.37$ nH; $L_s = 0.02$ nH; $L_d = 0.23$ nH. Courtesy of Matcom [17].

this figure, g_m is the transconductance. This power FET has 10 unit transistors which are monolithically connected in parallel [17]. Therefore the transconductance per unit FET is 6.5 mS. τ is the electron transit time in the channel; $1/\tau = 1/5.3 \times 10^{-12} = 189 \times 10^9$ Hz is the cutoff frequency of this transistor. The drain–gate capacitance C_{dg} and the gate source capacitance C_{gs} are indicative that the gate is insulated. Resistances R_{dg} and R_{gs} represent the small conduction loss and dielectric loss in the reverse biased gate. The microwave power loss within the channel is mainly represented by R_{ds}. In practice, an external input circuit will be connected between the gate and the source terminal and an external output circuit will be connected between the drain terminal and the source terminal in Fig. 2.7.

2.5 Microstrip Circuit with Microwave Transistors

Depending on the types of microwave transistor, the selection of the Q-point, the dc bias circuit, and the polarity may differ, yet the configurations of microwave circuit patterns are similar to each other. Therefore, as an example, a microwave FET circuit is shown in this section. The same idea will apply to microwave BJTs, JFETs, MESFETs, and HEMTs. A schematic circuit diagram of a conventional radio frequency amplifier circuit using an FET is shown in Fig. 2.8. In this diagram, the gate and the drain resonators are represented by parallel LC circuits. Capacitors C_c couple to the input or output rf circuit outside. These are also used for blocking the dc gate or drain voltage to the rf circuits. The radio frequency choke coils are represented by RFC. The RFC prevents escaping microwave energy through lead wires for dc biasing. Some residual rf signal is further grounded by the feedthrough capacitors, labeled FTC. Bypassing capacitors C_b are used to ground one end of the resonator for rf while keeping the dc bias voltage from grounding.

When the rf input voltage V_{in} is fed across the input terminal, the input line is properly tapped to the input resonator coil to match the input circuit impedance. Then the input resonator is sufficiently excited to produce a good control voltage at the gate

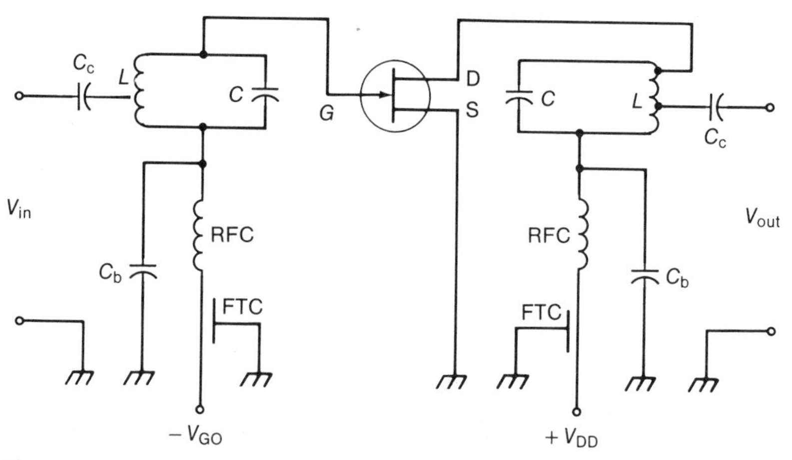

Fig. 2.8 Schematic diagram of a radio frequency amplifier.

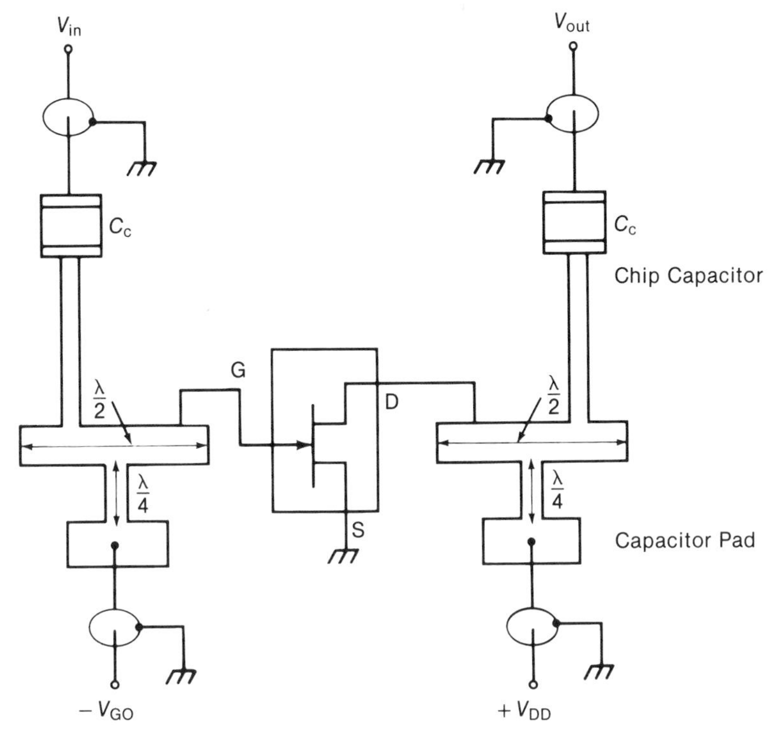

Fig. 2.9 Schematic diagram of a microstripline version of a GaAs MESFET microwave amplifier.

of the FET. The gate voltage is amplified at the drain, and the amplified rf voltage appears across the drain resonator circuit. The drain is tapped to match the drain impedance to the drain resonator impedance. The output is coupled through a proper tap to the resonator coil to match the impedance of the circuit outside.

The circuit analogous to Fig. 2.8 is built for high microwave frequencies utilizing microstripline technology. A schematic diagram of the microstripline version of Fig. 2.8 is shown in Fig. 2.9. In this figure, both the gate and drain resonators are replaced by half-wavelength resonators. The rf choke coils are replaced by quarter-wavelength lines, and the feedthrough capacitors are replaced by capacitor pads. The coupling capacitors are replaced by chip capacitors. The stripline resonator is tapped at a proper location to match the impedance of the tapping line to the impedance of the stripline resonator at the tapping point [12,13,18,19]. According to transmission line theory, the impedance of both ends of a half-wavelength resonator is maximum, and the impedance at the middle is minimum [4] (Appendix 1). Therefore,

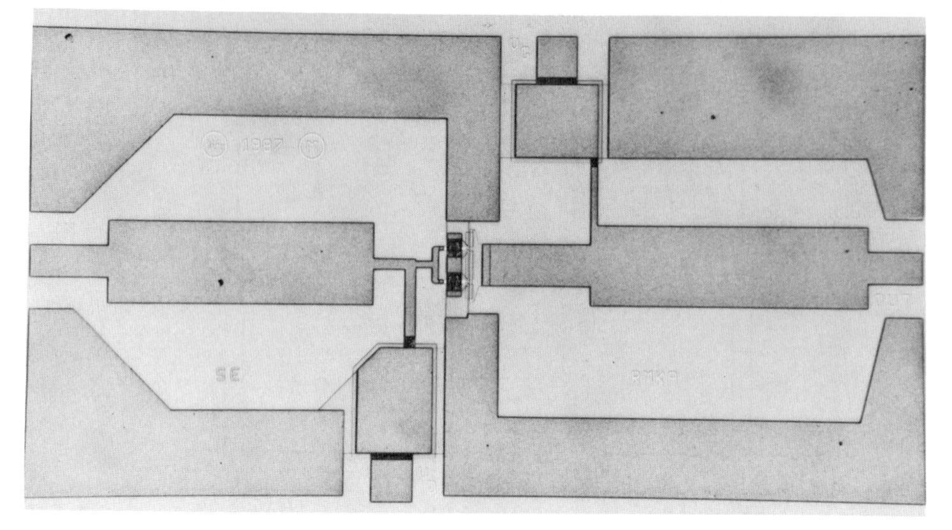

Fig. 2.10 Photograph of a monolithic 20–40 GHz amplifier. Chip size is 2.2 mm × 1.1 mm. Courtesy of Varian Associates [19].

the dc bias lines for the gate and the drain are tapped in the middle of the resonator with the quarter-wavelength line with the capacitor pads. The transmission line theory shows that the impedance of a point one quarter-wavelength (or an odd number multiple of quarter-wavelengths) from a capacitively-short-circuited point is infinite [4]. Therefore very little microwave energy will enter into the quarter-wavelength line in Fig. 2.9. Usually, microstriplines used as bias lines are made as narrow as practical. Narrow microstriplines have higher characteristic impedance than wider microstriplines [20]. This helps reduce the coupling of microwaves from resonators to the bias lines. A single stage amplifier, as shown in Fig. 2.9, can be monolithically integrated with either a Si or GaAs substrate to the size of 2.2 mm × 1.1 mm chip at the 20–40 GHz range [19]. An example is shown in Fig. 2.10. In Fig. 2.10, the active device is a dual gate HEMT [19]. Microwaves of 20 GHz are fed to the device from the left. The step microstrip near the input is the impedance-matching design. The input line is connected to the gate of the HEMT. The gate is biased from the lower side where the capacitor pad and a $\lambda_l/4$ thin bias line are shown. The short thin line section which is collinear with the input line is the input impedance-matching section between the input line and the gate of the HEMT. A wide area of metallization along the edge of this IC chip is the grounding metallization. The source of this HEMT is connected to this grounding metallization as seen in the middle of Fig. 2.10. The output step microstripline is connected to the drain at the right. A thin $\lambda_l/4$ bias line and a capacitor pad are seen in the upper part of Fig. 2.10. The reason for stepping the microstripline is to match impedance [18]. This amplifier can operate up to 10 dBm output with 1 dB output power compression at 30 GHz, a gain of 6 dB, and a noise figure of 5 dB [19].

If the operating frequency is in the low microwave frequencies (e.g. 1.6–2.3 GHz), the power is high (e.g. 20 W), and a packaged surface mounted microwave transistor (NE 345L–20B) and low permittivity substrate are used, the microstripline circuit board can be as large as 6 cm × 6 cm [24].

2.6 Microwave Cavity Circuit with Transistors

The microwave cavity resonator version of a microwave transistor amplifier as shown in Fig. 2.8 is schematically shown in Fig. 2.11.

The FET chip is packaged in a coaxial structure as shown. Both the gate resonator and the drain resonator are employed as semicoaxial resonators. As seen from Fig. 2.11, the source is grounded to the center wall of the cavity resonator. The length of the center stud of the gate cavity is three quarters of the wavelength. One end of the stud is grounded to the cavity wall. According to coaxial line theory, (Appendix 2) the radial electric field at a point one half wavelength from the shorting end is zero. Therefore, a bias lead for the gate is attached there through a

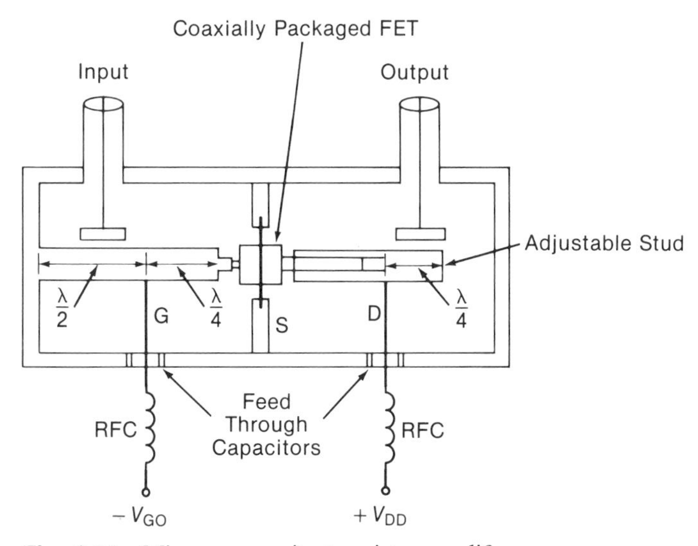

Fig. 2.11 Microwave cavity transistor amplifier.

feedthrough capacitor. The gate is attached approximately one quarter-wavelength from the bias feeding point as seen from Fig. 2.11. At this point, both the voltage and the impedance are considered to be maximum. The microwave input signals are fed through a coaxial capacitive coupler at a matching location along the stud [5].

The drain cavity resonator is structured by the same reasoning, but the length of the center stud is quite different from three quarters of a wavelength [5]. This is to accommodate the impedance-matching at the drain. The drain has a finite impedance. Adjusting the stud length produces an analogous effect to tapping the resonator coil in Fig. 2.8. Both the drain and the gate resonators usually resonate at the same frequency.

Thus, the coaxially packaged FET is properly biased. If the input microwaves are fed through the inpout coaxial cable, the center stud gives control signals to the gate and the amplified voltage appears on the center stud of the drain cavity. The amplified microwaves are coupled from the output coupling capacitor and the coaxial cable. For example, when a MSC85856 coaxial transistor is operated in the range of 2.5–3.5 GHz, the total length of the two cavities would be 5.7 cm. The center stud has dimensions of 3.8 cm cavity wall i.d. and 0.28 cm o.d. The operating output power is about 600 mW [6].

2.7 Features of Microwave Transistors

To be useful as a microwave transistor, the following conditions, among others, must be met.

1. Electron mobility must be high in the channel.
2. Thermal conductivity for cooling must be high.
3. Interelectrode capacitance must be small.
4. Lead inductance must be small.

As studied in this chapter, small size bipolar junction transistors, junction field effect transistors with n-type GaAs channels, and high electron mobility transistors meet all the conditions cited above. Most microwave transistors are inherently wide band devices. The frequency bandwidth is determined primarily by the bandwidth of the circuit with which it is to be connected.

The cutoff frequency of a microwave transistor calculated from the reciprocal of the electron transit time between the emitter and the collector, or between the source and the drain, will produce a far greater value than the actual upper limit of the operating frequency of the transistor. The upper limit of the operating frequency of a transistor is determined by:

1. electron transit time from the emitter to the collector or the source to the drain.
2. time constant of the transistor input and output circuits.

By adding positive feedback to an amplifier with $(2n + 1)\pi$ phase difference between the base and the collector microwave voltages, or the gate and the drain microwave voltages, the amplifiers turn into oscillators, where n is zero or an integral number. The simplest method of designing a feedback loop is connecting the base and the collector or the gate and the drain with a properly undercoupled $(2n + 1)\lambda_l/2$ transmission line, where λ_l is the wavelength of the transmission line. Because of the nature of practical microwave circuits, most microwave oscillators are Colpitts oscillators.

Depending on the selection of the Q-point, a microwave transistor can be highly nonlinear. If operated in this nonlinear mode, the transistors can be used as harmonic generators, mixers, demodulators, and mixer-local oscillators [23].

Many microwave transistor circuit design files are available for computer aided design using a personal computer [21,22]. Use of commercial CAD software is an efficient way of designing a specific transistor circuit if it fits the needs. Otherwise, the design must be created following an established procedure [18].

Problems

2.1 List the disadvantageous points of bipolar transistors and junction field effect transistors for operation at high microwave frequencies.

2.2 State the reason why the small size is required for microwave transistors.

2.3 Assuming that there is a dc energy of $V_D = 4$ V, $I_D = 12$ mA concentrated in the volume of 1 μm^3 of the active region, estimate the temperature rise per second. For simplicity, assume the specific heat is unity.

2.4 Sketch the potential distribution within the depletion layer. Sketch the family of curves for various values of V_{GO}.

2.5 Sketch the width of the depletion layer w–V_{GO} relationship.

2.6 Sketch the I_d–V_i curve.

2.7 Find the quantitative effect of the gate width and the gate length on the amplifier gain.
2.8 When a GaAs MESFET is under a fixed bias, point out a parameter to determine the gain and the frequency bandwidth of the FET amplifier.
2.9 State the consequence of a large electron transit time effect.
2.10 Sketch a circuit diagram of a radio frequency FET oscillator and explain how it works.
2.11 Sketch a schematic diagram of a GaAs MESFET microstrip oscillator.
2.12 Sketch a schematic diagram of a GaAs MESFET cavity oscillator.
2.13 In a microstripline resonator of one half-wavelength at 24 GHz, if the relative permittivity of the substrate is 10, what is the length of the resonator?
2.14 List the necessary requirements for high microwave frequency transistors.
2.15 Give a possible explanation for why the resistivity of GaAs is greater than the resistivity of Si.
2.16 Describe a homojunction FET and a heterojunction FET.
2.17 Describe the differences in transistor current control mechanisms among BJTs, JFETs, MESFETs, and HEMTs.
2.18 Point out a mechanism common to output signals for all types of transistors.

References

1 A. Bar-Lev, "Semiconductors and Electronic Devices." Prentice-Hall, London, 1979.
2 A. van der Ziel, "Solid-State Physical Electronics," 2nd ed. Prentice-Hall, Englewood Cliffs, New Jersey, 1976.
3 E. S. Young, "Fundamentals of Semiconductor Devices." McGraw-Hill, New York, 1978.
4 T. K. Ishii, "Microwave Engineering." Ronald Press, New York, 1966.
5 G. A. Bowman and T. K. Ishii, Linearize FM transmitters over wide bandwidth. *Microwaves* **18**(6), 66–75 (1979).
6 C. A. Liechti, Microwave field effect transistors. *IEEE Trans. Microwave Theory Tech.* **MTT-24**(6), 279–300 (1976).
7 Hewlett Packard, "1977 Diode and Transistor Designer's Catalog," pp. 6–8. Hewlett Packard, Palo Alto, California, 1977.
8 J. M. Golio, Ultimate scaling limits for high-frequency GaAs MESFET's *IEEE Trans. Electron Devices* **ED-35**(7), 839–848 (1988).
9 G.-W. Wang, Y.-K. Chen, W. J. Schaft, and L. F. Eastman, A. 0.1-μm Gate $A\ell_{0.5}In_{0.5}As/Ga_{0.5}$as MODFET fabricated on GaAs substrate. *IEEE Trans. Electron Devices* **ED-35**(7), 818–823 (1988).

10 M. Nanba, T. Shiba, T. Nakamura, and T. Toyabe, An analytical and experimental investigation of the cutoff frequency f_T of high-speed bipolar transistors. *IEEE Trans. Electron Devices* **ED-35**(7), 1021–1028 (1988).

11 B. G. Streetman, "Solid-State Electronic Devices," 2nd ed. Prentice-Hall, Englewood Cliffs, New Jersey, 1980.

12 S. Y. Liao, "Microwave Solid-State Devices." Prentice-Hall, Englwood Cliffs, New Jersey, 1985.

13 F. Emori, Y. Saito, K. Ueda, and T. Noguchi, Designing an ultra wideband monolithic distributed amplifier. *Microwave Syst. News Commun. Tech.* **18**(7), 28–38 (1988).

14 Y. H. Lo, J. Harbison, J. H. Abeles, T. P. Lee, and R. E. Nahory, High-performance GaAs MESFET's fabricated on misoriented InP substrate by heteroepitaxy. *IEEE Electron Device Lett.* **9**(8), 383–384 (1988).

15 P. Saunier, R. J. Mastyi, and K. Bradshaw, A Double-heterojunction doped-channel pseudomorphic power HEMT with a power density of 0.85 W/mm at 55 GHz. *IEEE Electron Device Lett.* **9**(8), 397–398 (1988).

16 *Microwave J.* **13**(7), 10 (1988).

17 Matcom, "Toshiba Microwave GaAs FET S8851, JS8851-AS," specification sheet. Matcom, Palo Alto, California, June 1988.

18 G. Gonsalez, "Microwave Transistor Amplifiers Analysis and Design." Prentice-Hall, Englewood Cliffs, New Jersey, 1984.

19 C. Yuen, M. Riaziat, S. Bandy, and G. Zdasiuk, Application of HEMT devices to MMICs. *Microwave J.* **31**(8), 87–104 (1988).

20 D. A. Willems and I. J. Bahl, An octave bandwidth MMIC balanced amplifier. *Microwave J.* **31**(8), 106–115 (1988).

21 EEsof, Inc., "Touchstone," "Microwave Spice," "MiCAD II," and "MIC Mask." EEsof, Inc., Westlake Village, California, August 1988.

22 Compact Software, "CADEC 4," "SUPERCOMPACT PC," "GaS STATION," "COMPLEX MATCH II," "SONANTA," and "LINMIC + ". Compact Software, Paterson, New Jersey, August 1988.

23 S. Wang and T. K. Ishii, Microwave down converter circuit with dually operated field effect transistor. *Proc. Midwest Symp. Circuit Syst., 31st, St. Louis, Missouri* pp. 254–257 (August 1988).

24 K. Miyagaki, A 20-watt L-band class A power amplifier. *rf Design* pp. 41–47 (September 1988).

3

Microwave Tunnel Diodes

3.1 Tunnel Effect in Degenerate Semiconductor Junctions

When a p-type semiconductor is heavily doped, the Fermi energy level is in the valence band. On the other hand, if an n-type semiconductor is heavily doped, the Fermi energy level is in the conduction band. These heavily doped semiconductors are called degenerate semiconductors. If a degenerate p-type semiconductor and n-type semiconductor are joined to make a junction diode, such a diode is called an Esaki diode [1]. An electron energy diagram of the Esaki diode [2–5] is sketched in Fig. 3.1. As seen from this diagram, a thin layer of forbidden energy gap is created at the junction. If the diode is slightly forward biased, raising the Fermi energy level E_F of the n-type semiconductor, electrons in the low edge of the conduction band in the n-type semiconductor tunnel through the thin forbidden gap and recombine with holes in the upper edge of the valence band of the p-type semiconductor. In this bias voltage region, the diode current is carried along mainly by this tunneling process. This is the reason the diode is also called a tunnel diode. The tunneling speed is extremely high, which is an advantage in high frequency operation.

For example, if the p-n junction is formed by conventional nondegenerate semiconductor, then the carrier concentration is on the order of 10^{16} carriers/cm^3 [5]. In the tunnel diode that is

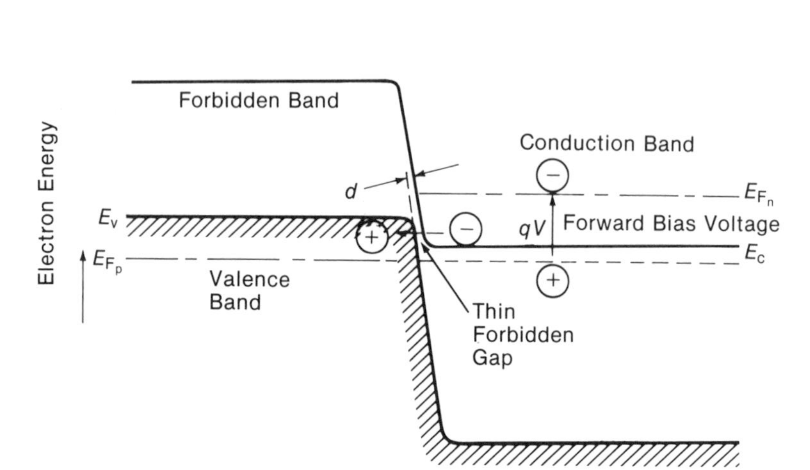

Fig. 3.1 Energy diagram of a degenerate semiconductor p–n junction.

formed by a degenerate semiconductor junction, the carrier concentration is on the order of 10^{17}–10^{20} carriers/cm^3 [11]. The forbidden band gap $E_c - E_v$ in Fig. 3.1 is known to be 1.1 eV for Si, 0.72 eV for Ge, 0.13 eV for GaAs, 0.18 eV for InSb, and 0.33 eV for InAs [5]. Depending on the carrier concentration, the relative location of the Fermi energy level $E_v - E_{F_p}$ or $E_{F_n} - E_c$ in Fig. 3.1 can be as high as $(E_c - E_v)/2$.

Consider a typical microwave tunnel diode made of a degenerate Ge semiconductor with the donor concentration $N_d = 10^{17}/\text{cm}^3$. The active p-n region has a size of 50 μm $\times$ 50 μm and a thickness of 10 μm. When forward biased to $V = 500$ mV, the thickness of the transitional forbidden band gap as shown in Fig. 3.1 becomes as narrow as $d = 0.3$ nm [11]. The electron mobility in degenerate Ge is known to be at least $\mu_n = 220$ cm^2/V-s [5]. The velocity of electrons tunneling through the narrow forbidden gap d is therefore $u = \mu_n V/d = (220 \times 10^{-4}$ m^2/V-s) (0.5 V/0.3 $\times 10^{-9}$ m $= 0.37 \times 10^8$ m/s. This is almost one tenth the velocity of light in a vacuum, and almost equal to the velocity of electrons in microwave vacuum traveling wave tubes. This tunneling speed is high compared with the electron velocity in conventional bulk semiconductor, which is on the order of 10^5 m/s.

In reality, the tunneling is done via the de Broglie wave function. Therefore, the upper bound of the tunneling speed approaches $c/\sqrt{\varepsilon_r}$ which is 3×10^8 m/s$/\sqrt{16} = 0.75 \times 10^8$ m.

The electron transit time for the narrow forbidden band gap is then $\tau = d/u = 0.3 \times 10^{-9}$ m/0.37 $\times 10^{8}$ m/s = 0.8 $\times 10^{-17}$ s = 8 μps. The possible upper limit of oscillation frequency is then $f_{max} = 1/10\tau = 1/(10 \cdot 0.08 \times 10^{-17}$ s) = 1.25 $\times 10^{16}$ Hz = 12.5 kTHz (kiloterahertz). In terms of the free space wavelength, $\lambda = c/f = (3 \times 10^{8}$ m/s)/1.25 $\times 10^{16}$ Hz = 24 nm. This is in the range of ultraviolet wavelengths [12]. Actually, this device is not capable of oscillating at 12.5 kTHz. The reason is that the tunnel diode has bulk resistance r and junction capacitance c. Before reaching f_{max}, the rc time constant will limit the operating frequency. For the Ge tunnel diode of 50 μm $\times$ 50 μm $\times$ 10 μm, the bulk resistance r is known to be on the order of 1 Ω and the junction capacitance is known to be about 2.7 pF [5]. Then the rc time constant is $rc = 1\ \Omega \times 2.7$ pF = 2.7 ps, and the frequency limit is $f_{limit} = 1/10rc$ = 37 GHz. At any rate, if the circuit is properly designed, the tunnel diode has a high frequency handling capability.

3.2 Diode Current Generated by the Tunnel Effect

Under slightly forward biased conditions, as shown in Fig. 3.1, the current is mainly created by the tunnel effect of electrons which tunnel through a thin forbidden gap at the junction. Let $f_c(E)$ be a Fermi distribution function that represents the probability of the energy level E being filled throughout the junction. $f_c(E)$ would lie near the low edge of the conduction band of the n-type semiconductor in Fig. 3.1. If $\rho_c(E)$ is the density of states per joule of energy band at the level, then $f_c(E)\rho_c(E)$ is the available number of states at the energy level E per joule of energy band for tunneling in the n-type semiconductor. Not all the electrons in these states actually tunnel through. In order to tunnel through, there must be some empty states in the valence band in the p-type semiconductor. If $f_v(E)$ represents the probability that energy level E is filled throughout the junction in the upper edge of the valence band of the p-type semiconductor, then $[1 - f_v(E)]$ is the probability that the level E is vacant and available for the electron to occupy the junction.

Therefore, if $\rho_v(E)$ is the density of states per joule of energy band at the energy level E, the number of empty states in the valence band of the p-type semiconductor is $[1 - f_v(E)]\rho_v(E)$ per joule of energy band. Therefore, the number of potentially tunnelable states from the conduction band of the n-type semiconductor to the valence band of the p-type semiconductor is $[1 - f_v(E)]\rho_v(E) \cdot f_c(E)\rho_c(E)$ per joule of energy state. If the tunneling rate is Z_{cv} electrons/s-m^2, for infinite availability, then the number of electrons actually tunneling through at energy level E is $Z_{cv}[1 - f_v(E)]\rho_v(E) \cdot f_c(E)\rho_c(E)$ electrons/s-m^2. Within a narrow energy bandwidth dE near the Fermi energy level, the number of electrons actually tunneling through per second is then $Z_{cv}[1 - f_v(E)]\rho_v(E)f_c(E)\rho_c(E)\,dE$ (electrons/s-m^2).

As seen from Fig. 3.1, the tunneling is possible within $E_c < E < E_v$, where E_c is the bottom of the conduction band of the n-type semiconductor, and E_v is the top of the valence band of the p-type semiconductor. Therefore

$$\int_{E_c}^{E_v} Z_{cv}[1 - f_v(E)]\,f_c(E)\rho_v(E)\,dE \text{ (electrons/s-m}^2\text{)}$$

is the total number of electrons which tunnel through the junction per unit of cross-sectional area. If the total cross section of the junction is A (m^2), and the magnitude of the charge of the electron is q (C), then the tunnel diode current is

$$I_d = qA\int_{E_c}^{E_v} Z_{cv}[1 - f_v(E)]\,f_c(E)\rho_v(E)\rho_c(E)\,dE \text{ (A)} \quad (3.2.1)$$

In this equation, q is known to be 1.603×10^{-19} C. A typical junction cross-sectional area for a microwave tunnel diode is $A = 50\ \mu\text{m} \times 50\ \mu\text{m} = 2500\ \mu\text{m}^2$. In Fig. 3.1, since the electronic energy level is relative, a location of zero electron-volt level can be chosen. For example, in the case of a degenerate Ge tunnel diode, if E_c is chosen to be $E_c = 0$ eV, then E_v can be about 0.03 eV when the diode has a forward bias $qV = 0.1$ eV. When the forward bias $qV = 0$ eV, as seen in Fig. 3.1, the value of $E_v - E_c$ is greater than in the case of $qV = 0.1$ eV. Since there is no forward bias, the value of Z_{cv} is zero or extremely small. In Eq. 3.2.1, the integral is basically an effective electron tunneling rate. When $qV = 0$, the integral range of E_c to E_v is large, as large as 0.15 eV, yet $Z_{cv} \approx 0$. The value of integral therefore is zero. When forward biased to $qV = 0.1$ eV, the integral range of E_c to E_v decreases to 0.03 eV but the effective tunneling rate or the value of the integral in Eq. 3.2.1 increases due to the increase in value of Z_{cv}. In a typical microwave Ge tunnel diode of junction area of 2500 μm^2, the diode current is on the order of

$I_d \approx 10^{-3}$ A [5]. Then the effective tunneling rate in this case is

$$\int_0^{0.03\text{ eV}} Z_{cv}[1 - f_v(E)]\, f_c(E)\rho_v(E)\rho_c(E)\,dE$$
$$= I_d/qA$$
$$= 10^{-3}\text{ A}/(1.602 \times 10^{-19}\text{ C})$$
$$\cdot(2500 \times 10^{-12}\text{ m}^2)$$
$$= 2.5 \times 10^{24}\text{ electrons/s-m}^2.$$

This is 2.5×10^{15} electrons/ns-m^2 or 2.5×10^{12} electrons/ps-m^2. Since the actual junction area is $A = 2500 \times 10^{-12}$ m, the actual tunneling rate across the junction is 2.5×10^{24} electrons/s-m^2 $\times$ 2500×10^{-12} m^2 $= 6.25 \times 10^9$ electrons/s. For example, in a Ge tunnel diode, when the bias $qV = 0$, $E_v - E_c \approx 0.15$ eV. When qV increases, $E_v - E_c$ decreases as seen from Fig. 3.1, but Z_{cv} increases. As a result, the tunnel current increases until $qV =$ 0.08 eV, reaching $I_d = 5$ mA [5]. Beyond this point $E_v - E_c$ gets smaller and smaller and for this reason, Z_{cv} gets smaller. Tunneling is possible only through the range of E_c and E_v.

When $qV \approx 0.15$ eV, both E_c and E_v line up, or $E_v - E_c = 0$. At this point Z_{cv} is getting small as is I_d. I_d is decreased as low as 2.5 mA [5]. If qV is further increased, then $E_c > E_v$. By the time qV reaches 0.3 eV, Z_{cv} is zero and tunneling ceases. From this point on, the transition of electrons from the conduction band of the n-type Ge to the conduction band of the p-type Ge gets started with small probability. This current created by conduction band-to-conduction band transfer is called the drift current. Therefore, with a value of $qV = 0.3$ eV, the tunnel current stops, and drift current starts to flow. At this point, I_d is as small as 1 mA [5]. As qV is increased further, the transition probability increases and the drift current increases.

When reverse biased, $E_v > E_c$ always. If $|qV|$ increases, $E_v - E_c$ increases. So does Z_{cv}. The tunnel current increases in the reverse direction as reverse bias $|qV|$ increases. Further investigation of the integrand in Eq. 3.2.1 is presented in the next section.

3.3 Concept of Negative Resistance

In Eq. 3.2.1, the energy level of a tunneling electron, E, is $E_c \le E \le E_v$, as seen from Fig. 3.1. When the diode is forward biased V volts, raising E_{F_n} with respect to E_{F_p} decreases the value of $(E_v - E_c)$. At a certain bias voltage $V = V_v$, E_c becomes equal to E_v, and the integral Eq. 3.2.1, or the tunnel current, becomes zero. Therefore $(E_v - E_c)$ is determined by the bias voltage under the biased condition, and the diode current I_d is also determined by the bias voltage. In the tunnel diode studied in Section 3.2, the tunnel current is zero at $qV_v = 0.3$ eV. This suggests that $(E_v - E_{F_p}) + (E_{F_n} - E_c) = 0.3$ eV.

It is an accepted assumption that

$$\rho_v(E) = k_v(E_v - E)^{1/2} \tag{3.3.1}$$

$$\rho_c(E) = k_c(E - E_c)^{1/2} \tag{3.3.2}$$

$$f_c(E) \simeq 1/2 - \frac{E - E_{F_n}}{4kT} \tag{3.3.3}$$

$$f_v(E) \simeq 1/2 + \frac{E_{F_p} - E}{4kT} \tag{3.3.4}$$

$$Z_{cv} = k_{cv}|v|^{1/2} \tag{3.3.5}$$

where k_v, k_c, and k_{cv} are proportionality constants, and k is the Boltzmann constant [5]. Substituting Eq. 3.3.1 through 3.3.5 in Eq. 3.2.1,

$$\begin{aligned} I_d &= qAk_ck_vk_{cv}|V|^{1/2}\int_{E_c}^{E_v}\left(1/2 - \frac{E - E_{F_p}}{4kT}\right)\left(1/2 - \frac{E - E_{F_n}}{4kT}\right) \\ &\quad \times (E_v - E)^{1/2}(E - E_c)^{1/2}\,dE \\ &= \frac{qAk_ck_vk_{cv}|V|^{1/2}}{4}\int_{E_c}^{E_v}\left(1 - \frac{E - E_{F_p}}{2kT}\right)\left(1 - \frac{E - E_{F_n}}{2kT}\right) \\ &\quad \times (E_v - E)^{1/2}(E - E_c)^{1/2}\,dE \end{aligned} \tag{3.3.6}$$

Under the forward biased condition in Fig. 3.1, noting that V is negative,

$$E_v = E_c = \left(E_{F_n} - E_c\right) + qV + \left(E_v - E_{F_p}\right). \tag{3.3.7}$$

Note in this equation that when $qV = 0$, $E_{F_n} = E_{F_p}$ and $E_v - E_c = (E_{F_n} - E_c) + (E_v - E_{F_p})$ is verified. The value of $(E_v - E_c)$ decreases when qV is negative or forward biased. The value of $(E_v - E_c)$ increases when qV is positive or reverse biased. Let the constant values be defined as follows:

$$E_{F_n} - E_c = \Delta E_n \tag{3.3.8}$$

$$E_v - E_{F_p} = \Delta E_p \tag{3.3.9}$$

Under the forward biased conditions from Eq. 3.3.7,

$$E_v = E_c + \Delta E_n + \Delta E_p + qV \tag{3.3.10}$$

with Eq. 3.3.6 and Eq. 3.3.10,

$$(E_c + \Delta E_n + \Delta E_p + qV)$$

$$I_d = \frac{qA\mathrm{k}_c\mathrm{k}_v\mathrm{k}_{cv}|V|^{1/2}}{4} \int_{E_c}^{E_v} \left(1 - \frac{E - E_{F_p}}{2\mathrm{k}T}\right)\left(1 - \frac{E - E_{F_n}}{2\mathrm{k}T}\right)$$

$$\times (E_v - E)^{1/2}(E - E_c)^{1/2}\,\mathrm{d}E \tag{3.3.11}$$

If Eq. 3.3.11 is numerically integrated for a specific tunnel diode, then the current–voltage characteristics, or the relation between the diode current and the bias voltage, is generally found to be a curve represented by a solid line shown in Fig. 3.2. When the bias voltage is zero, the tunneling rate is zero, and so is the diode current. At $V = V_v$, $E_c = E_v$, the integral of Eq. 3.2.1 will be zero, as is the diode current. Beyond $V \geq V_v$, Eq. 3.2.1 loses its meaning and does not apply. Beyond this point, there will be no more tunnel current and the drift current due to electron

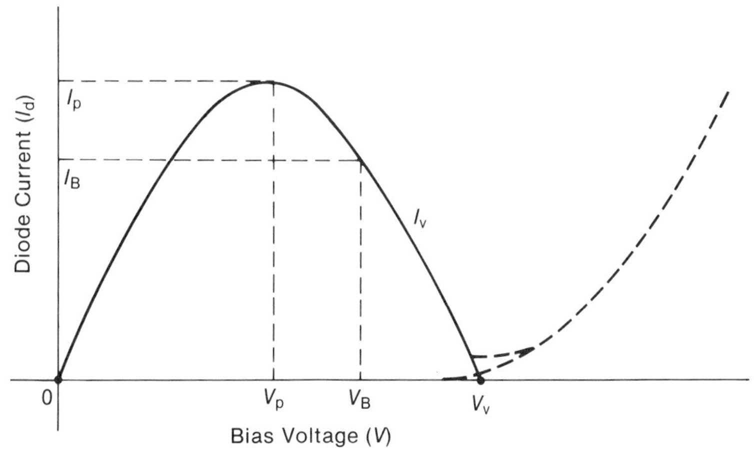

Fig. 3.2 Current–voltage characteristics of a tunnel diode. The tunnel current I_v is indicated by the solid curve. The drift current is indicated by the dashed curve.

transition from the conduction band of the n-type semiconductor to the conduction band of the p-type semiconductor takes over. This situation is indicated in Fig. 3.2 by a dashed line. The Ge tunnel diode chip of 50 μm $\times$ 50 μm is studied in Section 3.2, $V_v = 0.3$ V with $I_d = 1$ mA. The peak current of $I_d = 6$ mA is reached with the forward bias voltage $V_p = 0.08$ V [5].

When the bias voltage is reversed, as seen in Fig. 3.2 and Eq. 3.3.6, the value of $E_v - E_c$ gets larger than qV gets larger as seen in Fig. 3.2. Therefore the tunnel current increases in reverse direction as the reverse bias qV increases.

Between the two zeros of the tunnel current, the tunnel current reaches its peak at $V = V_p$. In a range of $V_p < V < V_v$,

$$\frac{\mathrm{dI_d}}{\mathrm{d}V} < 0 \tag{3.3.12}$$

If the differential conductance is represented by g_t, then

$$g_t = \frac{\mathrm{d}I_d}{\mathrm{d}V} < 0 \tag{3.3.13}$$

When a tunnel diode is operated as an amplifier or an oscillator, the bias voltage is always

$$V_p < V < V_v \tag{3.3.14}$$

and the differential conductance of the diode is always negative. When a microwave signal is superimposed on the dc bias voltage V, the bias voltage is perturbed by the microwave voltage ΔV, and the diode current is perturbed by ΔI. Therefore the microwave signals interact with the differential negative conductance $g_t = \Delta I/\Delta V$, which is negative. At this point, it should be noted that the static conductance $G = I/V$ is always positive no matter what the voltage is. For example, in the Ge tunnel diode presented earlier, the maximum $g_t = \Delta I/\Delta V$ occurred at the bias voltage $V_B = 0.12$ V with the bias current $I_B = 2.5$ mA [5]. At this point, $g_t = \Delta I/\Delta V = -3$ mA/0.30 V $= -100$ mS and $G_t = I_B/V_B = 2.5$ mA/0.12 V $= 21$ mS.

3.4 Dynamic Negative Conductance

The differential conductance g_t contrasts with the static conductance G_t. The differential conductance g_t is termed the dynamic negative conductance with respect to microwave signals for the bias voltage within a range of $V_p < V < V_v$. After being properly

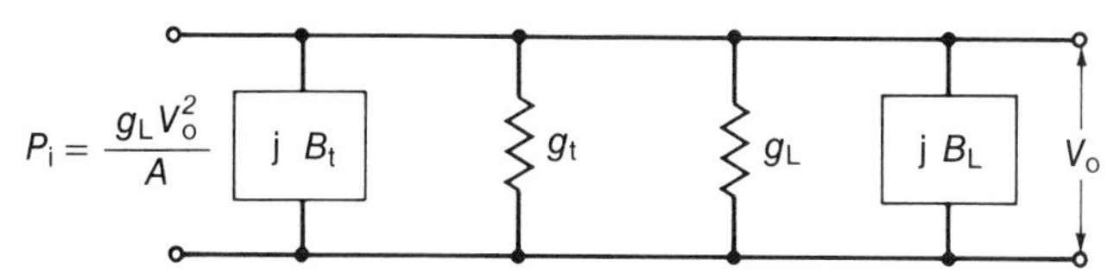

Fig. 3.3 Equivalent circuit of a tunnel diode amplifier operating at microwave frequency.

biased, only g_t becomes important for interaction with microwave signals. The microwave load is connected in parallel to the tunnel diode as shown in Fig. 3.3. Usually both the tunnel diode and the load are not purely conductive. There are some susceptances: B_t for the diode and B_L for the load. For best results, B_L is usually adjusted to tune out B_t.

$$\mathrm{B_L} + \mathrm{B_t} = 0 \quad \text{or} \quad \mathrm{B_L} = -\mathrm{B_t} \tag{3.4.1}$$

Under this condition, the law of conservation of energy at equilibrium requires that

$$|g_t|V_o^2 + \frac{1}{A} g_L V_o^2 = g_L V_o^2 \tag{3.4.2}$$

where V_o is the microwave voltage across the diode as shown in Fig. 3.3. The first term is the power generated by the negative conductance g_t. Since g_t is negative, it generates microwave power instead of consuming power [6]. The term $g_L V_o^2$ is the microwave output power to the load g_L. Then $g_L V_o^2/A$, where A is the power gain, is the input power to the amplifier. Eq. 3.4.2 states that, after tuning out the susceptances, the sum of the power generated by the dynamic negative conductance g_t and the input microwave power is equal to the microwave power consumed by the load conductance g_L. Then, solving Eq. 3.4.2 for the power gain A,

$$A = \frac{g_L}{g_L - |g_t|} \tag{3.4.3}$$

It should be noted that g_t is a negative quantity. If g_L is adjusted so that

$$g_L - |g_t| = 0 \tag{3.4.4}$$

with the condition from Eq. 3.4.1, then

$$A \to \infty \tag{3.4.5}$$

This is an oscillator.

In practical tunnel diode oscillators, both Eqs. 3.4.1 and 3.4.4 are complicated functions of the operating frequency. They are usually transcendental equations of the operating frequency ω.

The simultaneous solution of both Eqs. 3.4.1 and 3.4.4 for the operating frequency is the oscillation frequency [7,10].

For example, with regard to the microwave Ge tunnel diode studied in Section 3.3, $g_t = -100$ mS. For oscillation, the load conductance $g_L = -g_t = 100$ mS, or the shunt load resistance across the diode is 10 Ω. The tunnel diode is generally a low impedance device. In Eq. 3.4.4, if $g_L - |g_t| = 1$ mS, then from Eq. 3.4.3, the power gain $A = g_L/(g_L - |g_t|) = 100$ mS/1 mS $=$ 100 or 20 dB.

Generally, a microwave tunnel diode chip is packaged. If an equivalent tunnel diode capacitance, including the package capacitance and the junction capacitance, is $C_t = 5$ pF, then the admittance at 10 GHz is $B_t = \omega C_t = 2\pi \times 10 \times 10^9 \times 5 \times 10^{-12} =$ 0.314 S. So, to meet one of the oscillation conditions in Eq. 3.4.1, the load susceptance B_L must be inductive. $B_L = 1/\omega L_L = B_t = \omega C_t = 0.314$ S. Then $L_L = 1/(2\pi f B_t) = 1/(2\pi \times 10 \times 10^9 \times 0.314) = 51 \times 10^{-12}$ H $= 51$ pH. This small shunt inductance is difficult, if not impossible, to obtain in a lumped circuit component. However, when the diode is mounted on a distribution parameter circuit such as a waveguide, a coaxial line, or a microstripline, realization of small equivalent inductance is not so difficult [6,7].

For example, if the packaged tunnel diode is mounted on a microstripline of characteristic impedance $Z_o = 50$ ohms, the characteristic admittance is then $Y_o = 1/Z_o = 1/50 = 0.02$ S. The normalized susceptance $\tilde{B}_L$ to be realized is therefore $\tilde{B}_L \equiv B_L/Y_o = 0.314/0.02 = 15.7$. If the open microstripline has a length l, then $\tilde{B}_L = \tan(2\pi l/\lambda_l)$ or $(l/\lambda_l) = (1/2\pi)\tan^{-1} \tilde{B}_L =$ $(1/2\pi)\tan^{-1} 15.7 = 0.24$. This means that the package capacitance of the tunnel diode can be tuned out by attaching a 0.24 wavelength microstripline. At 10 GHz, the free space wavelength is 3 cm. So 0.24 wavelength is 7.2 mm in free space. If the microstripline is formed on a ceramic substrate of $\varepsilon_r = 4$, then the effective transmission line wavelength gets shorter and $\lambda_l \approx \lambda_o/\sqrt{\varepsilon_r} = \lambda_o/2 = 15$ mm. This means that 0.24 wavelength is 3.6 mm. At any rate, it is not difficult to realize one of the oscillation conditions in Eq. 3.4.1 using a distribution parameter approach.

3.5 Diode Packaging and Equivalent Circuit

Tunnel diodes operating at microwave frequencies are packaged in various configurations [13]. Only two representative configurations

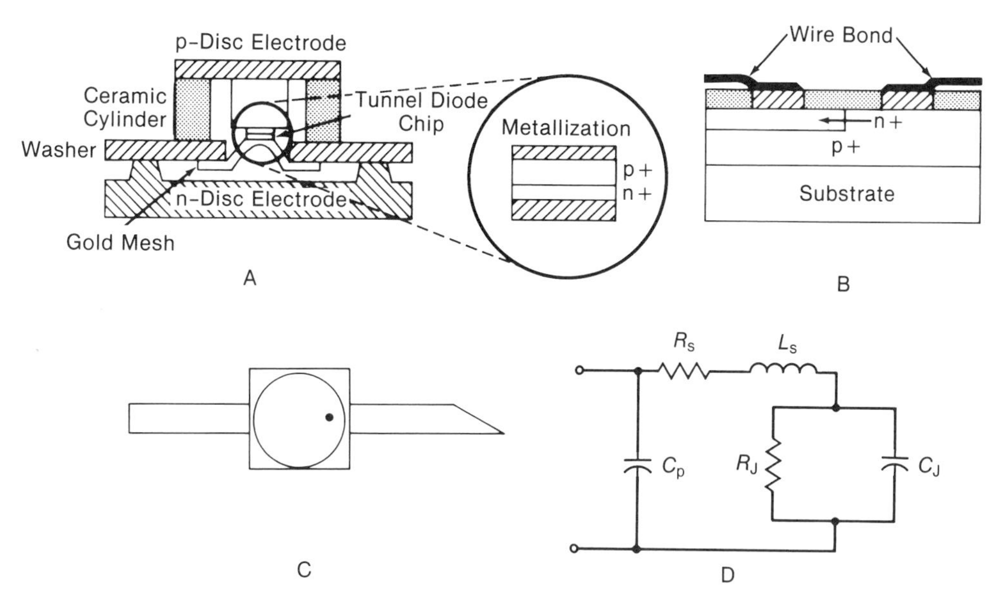

Fig. 3.4 Packaged tunnel diodes and equivalent circuit. A. Pill type packaging. B. Planar chip configuration. C. Drop-in type package. D. Equivalent circuit.

are shown in Fig. 3.4. These configurations are pill type packaging, Fig. 3.4 A, and the drop-in type packaging, Fig. 3.4 C. The pill type packages are used primarily for the cavity, waveguide, coaxial line, and microstripline applications. The drop-in type packages are for microstriplines. In this chapter, the packaged tunnel diodes are discussed, but the same configurations are used for various types of microwave diodes which will be presented in the following chapters. The microwave diodes include Gunn diodes,IMPATT diodes, Schottky barrier diodes, and PIN diodes.

A cross-sectional view of a pill type packaged tunnel diode is shown in Fig. 3.4 A. A tunnel diode chip of approximately 50 μm diameter and 10 μm thick is mounted between the upside-down metallic post and a gold wire mesh. The disk-shaped p^+–n^+ junction tunnel diode wafer is enlarged in Fig. 3.4 A. The gold wire mesh is bonded to the n^+ side of the tunnel diode chip. The wire mesh protects the diode chip from excessive mechanical pressure and accidental shock. The gold wire mesh is bonded on a metallic washer such as Kovar, and the washer is bonded to the bottom metallic disk. The top disk–post metallic structure is therefore for the p-side of the diode and the bottom metallic disk is for the n-side of the diode. The p-electrode and the n-washer are separated by an insulating ceramic cylinder as shown in Fig. 3.4 A.

Though the diode chip can be as small as 50 μm in diameter and 10 μm thick, the pill type packaged tunnel diode can be 1300–3230 μm in diameter and 1020–2460 μm in height [13].

The diode chip can be a planar structure as shown in Fig. 3.4 B. This type of chip can also be 50 μm $\times$ 50 μm and 10 μm thick, yet the active part can be as small as 10 μm $\times$ 10 μm.

Both the disk type diode chip and the planar diode chip can be packaged in a drop-in type package as shown in Fig. 3.4 C. The dotted side of the lead microstrip is the anode. This type of package can be 2390–2590 μm in diameter and 790–1120 μm thick [13].

As seen from these diode chip mounting configurations, the packaging is designed to minimize stray inductance and capacitance so that the diodes can be used at high microwave frequencies. A widely accepted equivalent circuit of a packaged microwave diode is shown in Fig. 3.4 D. In this figure, R_s is the series resistance which includes the resistance of the bulk parts of the semiconductor (also called the spreading resistance of the semiconductor) and the ohmic resistance of the metallic leading conductor in series with the diode chip. L_S is the lead inductance. R_J is the junction resistance and C_J is the junction capacitance. C_P is the package capacitance. Theoretically L_S, which is the inductance associated with the packaging of the diode chip, is in series with R_S and should be considered. Yet in reality, the inductive reactance associated with L_S is usually negligibly small in comparison with the rest of the impedance. It is therefore often omitted.

In most activated tunnel diodes, R_S is on the order of 1 Ω, L_S is on the order of 10^{-12} H, R_J is the differential negative resistance on the order of 10 Ω, C_J is on the order of 10^{-1} pF, and C_P is on the order of 1 pF. Tunnel diodes are generally considered low impedance devices.

The input impedance of the equivalent circuit of a microwave diode as shown in Fig. 3.4 D is

$$\dot{Z} = \frac{\mathrm{R}_T + j\{X_T(1 - \omega C_P X_T) - \omega C_P R_T^2\}}{(1 - \omega C_P X_T)^2 + \omega^2 C_P R_T^2} \tag{3.5.1}$$

where R_T and X_T are the resistance and reactance of the packaged diode without C_P [7–10]. Also,

$$R_T = R_S - \frac{-R_J}{1 + \omega^2 C_J^2 R_J^2} \tag{3.5.2}$$

and

$$X_T = \omega L_S - \frac{\omega C_J R_J^2}{1 + \omega^2 C_J^2 R_J^2} \tag{3.5.3}$$

The frequency at which the real part of $\dot{Z}$ becomes zero is called the resistive cutoff frequency, $f_{sc} = \omega_{sc}/2\pi$. When $\omega \leq \omega_{sc}$, then $Re\,\dot{Z} \leq 0$. The packaged tunnel diode is a negative resistance device. When $\omega \geq \omega_{sc}$, then $Re\,\dot{Z} \geq 0$. The packaged tunnel diode is no longer a negative resistance device in these frequencies.

So, letting $Re\,\dot{Z} = 0$ and solving for ω,

$$\omega_{sc} = \frac{1}{C_J(-R_J)} \sqrt{\frac{(-R_J)}{R_S} - 1} \tag{3.5.4}$$

Note that $R_J < 0$ in an activated tunnel diode.

It is of interest to know that both C_P, the package capacitance, and L_S, the lead inductance, do not contribute to the resistive cutoff frequency. For example, a typical 10 GHz range tunnel diode [7–10] has $-R_J = 65\ \Omega$, $L_S = 25 \times 10^{-12}$ H, $C_J = 0.9 \times 10^{-12}$ F and $R_S = 3.5\ \Omega$. Substituting these parameters in Eq. 3.5.4, the resistive cutoff frequency $f_{sc} = \omega_{sc}/2\pi = 11.35$ GHz. Even if the tunnel diode has a potentially high operating frequency, as calculated from the tunneling electron transit time across the junction, the reality is that the packaging and bulk semiconductor parameters R_S and C_J control the resistive cutoff frequency together with R_J.

A frequency at which the imaginary part of $\dot{Z}$ becomes zero is termed the self-resonance frequency of the diode. In Eq. 3.5.1, solving $Im\,\dot{Z} = 0$ for ω, it is found that

$$\omega_{sr} = \frac{X_T}{C_p(R_T^2 + X_T^2)} \tag{3.5.5}$$

This is actually a transcendental equation because both X_T and R_T are functions of ω_{sr}. Nonetheless, it clearly shows that the larger the C_P, the lower the ω_{sr}.

If C_P can be neglected, then solving Eq. 3.5.3 for ω by letting $X_T = 0$ [7–10],

$$\omega_{sr} = \frac{1}{C_J(-R_J)} \sqrt{\frac{C_J(-R_J)}{L_S} - 1} \tag{3.5.6}$$

For a typical 10 GHz tunnel diode of $-R_J = 65\ \Omega$, $L_S = 25$ pH and $C_J = 0.9$ pF, the self-resonance frequency without package capacitance is found to be $f_{sr} = \omega_{sr}/2\pi = 19.79$ GHz.

On the other hand, when an input admittance of the diode equivalent circuit as shown in Fig. 3.4 A is considered,

$$\dot{Y} = \frac{1}{\dot{Z}} = \frac{R_T}{R_T^2 + X_T^2} + j\left(\omega C_p - \frac{X_T}{R_T^2 + Y_T^2}\right) \qquad (3.5.7)$$

The condition of $Re\ \dot{Y} = 0$ (when the input conductance is zero) is fulfilled by $R_T = 0$. Therefore, the resistive cutoff frequency Eq. 3.5.4 also represents the conductive cutoff frequency. The condition $Im\ \dot{Y} = 0$ results in the same conclusion as shown in Eqs. 3.5.5 and 3.5.6. The susceptive resonance frequency is therefore equal to the reactive self-resonance frequency Eq. 3.5.6.

3.6 Microstripline Circuit

A schematic diagram of a microwave tunnel diode amplifier–oscillator using a microstripline configuration is shown in Fig. 3.5. A pill type packaged tunnel diode is mounted at a point one half-wavelength away from an open end of the stripline. At this point, transmission line theory states that the impedance of the stripline itself is infinity when viewed toward the open end. The dc bias to the diode is brought in through a bias feeder line of one quarter-wavelength, which is connected at a point one quarter-wavelength from the open end on the main microstripline. At this point, the impedance is zero. The quarter-wavelength dc bias

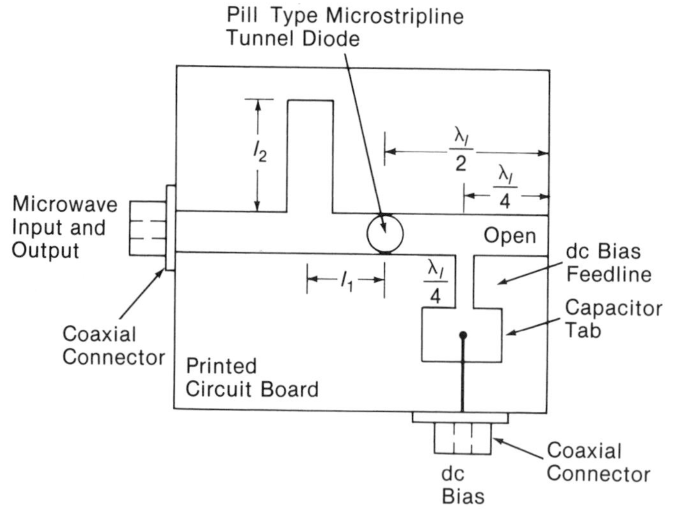

Fig. 3.5 Schematic diagram of a microwave microstrip circuit for pill type tunnel diodes.

feeder line acts as a radio frequency choke along with the capacitor tab, which presents a capacitive short to the ground plane and a bonding tab for the biasing lead wire.

Microwave signals are fed from the coaxial input to the microstripline at left in Fig. 3.5. The line length l_1 and l_2 are adjusted to satisfy the impedance matching as indicated in both Eqs. 3.4.1 and 3.4.2 [6]. The incident signals are amplified and taken out from the same place as entered. The negative resistance amplifier is essentially a one-port amplifier. The input port and the output port are identical. Both l_1 and l_2 may be adjusted to satisfy Eq. 3.4.4 and Eq. 3.4.1. Then it functions as an oscillator.

Since the diode is the negative impedance device, a negative impedance $\dot{Z}_N$ is presented at the junction of the collinear microstripline and the impedance-matching strip of length l_2 in Fig. 3.5. If the characteristic impedance of the collinear microstripline is Z_o, the voltage reflection coefficient $\dot{\rho}$ at the junction is, according to transmission line theory, given by

$$\dot{\rho} = \frac{\dot{Z}_L - Z_o}{\dot{Z}_L + Z_o} \tag{3.6.1}$$

Note that $\dot{Z}_L$ is negative [6,10]. The voltage reflection coefficient $\dot{\rho}$ is, by definition, the ratio of reflected voltage to the incident voltage. In this case, the reflected voltage is the output voltage $\dot{V}_o$, and the incident voltage is the input voltage $\dot{V}_i$. Therefore,

$$\dot{\rho} = \frac{\dot{V}_o}{\dot{V}_i} = \frac{\dot{Z}_L - Z_o}{\dot{Z}_L + Z_o} \tag{3.6.2}$$

If the voltage gain of the amplifier is $\dot{A}_v$, then, by definition $\dot{A}_v \equiv \dot{V}_o/\dot{V}_i$ and

$$\dot{A}_v = \frac{\dot{Z}_L - Z_o}{\dot{Z}_L + Z_o} \tag{3.6.3}$$

By adjusting l_1 and l_2 properly, it is possible to tune so that $\dot{Z}_L$ is purely real. Then

$$\dot{Z}_L = R_L \tag{3.6.4}$$

and

$$A_v = \frac{R_L - Z_o}{R_L + Z_o} \tag{3.6.5}$$

For example, it is common that $Z_o = 50\ \Omega$ for microstriplines. By selecting the diode and adjusting l_1, and l_2, R_L can be $-51\ \Omega$. Then $A_v = (-51 - 50)/(-51 + 50) = 101 = 20 \log_{10} 101$ dB $\approx$ 40 dB. If $R_N = -50\ \Omega$, then $A_v = \infty$. This is the condition for oscillation.

If Eq. 3.6.5 is solved for R_L, then

$$R_L = -Z_o \frac{A_v + 1}{A_v - 1} \tag{3.6.6}$$

For example, if $A_v = 10 = 20 \log_{10} 10$ dB $= 20$ dB is desired, $R_L = -50(10 + 1)/(10 - 1) = -61.1\ \Omega$ for negative resistance must be created at the input of the amplifier.

As seen from Fig. 3.5, the size of the entire circuit board is at most $\lambda_l/2 \times \lambda_l/2$. For example, at 10 GHz, the free space wavelength is $\lambda_o = c/f = 3 \times 10^8$ m/s$/10 \times 10^9$ Hz $= 30$ mm. For example, if an alumina or a ceramic substrate is used for the circuit board, the relative permittivity of the commercial substrate is $\varepsilon_r = 4.5$–10 [15]. Then the wavelength in the dielectric shrinks to $\lambda_l = \lambda_o/\sqrt{\varepsilon_r}$. If $\varepsilon_r = 4.5$ then $\lambda_l = \lambda_o/\sqrt{4.5} = \lambda_o/2.12$. If $\varepsilon_r = 10$, then $\lambda_l = \lambda_o/\sqrt{10} = \lambda_o/3.16$. Therefore, for $f = 10$ GHz and $\lambda_o = 30$ mm, $\lambda_l = 14.15$ mm for $\varepsilon_r = 4.5$ substrates and $\lambda_l = 9.49$ mm for $\varepsilon_r = 10$ substates. The size of the entire circuit board is then 21.75 mm $\times$ 21.75 mm for $\varepsilon_r = 4.5$ substrates and 14.24 mm $\times$ 14.24 mm for $\varepsilon_r = 10$ substrates. The thickness of the commercial circuit board is 250–1250 μm [14]. A similar circuit can be monolithically developed on a Si or GaAs substrate.

The negative resistance diode amplifier described above can be made into an oscillator simply by changing the bias voltage to the diode. If the diode is biased so that $R_N = Z_o$, then, as seen from Eq. 3.6.5, $A_v \rightarrow \infty$ and the circuit becomes an oscillator.

3.7 Waveguide Circuit

A waveguide circuit utilizing a microwave tunnel diode package is schematically sketched in Fig. 3.6. The impedance is matched to satisfy Eq. 3.4.1, 3.4.2, or 3.4.4 by adjusting distance l_1, l_2, and the screw penetration depth into the waveguide. The input microwave signals are fed into the waveguide opening, and the amplified output is taken out from the same port. The dc bias supply is fed through a coaxial connector. The design and operating principles of this waveguide amplifier are exactly the same as the microstrip version described in Section 3.6. The amplifier circuit can be used as an oscillator simply by changing the bias voltage [15]. For

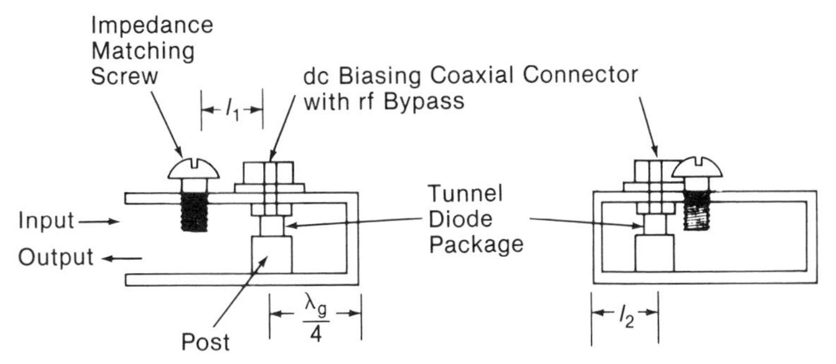

Fig. 3.6 Diagram of a waveguide mounted tunnel diode microwave amplifier–oscillator. Side view (*left*) and front view (*right*).

example, in Fig. 3.6 at 10 GHz, the internal dimension of the rectangular waveguide cross section is $a \times b = 22.9$ mm $\times$ 11.45 mm, where a is the width and b is the height of the waveguide. The waveguide wavelength λ_g is calculated to be

$$\lambda_g = \frac{\lambda_o}{\sqrt{1 - \left(\frac{\lambda_o}{2a}\right)^2}} \tag{3.7.1}$$

where λ_o is the free space wavelength. For $f = 10$ GHz, $\lambda_o = c/f = (3 \times 10^8 \text{ m/s})/10 \times 10^9$ Hz $= 3$ cm, substituting $\lambda_o = 3$ cm and $a = 2.29$ cm in Eq. 3.7.1, then $\lambda_g = 4$ cm. The length of the waveguide structure is at most one waveguide wavelength, 4 cm in this case. According to waveguide theory, the waveguide impedance is several hundred Ohms [6,20]. The tunnel diode impedance is on the order of several ten Ohms. Also, the low impedance point in a rectangular waveguide is located at an off-center position as shown in Fig. 3.6. The posts mounted at this location reduce the impedance further. At this point, a pill type packaged tunnel diode is mounted. The input impedance of the tunnel diode mounted waveguide is adjusted by the penetration length of the tuning screw protruding into the waveguide as shown in Fig. 3.6.

3.8 Tunnel Effect Microwave Amplification, Oscillation, Detection, Mixing, and Harmonic Generation

Tunnel diodes are made of Ge, Si, or GaAs. Usually V_v is less than 1.0 V. Both low voltage and low noise operations are possible. Because of low working voltages, high power operations cannot be expected. The transit time of the tunnel effect is much less than the electron transit time due to the drifting or diffusing of electrons. Therefore the high frequency limit of a tunnel diode itself exceeds comparable electron devices based on drift electrons, yet the associated circuit parameters prevent the realization of high frequency goals [7]. In summary, the tunnel effect creates negative conductance and the negative conductance amplifies microwaves. When the negative impedance of the diode equals the positive impedance of the circuit in magnitude, the diode goes into oscillation.

As seen in Fig. 3.2, the current–voltage curves of tunnel diodes are, in general, highly nonlinear [16–20]. Nonlinear devices are useful for microwave detection (Chapter 7), mixing (Chapter 8), and harmonic generation (Chapter 10) [21].

Problems

3.1 Define degenerate semiconductors.
3.2 Explain why the Fermi energy level exists within the conduction band of the n-type degenerate semiconductor.
3.3 Define the Fermi energy level.
3.4 Explain why the Fermi energy level exists within the valence band of p-type degenerate semiconductors.
3.5 Explain the energy band theory of semiconductor electronics.
3.6 Explain the tunnel effect.
3.7 Sketch a band structure of a tunnel diode.
3.8 Sketch a band structure of a nondegenerate p–n junction diode.
3.9 Explain why the fast process of tunneling is advantageous to microwave frequency operation.
3.10 Express the available number of electrons in n-type semiconductors in a tunnel diode in terms of the Fermi distribution function and the density of states.
3.11 Express the number of available states in the p-type semiconductor of a tunnel diode in terms of the Fermi distribution function and the density of states.
3.12 Express the number of vacant states in the valence bands of p-type semiconductors for a degenerate p–n junction.
3.13 Assuming that electrons are 100% available in an n-type semiconductor and vacant states in a p-type semiconductor

are also 100% available in a forward biased degenerate p–n junction, what determines the number of electrons in transit from the n-type to the p-type?

3.14 Derive an integral equation for the tunnel diode current.

3.15 Sketch the energy band diagram of a degenerate p–n junction as the forward bias gradually increases.

3.16 Sketch the energy band diagram of a degenerate p–n junction with the condition that there is zero diode current, in spite of a finite forward bias voltage V.

3.17 Explain why the forward bias of the p–n junction raises the Fermi energy level of an n-type semiconductor.

3.18 Define static negative conductance and dynamic negative conductance.

3.19 When a lead inductance L and a bulk resistance r are in series with a parallel combination of the junction conductance g_t and the junction capacitance C_J, the packaged tunnel diode's total conductance g_t' becomes a function of frequency. Obtain the expression for g_t' as a function of the operating frequency ω.

3.20 The packaged tunnel diode conductance g_t' in Problem 3.19 becomes zero at high frequency. This frequency is called the resistive cutoff frequency. Obtain an expression for the resistive cutoff frequency.

3.21 Differentiate the condition for the resistive cutoff frequency and the self-oscillation condition.

3.22 When the susceptance of a packaged tunnel diode becomes zero at a certain frequency, the frequency is called the self-resonance frequency. Differentiate the self-resonance frequency from the self-oscillation frequency stated in Problem 3.21.

3.23 In practice, the amplitude of the oscillation reaches a certain magnitude of equilibrium. For a microwave tunnel diode, list what determines the equilibrium amplitude of oscillation.

3.24 Applying transmission line theory, find the susceptance created by a piece of microstripline of length l.

References

1 L. Esaki, New phenomenon in narrow PN junctions. *Phys. Rev.* **109**, 602–603 (June 1958).

2 A. Bar-Lev, "Semiconductors and Electronic Devices." Prentice-Hall, London, 1979.

3 R. F. Soohoo, "Microwave Electronics." Addison-Wesley, Reading, Massachusetts, 1971.

4 A. van der Ziel, "Solid-State Physical Electronics," 2nd ed. Prentice-Hall, Englewood Cliffs, New Jersey, 1976.
5 K. K. N. Chang, "Parametric and Tunnel Diodes." Prentice-Hall, Englweood Cliffs, New Jersey, 1964.
6 T. K. Ishii, "Microwave Engineering." Ronald Press, New York, 1966.
7 C. C. Hoffins and K. Ishii, Conditions of oscillation for waveguide mounted tunnel diodes. *IEEE Trans. Microwave Theory Techniques* **MTT-12**(2), 176–183 (1964).
8 S. V. Jaskolski and K. Ishii, Millimeter wave generation employing a packaged microwave tunnel diode. *West. Electron. Show Conv. Rec.* **6.4** (1964).
9 K. Ishii, Theory of packaged millimeter-wave tunnel diode circuit. *Int. Conf. Microwave Circuit Theory Inf. Theory, Tokyo. Summary of Papers, Part 1: Microwaves* (1964).
10 T. K. Ishii, Equivalent circuit parameters of microwave tunnel diodes. *Proc. Colloq. Microwave Commun., 3rd Budapest, April 19–22, 1966* pp. 723–729 (1968).
11 S. Y. Liao, "Microwave Solid-State Devices." Prentice-Hall, Englewood Cliffs, New Jersey, 1985.
12 T. S. Laverghetta, "Practical Microwaves." Sams & Co., Indianapolis, Indiana, 1984.
13 For example, M/A-COM Semiconductor Product Operation, "Packaged Silicon PIN Diodes," Bull. 4325 and 4326. Burlington, Massachusetts, 1988.
14 Trans-Tech, "Trans-Strates Microwave Substrate." Trans-Tech, Adamstown, Maryland, 1988.
15 M. E. Hines, High frequency negative resistance circuit principles for Esaki diode applications. *Bell Syst. Tech. J.* **39**, 477–513 (May 1960).
16 J. M. Gering, T. J. Rudnick, and P. D. Coleman, Microwave detection using the resonant tunneling diode. *IEEE Trans. Microwave Theory Techniques* **MTT-36**(7), 1145–1150 (1988).
17 T. Tanoue, H. Mizuta, and S. Takahashi, A triple-well resonant tunneling diode for multiple-valued logic application. *IEEE Electron Device Lett.* **9**(8), 365–367 (1988).
18 S. Sen, F. Capasso, D. Sivco, and A. Y. Cho, New resonant-tunneling devices with multiple negative resistance regions and high room temperature peak-to-valley ratio. *IEEE Electron Device Lett.* **9**(8), 402–404 (1988).
19 N. C. Kluksdahl, A. M. Kriman, D. K. Ferry, and C. Ringhofer, Transient switching behavior of the resonant-tunneling diode. *IEEE Electron Device Lett.* **9**(9), 457–459 (1988).
20 S. Y. Liao, "Microwave Devices and Circuits," 2nd ed. Prentice-Hall, Englewood Cliffs, New Jersey, 1985.
21 FEI Microwave, Inc., Planar tunnel diode detectors with high power handling capability. *Microwaves RF* **27**(9), 13 (1988).

4

Microwave Avalanche Diodes

4.1 Avalanche Effect

In a depletion region of a reverse biased p–n junction, there will be a high field due to a high voltage drop across the depletion layer. Since there are not many charge carriers in the carrier depletion region, the resistivity of this region is higher than in the bulk portions of the semiconductors. Most of the applied reverse voltage will appear across the depletion region. When the reverse bias voltage increases to a certain point, the speed of residual electrons in the depletion region produces enough kinetic energy to ionize a neutral atom, thus producing a pair of electrons and a hole. The newly generated electron will be accelerated and collide with another neutral atom to produce another electron–hole pair. This process repeats itself rapidly, and a large number of carriers is suddenly created when the reverse bias voltage reaches a certain voltage, which is determined by the ionization potential of the atoms of the semiconductor material. This rapid production of electron–hole pairs by multiple impact ionization is termed the avalanche effect. The production and multiplication of carrier electrons in the depletion region resembles an avalanche in the snow covered mountains.

When the avalanche effect takes place, there will be a sudden burst of reverse current as shown in Fig. 4.1 A for a certain reverse bias voltage V_B. This voltage is termed the avalanche breakdown voltage. Depending on the ionization multiplication factor of the

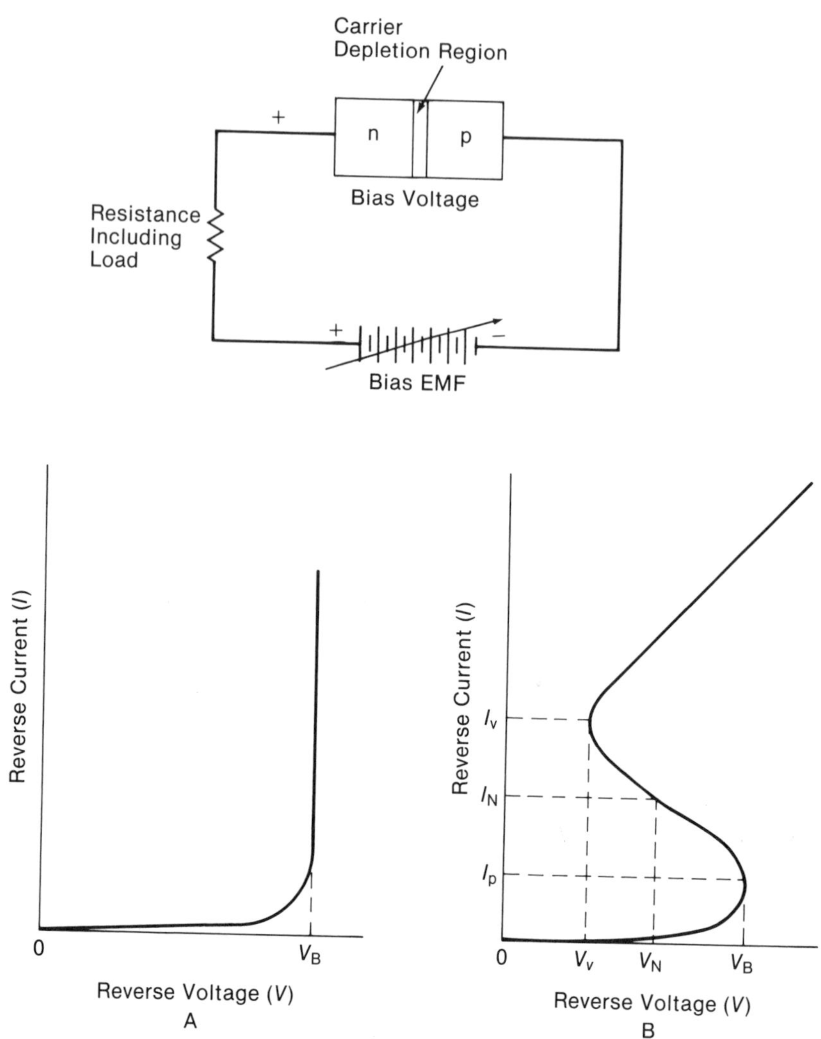

Fig. 4.1 Avalanche effect as seen in the current–voltage characteristic curve. A. Avalanche breakdown. B. Current controlled negative resistance.

material, if the carriers are rapidly multiplied and suddenly produce a large current, due to the finite resistance of biasing circuitry, the diode terminal voltage drops, as shown in Fig. 4.1 B. The avalanche effect current increases, but the terminal voltage drops in spite of an increase in the reverse bias electromotive force. If the reverse bias electromotive force is further increased, the depletion region gets wider due to the high resistivity of the depletion region. The ionization multiplication factor begins to saturate and the bias voltage begins to increase again with the steady increase of current, as seen from Fig. 4.1 B. The current–

voltage curve of Fig. 4.1 B is in great contrast to the current–voltage curve for tunnel diodes shown in Fig. 3.2. Both the tunnel diode and the avalanche diode of Fig. 4.1 B have a negative differential conductance or negative dynamic conductance. However, the negative dynamic conductance of the tunnel diode is controlled by the bias voltage. The negative dynamic conductance of the avalanche diode is controlled by the current. The tunnel diode is therefore termed a voltage controlled device, and the avalanche diode is termed a current controlled device.

In microwave electronics, characteristics shown in Fig. 4.1 A are desirable for microwave switching or momental carrier injection or generation. The characteristics shown in Fig. 4.1 B are suited for microwave amplification and oscillation by the use of negative conductance, as in the case of the microwave tunnel diode amplifiers and oscillators. An avalanche diode with characteristics shown in Fig. 4.1 A with a sharp "knee" is known as a Zener diode. It is widely used in voltage regulator circuits and clipper circuits.

4.2 Avalanche Effect Current

When a p–n junction diode is reverse biased, the reverse current due to the drifting of residual carriers in the depletion layer is given by

$$I = I_o(1 - e^{-qV/kT}) \tag{4.2.1}$$

where

q is the magnitude of electronic charge 1.6×10^{-19} C;
V is the magnitude of reverse bias voltage across the junction, in volts;
k is the Boltzmann constant 1.38044×10^{-23} J/K;
T is the absolute temperature of the junction K; and
I_o is the magnitude of the reverse saturation current, in amps [1–4].

The reverse saturation current I_o is related to the device parameter as follows

$$I_o = qA\left(\frac{D_p p_{no}}{L_p} + \frac{D_n n_{po}}{L_n}\right) \tag{4.2.2}$$

In this equation,

L_p is the hole diffusion length into the n-type region at the junction, in meters;

L_n is the electron diffusion length into the p-type region at the junction, in meters;
D_p is the hole diffusion constant in the n-type region, in m^2/sec;
D_n is the electron diffusion constant in the p-type region, in m^2/sec;
p_{no} is the hole density in the n-region of the junction under equilibrium condition, in m^{-3}; and
n_{po} is the electron density in the p-region of the junction under equilibrium condition, in m^{-3} [2].

The reverse current expressed by Eq. 4.2.1 is correct only if the diode voltage is far less than the avalanche breakdown voltage V_B. Near the avalanche breakdown voltage V_B, the carrier electrons multiply by impact ionization multiplication. Just before impact ionization multiplication sets in, the reverse current is usually saturated, and it is approximately I_o. If the width of the depletion layer is w at the avalanche breakdown voltage V_B, and if an original number of electrons is to be multiplied to N times the number of original electrons at the end of the traverse, then

$$N \equiv \int_0^w \alpha(V)\,dx \tag{4.2.3}$$

where x is the one dimensional coordinate in the direction parallel to the electric field, and $\alpha(V)$ is McKay's ionization rate per unit distance and is a function of the bias voltage V [5].

The magnitude of the reverse current under the avalanche breakdown is then

$$I(1 - N) = I_o \tag{4.2.4}$$

or

$$I = \frac{I_o}{1 - N}. \tag{4.2.5}$$

Therefore when $N \to 1$,

$$I \to \infty \tag{4.2.6}$$

This is the avalanche breakdown. Substituting Eq. 4.2.3 in Eq. 4.2.4

$$I = \frac{I_o}{1 - \int_0^{w(V)} \alpha(V)\,dx} \tag{4.2.7}$$

Note that the width of the depletion layer is also a function of the bias voltage V.

In Eq. 4.2.7,

$$\frac{1}{1 - \int_0^{w(V)} \alpha(V)\, dx} \equiv M \qquad (4.2.8)$$

M is termed the avalanche multiplication factor. If the integral Eq. 4.2.7 is numerically integrated, the current–voltage curve as shown in Fig. 4.1 will be obtained.

The voltage across the reverse biased diode, V, is related to the biasing electromotive force E_B and a combination of the load resistance and the bulk resistance R by

$$V = E_B - IR \qquad (4.2.9)$$

For a proper value of R, a sudden increase of current makes the voltage across the diode drop suddenly, creating the current–voltage characteristic curve as shown in Fig. 4.1 B. The characteristic curve is determined by the McKay's ionization rate α and the external resistance R.

For example, if the reverse saturation current of a microwave avalanche diode is $I_o = 100\ \mu A$, and it started breakdown at $V = 15$ V with reverse current of $I = 1000\ \mu A$, then from Eq. 4.2.4, $N = (I - I_o)/I = [-1000\ \mu A - (-100\ \mu A)]/-1000\ \mu A = 0.9$.

At $V_B = -15$ V, if $w = 5\ \mu m$, then according to Eq. 4.2.3, $0.9 = \int_0^w \alpha(-15\text{ V})\, dx = \alpha(-15\text{ V}) w = \alpha(-15\text{ V}) \cdot 5\ \mu m$. In this example, McKay's ionization rate per micrometer is therefore $\alpha(-15\text{ V}) = 0.9/5\ \mu m = 0.18\ \mu m^{-1}$.

The avalanche breakdown voltage V_B depends on donor concentration on the n-side. Generally, the more donors, the lower the breakdown voltage. If $V_B = -15$ V, then for Ge, $N_d \approx 2.5 \times 10^{16}\text{ cm}^{-3}$; for Si, $N_d \approx 7 \times 10^{16}\text{ cm}^{-3}$; and for GaAs, $N_d \approx 1.5 \times 10^{17}\text{ cm}^{-3}$ [6].

Combining Eqs. 4.2.7 and 4.2.8, $I = MI_o$. Therefore $M = I/I_o$. In this example, $M = -1000\ \mu A/-100\ \mu A = 10$. The avalanche multiplication factor is the ratio of the reverse current I to the reverse saturation current I_o.

In this example, the bias circuit has an internal resistance of 50 Ω, including the bulk resistance, contact resistance, and lead resistance of the avalanche diode. If the diode does not break down at $E_B = -14.9$ V, then the terminal voltage across the junction is, using Eq. 4.2.9, $V = E_B - IR = -14.9\text{ V} - (-100 \times 10^{-6}\text{ A} \cdot 50\ \Omega) \approx -14.9$ V. If the diode breaks down at $E_B = -15.1$ V and $I = 60$ mA, using Eq. 4.2.9, $V = E_B - IR =$

-15.1 V $-$ (-60×10^{-3} A $\cdot$ 50 Ω) $= -12.1$ V. So, the magnitude of voltage across the junction decreases after sufficient breakdown. The more avalanche current, the lower the voltage across the diode, as shown in Fig. 4.1 B. In this example, the avalanche multiplication factor is $M = I/I_o = -60 \times 10^{-3}$ A$/-100 \times 10^{-6}$ A $= 600$, and the N number is $N = (I - I_o)/I = [-60{,}000\ \mu A - (-100\ \mu A)]/-60{,}000\ \mu A = 59{,}900\ \mu A/60{,}000\ \mu A = 0.9983$. McKay's ionization rate per micrometer in this example is $\alpha(-12.1\ \text{V}) = N/w = 0.9983/5\ \mu m = 0.1997\ \mu m^{-1}$.

An alternative empirical expression for the avalanche multiplication factor is

$$M = \frac{1}{1 - (V/V_B)^n} \tag{4.2.10}$$

This is valid only if $|V| < |V_B|$ and $n = 3$ to 6, depending on the semiconductor, where V_B is the beginning voltage of substantial breakdown [6]. Equation 4.2.10 is therefore useful for an avalanche diode as in Fig. 4.1 A or before substantial breakdown of the avalanche diode of Fig. 4.1 B. After a substantial breakdown of the diode of Fig. 4.1 B, Eq. 4.2.10 does not apply.

Solving Eq. 4.2.10 for (V/V_B),

$$(V/V_B) = \sqrt[n]{1 - \frac{1}{M}} \tag{4.2.11}$$

$$V = V_B \sqrt[n]{1 - \frac{1}{M}} \tag{4.2.12}$$

If $n = 4$, $M = 10$ and $V_B = -15$ V, then $V = -14.6$ V. If $n = 4$, $M = 600$ and $V_B = -15$ V, then $V = -14.994$ V.

At any rate, Eq. 4.2.10 describes proximity to $V = V_B$ and $V < V_B$ for the first time.

After $V = V_B$ and beyond, Eqs. 4.2.7, 4.2.8, and 4.2.9 take over, though these equations are valid even when $V < V_B$ for the first time.

After all atoms are ionized, then the junction reaches a saturation resistance of R_J. The avalanche diode current is, using Eq. 4.2.9,

$$I = \frac{V}{R_J} = \frac{E_B - IR}{R_J} \tag{4.2.13}$$

$$I = \frac{E_B}{R_J + R} \tag{4.2.14}$$

Substituting Eq. 4.2.14 into Eq. 4.2.13,

$$I = \frac{E_B}{R_J}\left(1 - \frac{R}{R_J + R}\right) \tag{4.2.15}$$

or, using Eq. 4.2.9,

$$V = E_B - \frac{R}{R_J + R} E_B \tag{4.2.16}$$

Using ionization saturation, or when R_J becomes a constant value,

$$\Delta V = \Delta E_B - \frac{R}{R_J + R}\Delta E_B = \left(1 - \frac{R}{R_J + R}\right)\Delta E_B \tag{4.2.17}$$

If $\Delta E_B > 0$, then $\Delta V > 0$. From Eq. 4.2.13,

$$\Delta I = \frac{\Delta V}{R_J} \tag{4.2.18}$$

If $\Delta V > 0$, then $\Delta I > 0$. This is the reason that the current–voltage curve of Fig. 4.1 B bends back after reaching a point (V_v, I_v).

The avalanche diode of this example carries 15 V across a 5-μm wide depletion region. This means that the electric field across the depletion region is $E = 15\ \text{V}/5 \times 10^{-6}\ \text{m} = 3 \times 10^6\ \text{V/m} = 30\ \text{kV/cm} = 3\ \text{V}/\mu\text{m}$. It is of interest to note that the same field strength appears with different impressions depending on the unit employed.

4.3 Dynamic Negative Conductance

If the device parameters and the load resistance are adjusted to produce the current–voltage characteristics curve of a reverse biased p–n junction diode, as shown in Fig. 4.1 B, then by controlling the bias current,

$$I_p \ll I_N \ll I_v \tag{4.3.1}$$

$$\frac{dI_N}{dV_N} < 0 \tag{4.3.2}$$

where

I_p is the reverse diode current at the peak bias voltage V_B,
I_v is the reverse diode current at the valley bias voltage V_v,
I_N is the reverse bias current between I_p and I_v, and
V_N is the reverse bias voltage across the diode corresponding to I_N.

As seen from Fig. 4.1 B, at the bias current I_N, the differential conductivity

$$g_a \equiv \frac{dI_N}{dV_N} < 0 \tag{4.3.3}$$

For example, in Fig. 4.1 B, if $V_B = 15$ V, $I_p = 200$ μA, $V_v = 12.1$ V, and $I_v = 60{,}000$ μA, then,

$$g_a = \frac{dI_N}{dV_N} \approx \frac{I_P - I_v}{V_B - V_v} = \frac{200\ \mu\text{A} - 600{,}000\ \mu\text{A}}{15\ \text{V} - 12.1\ \text{V}}$$

$= -0.021$ S. The static conductances are at $V = V_B$, $G_B = I_p/V_B = 200$ μA/15 V $= 13.3$ μS and at $V = V_v$, $G_v = I_v/V_v = 60{,}000$ μA/12.1 V $= 0.00496$ S. In terms of resistances, the incremental resistance of the diode is $r_a = 1/g_a = -47.6\ \Omega$, $R_B = 1/G_B = 75$ kΩ, and $R_V = 1/G_V = 201.6\ \Omega$.

Equation 4.3.3 is analogous to Eq. 3.3.13 for tunnel diodes. Therefore, the principles of microwave amplification and oscillation and the concept of dynamic negative conductance for tunnel diodes are also applicable to avalanche diodes.

4.4 Further Application of the Avalanche Effect

Though the principles of amplification and oscillation of avalanche effect diodes are the same as the ones with the tunnel diode as far as the application of the negative conductance is concerned, the mechanism for producing the negative conductance is quite different as studied in previous sections. The avalanche effect requires at least one order of magnitude, and generally more orders of magnitude, of the bias voltage than tunnel diodes. The avalanche effect relies on impact ionization multiplication. This tends to generate a greater amount of noise than the tunnel effect. The avalanche effect takes more time than the tunnel effect. This will affect extremely high frequency operation in the microwave frequency range. The direct application of dynamic negative conductance in the area of amplification and oscillation is therefore not as popular compared to the tunnel diodes.

The avalanche effect, however, provides a useful carrier injection mechanism [7] for the IMPATT diode which will be studied in Chapter 6. The avalanche effect also provides an effective nondestructive microwave short circuit to the microwave transmission line when it is properly mounted and operated [8,9]. This is useful for microwave diplexing, which will be studied in Chapter 10.

Problems

4.1 Sketch the energy band structure of a nondegenerate p–n junction diode and describe the mechanism of formation of the depletion layer at the junction. Also, explain why the width of the depletion layer increases when the reverse bias voltage across the diode increases.

4.2 Sketch the voltage drop for all portions of a reverse biased p–n junction diode.

4.3 Identify the charge carriers in the depletion layer of a reverse biased p–n junction.

4.4 Knowing the ionization potential of the semiconductor material in the depletion region of a reverse biased p–n junction diode, try to formulate an equation to obtain the avalanche breakdown voltage.

4.5 State the reason avalanche effect devices are noisier than tunnel effect devices.

4.6 State the reason avalanche effect devices need higher operating voltages than tunnel diodes.

4.7 Explain the mechanism for the formation of negative dynamic conductance in reverse biased p–n junction diodes.

4.8 Differentiate current controlled dynamic negative conductance from voltage controlled dynamic negative conductance.

4.9 At room temperature, find the reverse bias voltage which produces 90% of the reverse saturation current in a reverse biased p–n junction diode.

4.10 Devise a procedure to numerically or graphically integrate the integral equation for the reverse current under avalanche breakdown conditions in order to obtain the current–voltage curve which contains a negative differential conductance.

4.11 Sketch a circuit diagram of a microwave amplifier using an avalanche diode in a microstripline configuration.

4.12 Derive an equation to obtain the voltage gain of an avalanche diode microwave amplifier.
4.13 Specify an oscillation condition for an avalanche diode microwave oscillator.
4.14 Find a formula to determine the oscillation frequency of a microwave avalanche diode oscillator.
4.15 Find a formula to estimate the probable maximum oscillation voltage and the power of an avalanche diode microwave oscillator.
4.16 Sketch a diagram of a waveguide mounted avalanche diode microwave amplifier.
4.17 List the difference in the mechanisms for producing dynamic negative conductance in avalanche diodes and tunnel diodes.
4.18 Quantitatively examine the electron transit time in avalanche effect devices and in tunnel effect devices.

References

1 J. F. White, "Semiconductor Control." Artech House, Dedham, Massachusetts, 1977.
2 E. S. Yang, "Fundamentals of Semiconductor Devices." McGraw-Hill, New York, 1978.
3 A. B. Philips, "Transistor Engineering." McGraw-Hill, New York, 1962.
4 H. A. Watson, "Microwave Semiconductor Devices and Their Circuit Applications." McGraw-Hill, New York, 1969.
5 K. G. McKay, Avalanche breakdown in silicon. *Phys. Rev.* **94**, 877–884 (May 1954).
6 B. G. Streetman, "Solid-State Electronic Devices," 2nd ed. Prentice-Hall, Englewood Cliffs, New Jersey, 1980.
7 Hughes Aircraft, We're turning up the volume on mm-wave diode production. *Microwave Syst. News Commun. Tech.* **18**(1), 33 (1988).
8 Metalics, Mesa beam lead PIN diodes. *Microwaves RF* **27**(9), 179 (1988).
9 Alpha Industries, Series shunt elements. *Microwave J.* **31**(9), 118–119 (1988).

5

Transferred Electron Devices

5.1 Electron Energy Transfer within Conduction Bands

This chapter is about Gunn diodes. The original Gunn diode was a simple bulk n-type GaAs semiconductor with metallic ohmic contacts at both ends [1]. In this bulk n-type semiconductor, when the minimum available total electron energy ε is plotted against the wave number κ of the wave function ψ of the Schrödinger equation, the plot takes various forms, depending on the material. Even with respect to similar materials, the plot takes a variety of forms depending on the direction of the wave vector $\boldsymbol{\kappa}$. In n-type GaAs, for example, the wave vector $\boldsymbol{\kappa}$ is placed parallel to the crystallographic axis [100] (Appendix 4); the total energy ε versus the wave number κ curve looks like Fig. 5.1 [2,3,6,8,13,14]. According to quantum theory, the wave number is proportional to the momentum p by the following relation

$$p = \frac{h}{2\pi}\kappa \tag{5.1.1}$$

where h is Planck's constant, 6.6256×10^{-34} J-s [2]. The wave vector $\boldsymbol{\kappa}$ is in the direction of propagation of the wave function ψ, which is the solution of Schrödinger's equation [4]. The wave of wave function ψ is the de Broglie wave; it is *not* the photon wave or electromagnetic wave [2,3]. Thus, the direction of the wave

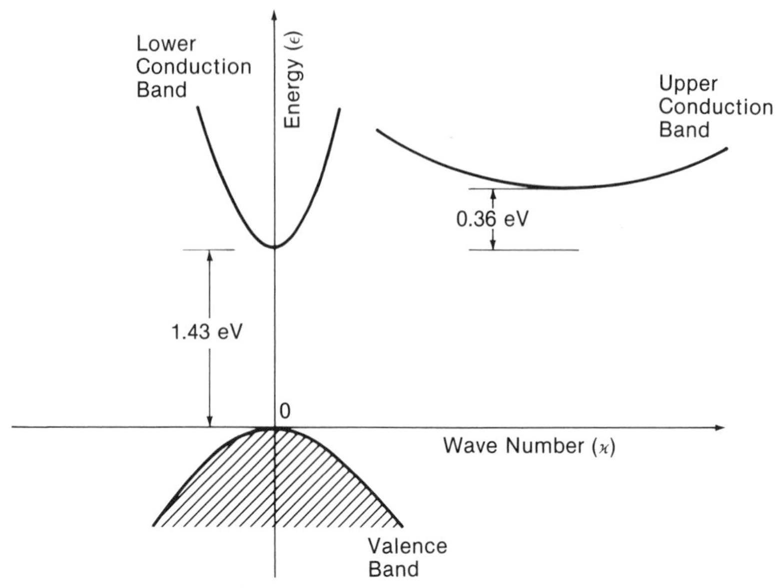

Fig. 5.1 Energy band structure of an n-type GaAs crystal in a particular orientation designated by the crystallographic direction (100) [1,2].

vector is in the direction of electron motion, or the momentum vector **p**. In order to produce an energy wave number diagram like the one shown in Fig. 5.1, the applied electric field which accelerates electrons in the n-type GaAs crystal must therefore be parallel to the [100] crystallographic axis. Since the wave number is proportional to the momentum, some people call the energy–wave number diagram the energy–momentum diagram, even though it is the ε–κ diagram.

Wave number κ is the same as the phase constant $2\pi/\lambda$ in electromagnetic field theory except that λ is *not* the wavelength of electromagnetic waves but the wavelength of de Broglie waves. A de Broglie wave is a wave function ψ which is a solution of Schrödinger's equation. The units of κ are radians/meter. A particle of total energy ε can be considered a de Broglie wave of wavelength λ. If the particle's mass is m (kg), the velocity is u (m/s), and the total energy W (J) is kinetic energy, then the momentum p is

$$p^2 = (mu)^2 = 2m \cdot \frac{1}{2}mu^2 = 2mW$$

The saturation velocity of an electron of nondegenerate n-type GaAs is known to be $u = 100$ km/s [6,8,12,14]. Then $p = mu =$

9.11×10^{-31} kg $\times$ 100 $\times$ 10^3 m/s = 9.11×10^{-26} kg-m/s. Therefore, from Eq. 5.1.1, the wave number $\kappa = 2\pi\, p/h = 2\pi \times 9.11 \times 10^{-26}$ kg-m/s/6.6256×10^{-34} J-s = 8.64×10^8 rad/m. This means that the de Broglie wave of an electron with momentum $p = 9.11 \times 10^{-26}$ kg-m/s and energy $W = p^2/2\mathrm{m} = (9.11 \times 10^{-26}\ \mathrm{kg\text{-}m/s})^2/(2 \times 9.11 \times 10^{-31}\ \mathrm{kg}) = 4.55 \times 10^{-21}$ J has the de Broglie wavelength of $\lambda = 2\pi/\kappa = 2\pi$ rad/8.64×10^8 rad/m = 0.727×10^{-8} m = 7.27 nm. This is the wavelength of the wave function ψ, not the wavelength of electromagnetic radiation from the moving electron nor the oscillation wavelength of the Gunn diode.

Starting with zero applied electric field, if an electron occupies the bottom of the lower conduction band in Fig. 5.1, it will gradually rise; it will gain kinetic energy as the applied electric field is increased. If the additional energy due to the applied electric field exceeds 0.35 eV, the electron will transfer to the upper conduction band. When this transfer occurs, the following things accompany the transfer.

1. The mass of the electron increases. If the mass of an electron in the lower conduction band is m_{L}, and the mass of an electron in the upper conduction band is m_{u},

$$m_{\mathrm{u}} \approx 17 m_{\mathrm{L}} = 1.2 m_{\mathrm{o}} \tag{5.1.2}$$

where m_{o} is the static mass of the electron [3].

2. The mobility of the electron decreases. As seen above from Eq. 5.1.2,

$$m_{\mathrm{L}} \approx 0.07 m_{\mathrm{o}}; \tag{5.1.3}$$

the mobility of the electrons in the lower conduction band μ_{L} is

$$\mu_{\mathrm{L}} = \frac{u_{\mathrm{dL}}}{E_{\mathrm{L}}} \gg \mu_{\mathrm{u}} = \frac{u_{\mathrm{du}}}{E_{\mathrm{u}}} \tag{5.1.4}$$

where u_{dL} is the drift velocity of electrons in the lower conduction band in m/s; u_{du} is the drift velocity of electrons in the upper conduction band; E is the applied electric field strength in V/m. The theoretical values of $\mu_{\mathrm{u}} = 0.010$–$0.018\ \mathrm{m^2/V\text{-}s}$ and $\mu_{\mathrm{L}} = 0.5$–$0.8\ \mathrm{m^2/V\text{-}s}$ are reported [6,8,13].

3. The electron velocity slows down in the upper conduction band.

$$u_{\mathrm{du}} < u_{\mathrm{dL}} \tag{5.1.5}$$

For example, if the actual electron transit time across the bulk portion of the Gunn diode is τ s, then the oscillation frequency f must be $f = 1/\tau$. If the device length is d, and the saturation velocity of the electron in the nondegenerate n-type GaAs is $u_s = 100$ km/s, then $\tau = d/u_s$, $f = u_s/d$ and therefore $d = u_s/f$. If $f = 10.525$ GHz then $d = 100 \times 10^3$ m/s/10.525 $\times 10^9$/s $= 9.5 \times 10^{-6}$ m $= 9.5$ μm. If the bias voltage is $V = 8$ V, then the electric field strength E must be $E = V/d = 8$ V/9.5 $\times 10^{-6}$ m $= 0.842 \times 10^6$ V/m $= 8.42$ kV/cm $= 842$ V/mm. Now if $u_s = u_{du}$, then the mobility of electrons in the upper conduction band is $\mu_u = u_{du}/E = 100 \times 10^3$ m/s/0.842 $\times 10^6$ V/m $= 0.119$ m^2/V-s.

According to Eqs. 5.1.2 and 5.1.3, even if the same kinetic energy W is given to the electron in the lower conduction band as in the upper conduction band, then

$$W = \frac{1}{2} m_L u_{dL}^2 = \frac{1}{2} m_u u_{du}^2 \tag{5.1.6}$$

Therefore

$$\frac{u_{dL}}{u_{du}} = \sqrt{\frac{m_u}{m_L}} = \sqrt{\frac{1.2 m_o}{0.07 m_o}} = 4.14 \tag{5.1.7}$$

In practice, in Eq. 5.1.4, $E_L < E_u$. The ratio of E_L/E_u can be any value between zero and one. So if $E_L/E_u = 0.8$, then

$$\frac{\mu_L}{\mu_u} = \frac{u_{dL}}{u_{du}} \frac{E_u}{E_L} = 4.14 \times \frac{1}{0.8} = 5.2 \tag{5.1.8}$$

According to published theory, E_L/E_u can be as low as 0.1 [6,8,13]. This argument fails when E_L/E_u is too small. If E_L/E_u is too small, then Eq. 5.1.6 does not apply and therefore Eq. 5.1.7 does not apply either. What is of importance is the fact that the mobility of the low field electron is greater than the mobility of the electron in the high field region. Now, according to Eq. 5.1.7, if $u_{du} = u_s = 100$ km/s, then $u_{dL} = 4.14 \times u_{du} = 4.14 \times 100$ km/s $= 414$ km/s. The mobility of electrons in the lower conduction band is then $\mu_L = 5.2\mu_u = 5.2 \times 0.119$ m^2/V-s $= 0.62$ m^2/V-s. This is at $E_L = 0.8E_u = 0.8 \times 0.842$ V/m $= 0.674 \times 10^6$ V/m and is equivalent to device voltage of $V = E_L d = 0.674 \times 10^6$ V/m $\times 9.5 \times 10^{-6}$ m $= 6.403$ V.

5.2 Current – Voltage Curve of Transferred Electron Devices

If the electron density of the lower conduction band is n_L (m^{-3}), and the electron density of the upper conduction band is n_u, the current density in the transferred electron device (TED) under a given electric field strength E (V/m) is

$$J = qn_L u_L + qn_u u_u$$

$$= qn_L \mu_L E + qn_u \mu_u E \quad (A/m^2) \tag{5.2.1}$$

where q is the magnitude of electronic charge in coulombs.

When the applied voltage V is low and E is small, most electrons are in the lower conduction band where $n_L \gg n_u$ and $\mu_L > \mu_u$ all the time. The first term, $qn_L \mu_L E$, dominates in Eq. 5.2.1. In this low voltage region

$$E \approx \frac{V}{d} \quad (V/m) \tag{5.2.2}$$

where d is the device length along the applied electric field,

$$J \approx qn_L \mu_L \frac{V}{d} + qn_u \mu_u \frac{V}{d} \quad (A/m^2) \tag{5.2.3}$$

If the cross sectional area of the TED is A (m^2), the device current I is

$$I = JA = qAn_L \mu_L \frac{V}{d} + qAn_u \mu_u \frac{V}{d} \quad (A) \tag{5.2.4}$$

In the low voltage region, the device current increases almost linearly with the applied voltage as shown in Fig. 5.2.

When the applied voltage V approaches the threshold voltage V_{TH} for the electron transfer, some of the electrons begin the transfer from the lower conduction band to the upper conduction band. At this point the magnitude of the first term of Eq. 5.2.4 decreases and the magnitude of the second term increases. With a sufficiently high voltage, most electrons are transferred to the upper conduction band and the lower conduction band is almost empty. In the high voltage range, the second term of Eq. 5.2.4 therefore dominates. In the second term, there is slow drift velocity, a heavy effective electron mass, and small electron mobility. Consequently, in spite of a high applied voltage, the device current does not flow much, as seen from Fig. 5.2. Between the low voltage V_L and the high voltage V_H in Fig. 5.2, there is a

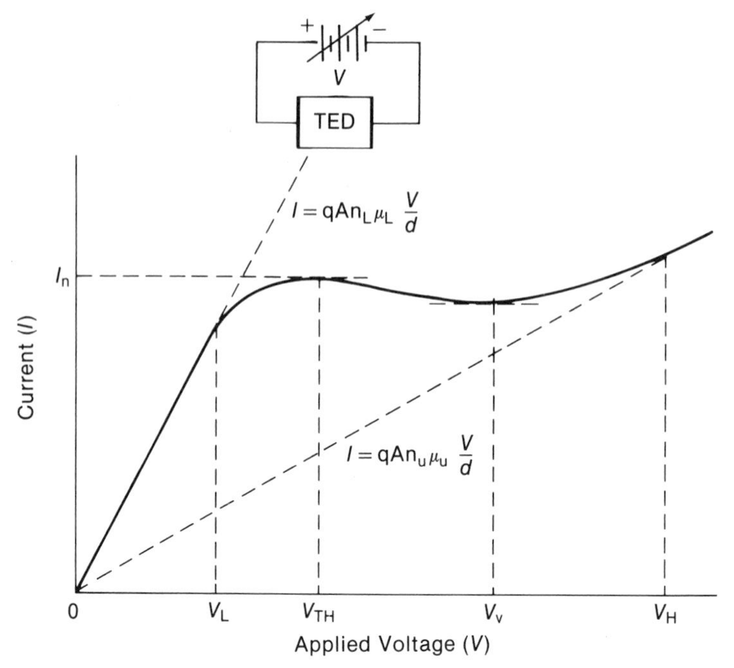

Fig. 5.2 Current–voltage characteristics of a transferred electron device (TED).

possibility that the current dips to form a negative differential conductivity.

For the 10.525 GHz Gunn diode discussed in Section 5.1, if the diode dc current is 30 mA and at a dc bias of 8 V, then the diode is consuming 8 V × 30 mA = 240 mW of dc power. If the microwave power emission is 80 mW, then the oscillation efficiency is (80 mW/240 mW) × 100% = 33.2%. In Fig. 5.2, a typical value for V_H is 12 V, then $I = qAn_u\mu_u V_H/d$. Then $A = Id/qn_u\mu_u V_H = 30 \times 10^{-3}\text{A} \times 9.5 \times 10^{-6}$ m/[(1.6 × 10^{-19} C) (10^{21} m^{-3})(0.119 m^2/V-s)(12 V)] = 1247.3 × 10^{-12} m^2 = (35.3 × 10^{-6} m)^2 = (35.3 μm)^2.

A GaAs of cross-sectional area $A = 1247.3 \times 10^{-12}$ m^2 and thickness $d = 9.5 \times 10^{-6}$ m can expect the capacitance of $C_d = \varepsilon_r\varepsilon_o A/d = 13.2 \times 8.854 \times 10^{-12}$ F/m × 1247.3 × 10^{-12} m^2/9.5 × 10^{-6} m = 0.153 ×10^{-12} F.

An emergent diode structure here is an n-type GaAs chip of donor concentration $n_u = 10^{15}$ $\text{cm}^{-3} = 10^{21}$ m^{-3} of area A = 35.3 μm × 35.3 μm and thickness $d = 9.5 \times 10^{-6}$ m. The actual device is much thicker however. The reason is that the above mentioned thickness $d = 9.5 \times 10^{-6}$ m is only the active part of the diode structure. The diode must have good ohmic contacts on both sides of the surfaces of the active region and the active region must be developed on a substrate. This active region is therefore usually grown epitaxially on top of a good conducting, heavily doped n^+ GaAs substrate 100 μm thick. On top of this n^+-substrate, a 9.5 μm thick n-layer, which is the active layer, is epitaxially grown. On top of this active layer, a thin layer of heavily doped n^{++},-layer is epitaxially grown for good metallic contact. The exterior of this thin n^{++}-layer and thick n^+-substrate is metallized by evaporating a thin layer of AuSn or AuGe for lead structure bonding. The diodes are packaged in a similar way as described in Chapter 3. Usually, the thin n^{++}-side is the cathode and the thick n^+-side is the anode. The thin n^{++}-side is directly attached to a heavy heat sink [5,7,9,10,11,13,15,16].

If all electrons are transferred to the upper conduction band then, as seen from Fig. 5.2,

$$\begin{aligned} I &= qAn_u\mu_u V/d \\ &= (1.6 \times 10^{-19}\ \text{C})(1247.3 \times 10^{-12}\ \text{m}^2)(10^{21}\ \text{m}^{-3}) \\ &\quad (0.119\ \text{m}^2/\text{V-s})V/9.5 \times 10^{-6}\ \text{m} \\ &= (0.002475\ \text{S})V \quad (\text{A}) \end{aligned} \tag{5.2.5}$$

$$V = I/0.002475 \quad (\text{V}) \tag{5.2.6}$$

The saturation current of 30 mA is therefore reached at $V_H = 0.03\ \text{A}/0.002475\ \text{S} = 12\ \text{V}$.

If all electrons are in the lower conduction band, as seen from Fig. 5.2

$$\begin{aligned} I &= qAn_L\mu_L \frac{V}{d} \\ &= (1.6 \times 10^{-19}\ \text{C})(1247.3 \times 10^{-12}\ \text{m}^2)(10^{21}\ \text{m}^{-3}) \\ &\quad (0.62\ \text{m}^2/\text{V-s})V/9.5 \times 10^{-6}\ \text{m} \\ &= (0.01287\ \text{S})V \quad (\text{A}) \end{aligned} \tag{5.2.7}$$

or

$$V = I/0.01287 \quad (\text{V}) \tag{5.2.8}$$

This means that in order to reach the saturation current of 30 mA, V must be $V_L = 30 \times 10^{-3}\text{A}/0.01283\ \text{S} = 2.34\ \text{V}$. As seen from Eq. 5.2.5, the high field static conductance of the Gunn diode G_H is

$$G_H = qAn_u\mu_u/d \quad (\text{S}) \tag{5.2.9}$$

which is 0.002475 S in this example.

The Gunn diode's low field static conductance G_L is, according to Eq. 5.2.7,

$$G_L = qAn_L\mu_L/d \quad \text{(S)} \tag{5.2.10}$$

which is 0.01283 S in this example.

Between $V = V_L$ and $V = V_H$, the available electrons change their distribution. Some electrons are in the lower conduction band and the rest of them are in the upper conduction band. In the region where $V_L < V < V_H$, $n_L + n_u =$ constant but the distribution changes and Eq. 5.2.4 stands.

For example, if the diode reaches saturation current I_s for a certain voltage in the range $V_L < V < V_H$, then from Eq. 5.2.4,

$$I_s = qA(n_L\mu_L + n_u\mu_u)\frac{V}{d} \tag{5.2.11}$$

then

$$n_L\mu_L + n_u\mu_u = I_s d/qAV \tag{5.2.12}$$

If the total carrier concentration is n (m^{-3}), then

$$n_L + n_u = n \tag{5.2.13}$$

Solving Eqs. 5.2.12 and 5.2.13 simultaneously,

$$n_u = \frac{n\mu_L - (I_s d/qAV)}{\mu_L - \mu_u} \tag{5.2.14}$$

$$n_L = n - n_u \tag{5.2.15}$$

For the 10.525 GHz Gunn diode in question, at $V = 8$ V and $I_s = 30 \times 10^{-3}$ A,

$$n_u = \frac{(10^{21}\ \text{m}^{-3})(0.62\ \text{m}^2/\text{V-s})}{(0.62\ \text{m}^2/\text{V-s}) - (0.119\ \text{m}^2/\text{V-s})} - \frac{(30 \times 10^{-3}\ \text{A})(9.5 \times 10^{-6}\ \text{m})/(1.6 \times 10^{-19}\ \text{C})(1247.3 \times 10^{-12}\ \text{m}^2)(8\ \text{V})}{(0.62\ \text{m}^2/\text{V-s}) - (0.119\ \text{m}^2/\text{V-s})}$$

$$= 0.88 \times 10^{21}\ \text{m}^{-3}$$

and $n_L = n - n_u = 10^{21}\ \text{m}^{-3} - 0.88 \times 10^{21}\ \text{m}^{-3} = 0.12 \times 10^{21}\ \text{m}^{-3}$. So at $V = 8$ V, n_u: $n_L = 0.88{:}0.12$. Similarly, at $V = 10$ V, $n_u{:}n_L = 0.95{:}0.05$. At $V = 4$ V, $n_u{:}n_L = 0.52{:}0.48$. This means that between V_L and V_H in Fig. 5.2, electrons are gradually transferred from the lower conduction band to the upper conduction band as the bias voltage V increases.

5.3 Negative Differential Conductivity

In the transition from the low energy state to the high energy state in Eq. 5.2.4, in an exact concept, all of n_{L}, μ_{L}, n_{u} and μ_{u} are functions of the bias voltage V. If the differential of both sides of Eq. 5.2.4 is taken, then by first approximation

$$\mathrm{d}I = \frac{qA}{d}\left[\left(\frac{\mathrm{d}n_{\mathrm{L}}}{\mathrm{d}V}\mu_{\mathrm{L}} + n_{\mathrm{L}}\frac{\mathrm{d}\mu_{\mathrm{L}}}{\mathrm{d}V}\right) + \left(\frac{\mathrm{d}n_{\mathrm{u}}}{\mathrm{d}V}\mu_{\mathrm{u}} + n_{\mathrm{u}}\frac{d\mu_{\mathrm{u}}}{\mathrm{d}V}\right)\right]\mathrm{d}V \tag{5.3.1}$$

So the differential conductivity is

$$g_{\mathrm{e}} \equiv \frac{\mathrm{d}I}{\mathrm{d}V} = \frac{qA}{d}\left[\frac{\mathrm{d}n_{\mathrm{L}}}{\mathrm{d}V}\mu_{\mathrm{L}} + n_{\mathrm{L}}\frac{\mathrm{d}\mu_{\mathrm{L}}}{\mathrm{d}V} + \frac{\mathrm{d}n_{\mathrm{u}}}{\mathrm{d}V}\mu_{\mathrm{u}} + n_{\mathrm{u}}\frac{\mathrm{d}\mu_{\mathrm{u}}}{\mathrm{d}V}\right] \tag{5.3.2}$$

when the bias V is in the transitional state, or

$$V_{\mathrm{TH}} < V < V_{\mathrm{H}} \tag{5.3.3}$$

In Fig. 5.2, n_{L} is decreasing and n_{u} is increasing.

$$\frac{\mathrm{d}n_{\mathrm{L}}}{\mathrm{d}V} < 0 \tag{5.3.4}$$

$$\frac{\mathrm{d}n_{\mathrm{u}}}{\mathrm{d}V} > 0 \tag{5.3.5}$$

In the transition region, electron mobility is always slightly smaller than in a higher voltage region. Therefore,

$$\frac{\mathrm{d}\mu_{\mathrm{L}}}{\mathrm{d}V} < 0 \tag{5.3.6}$$

$$\frac{\mathrm{d}\mu_{\mathrm{u}}}{\mathrm{d}V} < 0 \tag{5.3.7}$$

So, in Eq. 5.3.2, when

$$\left|\frac{\mathrm{d}n_{\mathrm{u}}}{\mathrm{d}V}\mu_{\mathrm{u}}\right| < \left|\mu_{\mathrm{L}}\frac{\mathrm{d}n_{\mathrm{L}}}{\mathrm{d}V} + n_{\mathrm{L}}\frac{\mathrm{d}\mu_{\mathrm{L}}}{\mathrm{d}V} + n_{\mathrm{u}}\frac{\mathrm{d}\mu_{\mathrm{u}}}{\mathrm{d}V}\right| \tag{5.3.8}$$

$$g_{\mathrm{e}} < 0 \tag{5.3.9}$$

It is well established that

$$\mu_{\mathrm{u}} < \mu_{\mathrm{L}} \tag{5.3.10}$$

and

$$\left|\frac{dn_u}{dV}\right| \approx \left|\frac{dn_L}{dV}\right| \qquad (5.3.11)$$

Eq. 5.3.8 is therefore well satisfied by the first term of the right hand side of the equation.

For the Gunn diode of 10.525 GHz (the example discussed in Section 5.2), it is assumed that μ_L and μ_u are constant. $\mu_L = 0.62\ m^2/V\text{-}s$ and $\mu_u = 0.119\ m^2/V\text{-}s$. Therefore, in Eq. 5.3.2, $d\mu_L/dV = 0$ and $d\mu_u/dV = 0$. Then

$$g_e \approx \frac{qA}{d}\left(\frac{dn_L}{dV}\mu_L + \frac{dn_u}{dV}\mu_u\right) \quad (S) \qquad (5.3.12)$$

Using values of n_L and n_u at $V = 4$ V and $V = 8$ V, respectively,

$$\left.\frac{dn_L}{dV}\right|_{V=8\,V} = \frac{0.12 - 0.48}{8\ V - 4\ V} \times 10^{21}\ m^{-3}$$

$$= -0.09 \times 10^{21}\ m^{-3}/V \qquad (5.3.13)$$

$$\left.\frac{dn_u}{dV}\right|_{V=8\,V} = \frac{0.88 - 0.52}{8\ V - 4\ V} \times 10^{21}\ m^{-3}$$

$$= 0.09 \times 10^{21}\ m^{-3}/V \qquad (5.3.14)$$

Eq. 5.3.11 is therefore verified.

Then from Eqs. 5.3.11 and 5.3.12,

$$g_e \approx \frac{qA}{d}\frac{dn_L}{dV}(\mu_L - \mu_u) \quad (S) \qquad (5.3.15)$$

Therefore, in the Gunn diode of this example at V = 8 V,

$$g_e \approx \frac{(1.6 \times 10^{-19}\ C)(1247.3 \times 10^{-12}\ m^2)}{9.5 \times 10^{-6}\ m}$$

$$\cdot(-0.09 \times 10^{21}\ m^{-3})\left(0.62\frac{m^2}{V\text{-}s} - 0.119\frac{m^2}{V\text{-}s}\right)$$

$$= -9.47 \times 10^{-4}\ S$$

This is not much dynamic negative conductance at $V = 8$ V. Tunnel diodes and avalanche diodes have higher dynamic conductance. On the other hand, the static conductance of this Gunn diode at the bias voltage $V = 8$ V is

$$G_e = \frac{I_s}{V} = \frac{30 \text{ mA}}{8 \text{ V}} = 3.75 \times 10^{-3} \text{ S}$$

The dynamic conductance and static conductance are different from each other in both magnitude and signs.

Once negative conductance is established, the effect of dynamic negative conductance is identical to the one discussed in Section 3.4 for tunnel diodes and in Section 4.3 for avalanche diodes. Actual diode packages and circuit design are also identical to those in Chapters 3 and 4. Therefore the effect of negative conductance on the amplification, the oscillation, packaging, and circuitry are not repeated here.

5.4 High Field Domain and the Gunn Effect

When a bulk n-type GaAs semiconductor is sandwiched between a pair of ohmic contact conducting parallel electrodes as shown in Fig. 5.2, the electric potential distribution and carrier electron charge distribution are as sketched in Fig. 5.3.

In the beginning, at t = 0, assuming that the bias voltage is beyond the threshold voltage V_{TH} and that it is in the negative

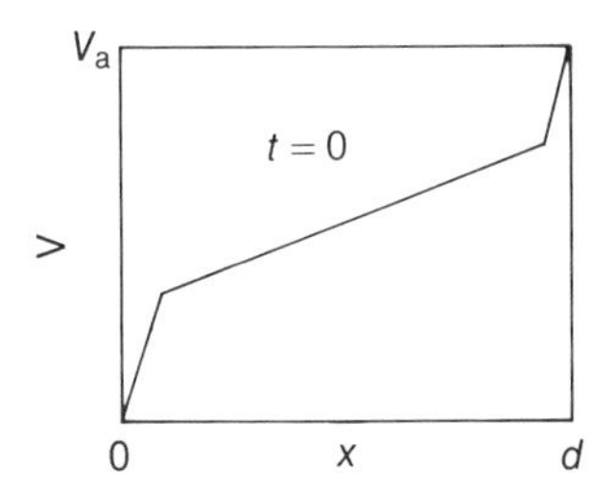

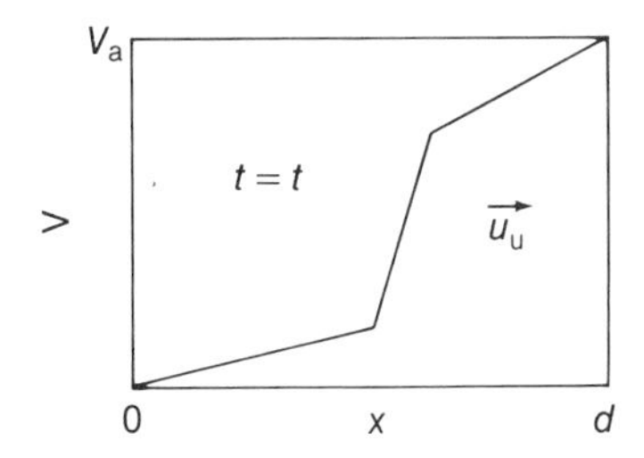

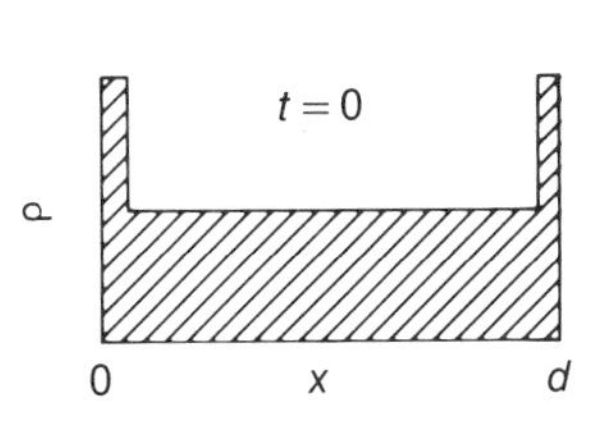

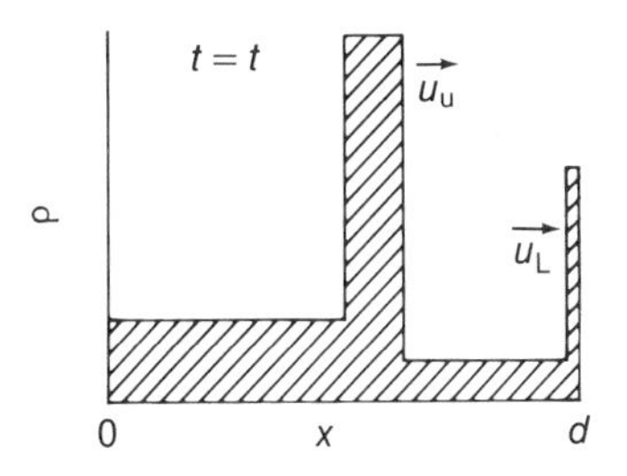

Fig. 5.3 High field domain formation in a TED. $V_{th} \ll V_a \ll V_H$.

slope region of Fig. 5.2, a high field region will appear at the contacts in both the cathode and the anode. Contact potential and contact resistance are the reason for the high field. Beyond the threshold voltage, the drift velocity of electrons at the high field region, u_u, is slower than the drift velocity of electrons at the low field region, u_L, or $u_L \gg u_u$. As a result, electrons tend to accumulate near the high field domain. Electrons near the high field region at the anode are discharged almost immediately, but electrons at the cathode high field domain must drift across the device thickness d (meters).

As sketched in Fig. 5.3, at time t the high field domain and the associated carrier electron accumulation region drift across the device thickness d with the upper band drift velocity u_u. Since $u_L \gg u_u$, the charge accumulation grows as it travels across the device thickness d. At $x = d$, the accumulated electron charge is discharged at the anode, and the initial condition at $t = 0$ is again established. The phenomena will repeat itself again and again until the bias voltage is shut off. The oscillation frequency must therefore be

$$f = \frac{u_u}{d} \tag{5.4.1}$$

This effect was found experimentally by J. B. Gunn [1], and is called the Gunn effect. Note that, in order to observe the Gunn effect, the bias voltage must be beyond threshold, and it must be placed in the negative differential conductance region.

In most commercial Gunn diodes, $u_u \approx 100$ km/s. By knowing the oscillation frequency, the effective device thickness is known or vice versa. For example, if $f = 26$ GHz, then $d = 10^5$ m/s/26 $\times 10^9$ s^{-1} $= 3.8 \times 10^{-6}$ m. The effective device thickness can be smaller than the actual device thickness. Electron charge in the high field domain discharges and the domain collapses before reaching the anode metallization. This is the case of the limited space–charge accumulation mode (LSA) operation [6,8,14].

5.5 Oscillation by Transferred Electrons

Oscillation by transferred electrons is distinctly different from the oscillation due to the tunnel effect or the avalanche effect. In both the tunnel effect oscillator and the avalanche effect oscillator, the negative differential mobility of carriers is not required to form a negative differential conductance. In the tunnel effect oscillator, the carrier mobility through the junction is not related at all. After all, the tunnel effect current is the quantum mechanical current carried by de Broglie waves; it is not the drift current. The avalanche effect current is the drift current, yet the differential mobility of the carrier electron is always positive. For both tunnel diodes and avalanche diodes, the negative differential conductivity is related to the magnitude of the quantity of the carrier transportation rather than the mobility of the carriers. For transferred electron devices (TED), the negative differential mobility due to the energy transfer from the lower conduction band to the upper conduction band is important in order to form both the negative differential conductance, as explained in Section 5.3, and the high field domain, as explained in Section 5.4. Due to the domain formation, the oscillation frequency is determined by Eq. 5.4.1. In this equation, the average drift velocity of the high field domain u_u is not uniquely determined by the bias voltage for a given device only. For *optimum* oscillation, both Eqs. 3.4.1. and 3.4.4 must be satisfied after adaptation to the TED. Usually the simultaneous solution of Eqs. 3.4.1 and 3.4.4 with respect to the oscillation frequency is not analytically easy, for these equations are likely to be transcendental. In addition, the device is likely to be degenerate. In other words, not all of the roots of the transcendental equation are useful to explain the experimentally observed oscillation frequency or frequencies. But one of the oscillation frequencies must fit with Eq. 5.4.1. If it does, the device satisfies Gunn effect oscillation. The domain velocity u_u is therefore influenced by external circuit parameters such as g_L and B_L, beside G_{TED} and B_{TED}. For this particular Gunn diode, it is sometimes possible to oscillate with Eqs. 3.4.1 and 3.4.4 but not with Eq. 5.4.1. In such a case, it is not Gunn effect oscillation, even though the device is a Gunn diode. Practical configurations of the oscillator circuits [5,7,9,11,15,16,18,19] are similar to the ones shown in Fig. 3.5 and Fig. 3.6. For example, in the Gunn diode of 10.525 GHz presented in Section 5.3, by calculating $g_e = -9.47 \times 10^{-4}$ S, the circuit conductance should be $g_L = 9.47 \times 10^{-4}$ S. This is small. The Gunn diode must be mounted at the maximum electric field location of a high-Q cavity resonator. In actual commercially packaged Gunn diodes, g_e is on the order of 10^{-4} S.

5.6 Amplification by Transferred Electrons

TED amplifiers are not as popular as TED oscillators, but can be achieved if necessary as in the case of the tunnel diode using circuits similar to those shown in Figs. 3.5 and 3.6 [5,7,9,11,15,16,18,19]. The gain of the TED amplifiers is given by Eq. 3.4.3, provided that the input signal frequency approximately satisfies Eq. 5.4.1. It is not necessary to satisfy Eq. 3.4.1 for the case of amplification. For example, if the Gunn diode of 10.525 GHz, discussed in Section 5.5, is to make a 10 dB power gain amplifier, then $A = 10$ dB $= g_L/(g_L + g_e)$, or $g_L = -Ag_e/(A-1) = -10(-9.47 \times 10^{-4}\ \mathrm{S})/(10-1) = 10.52 \times 10^{-4}$ S.

5.7 Transferred Electron Devices

When a properly doped n-type GaAs is biased beyond the threshold voltage, electrons in the lower conduction band transfer to the upper conduction band. The transferred electrons have less mobility than the electrons in the lower conduction band. As a result, negative differential conductivity is created and the high field domain is formed. The oscillation and amplification of the transferred electron devices are determined by the dynamic negative conductivities of the device, the circuit admittance, and the device susceptance. If the oscillation frequency of the device is equal to the reciprocal of the electron transit time of the high field domain crossing the device, then the oscillation is the Gunn effect oscillation.

n-type GaAs is the original material for the TED. But for higher frequency applications, InP is widely used. The high frequency limitation is not due to the device transit time but the packaging RC time constant. By compensating the diode capacitance C electronically, the upper frequency limit of a TED is extended to several hundred GHz with appreciable output power [17].

Problems

5.1 Write down Schrödinger's equation and identify each term.

5.2 In a picture tube of a television set, electrons are accelerated through 20,000 V. Find the wave number of one of the electrons.

5.3 Define a wave function.

5.4 State what makes electrons transfer from the lower conduction band to the upper conduction band.

5.5 List the results of the electron transfer.
5.6 Explain the current–voltage curve of a TED.
5.7 Explain the mechanism of the dynamic negative differential conductivity of a TED.
5.8 Explain, step by step, the process of the high field domain formation in the TED.
5.9 Explain the process of the charge accumulation layer development in the high field domain in the TED.
5.10 Explain how the negative conductance of TED differs from the negative conductance of tunnel diodes and avalanche diodes.
5.11 Explain how the mechanism of negative conductance formation in a TED differs from negative conductance formation in a tunnel diode and an avalanche diode.
5.12 List the factors which influence the average drift velocity of the high field domain in a TED.
5.13 List the factors which determine the oscillation frequency of a TED.
5.14 In an experiment involving a Gunn diode, simultaneous oscillation at 10.525 GHz and 10.723 GHz is observed. Offer a plausible explanation for this.
5.15 Explain why TED amplifiers are not as popular as TED oscillators.

References

1 J. B. Gunn, "Instabilities of current and of potential distribution in GaAs and InP. *In* J. Bok, ed., "Plasma Effects in Solids." Dunod, London, 1965.
2 E. S. Yang, "Fundamentals of Semiconductor Devices." McGraw-Hill, New York, 1978.
3 B. G. Streetman, "Solid-State Electronic Devices," 2nd ed. Prentice-Hall, Englewood Cliffs, New Jersey, 1980.
4 R. F. Soohoo, "Microwave Electronics." Addison-Wesley, Reading, Massachusetts, 1971.
5 T. S. Laverghetta, "Practical Microwaves." Sams & Co., Indianapolis, Indiana, 1984.
6 S. Y. Liao, "Microwave Solid-State Devices." Prentice-Hall, Englewood Cliffs, New Jersey, 1985.
7 D. Roddy, "Microwave Technology." Prentice-Hall, Englewood Cliffs, New Jersey, 1986.
8 S. Y. Liao, "Microwave Devices and Circuits." Prentice-Hall, Englewood Cliffs, New Jersey, 1985.
9 T. C. Edwards, "Introduction to Microwave Electronics." Arnold, London, 1984.
10 A. J. B. Fuller, "Microwaves," 2nd ed. Pergamon, Oxford, England, 1979.

11 N. P. Cook, "Microwave Principles and Systems." Prentice-Hall, Englewood Cliffs, New Jersey, 1986.
12 J. T. Coleman, "Microwave Devices." Reston Publ., Reston, Virginia, 1982.
13 A. van der Ziel, "Solid-State Physical Electronics," 2nd ed. Prentice-Hall, Englewood Cliffs, New Jersey, 1976.
14 A. Bar-Lev, "Semiconductors and Electronic Devices." Prentice-Hall, London, 1979.
15 H. E. Thomas, "Handbook of Microwave Techniques and Equipment." Prentice-Hall, Englewood Cliffs, New Jersey, 1972.
16 J. A. Seeger, "Microwave Theory, Components and Devices." Prentice-Hall, Englewood Cliffs, New Jersey, 1986.
17 I. Song and D.-S. Pan, "Analysis and simulation of the quantum well injection transit time diode." *IEEE Trans. Electron Devices* ED-**35**(12), 2315–2322 (1988).
18 R. A. Strangeway, T. K. Ishii, and J. S. Hyde, Design and fabrication of a 35 GHz, 100 mW low phase noise Gunn diode oscillator. *Microwave J.*, **31**(7), 107–111 (1988).
19 T. Cutsinger, Millimeter-wave cavity-stabilized GaAs Gunn oscillator. *Microwave Syst. News Commun. Tech.* **17**(12), 8–16 (1987).

6

Impact Avalanche Transit Time Diodes

6.1 Carrier Injection by Avalanche Effect

The impact avalanche transit time effect takes place when a p–n$^+$ junction diode is reverse biased through a microwave resonator, as shown in Fig. 6.1 [1–3]. The dc bias voltage V_o is adjusted to the verge of avalanche breakdown at the p–n$^+$ junction. The n-type semiconductor, usually silicon, is doped enough to short circuit the electric field inside the n$^+$ semiconductor as seen from the field plot of E as a function of x diagram in Fig. 6.1 [7,12,13,17]. There is a high field at the junction due to carrier depletion caused by the reverse bias. The doping in the p-side is graded and the doping density is greater toward the cathode. The electric field distribution in the p-type semiconductor is therefore large at the junction, but decreases in magnitude toward the cathode as sketched in Fig. 6.1. At $x = 0$, the electric field junction is at the maximum, E_o (V/m). Toward the cathode, the field strength decreases. At the cathode, $x = d$, and the field strength is zero as it reaches the conducting cathode. Under this condition, if a small amount of microwave bias field e (V/m) is superimposed on E_o, as shown in Fig. 6.1, then $E_o + e$ momentarily exceeds the avalanche breakdown field strength E_a,

$$(E_o + e) > E_a \qquad (6.1.1)$$

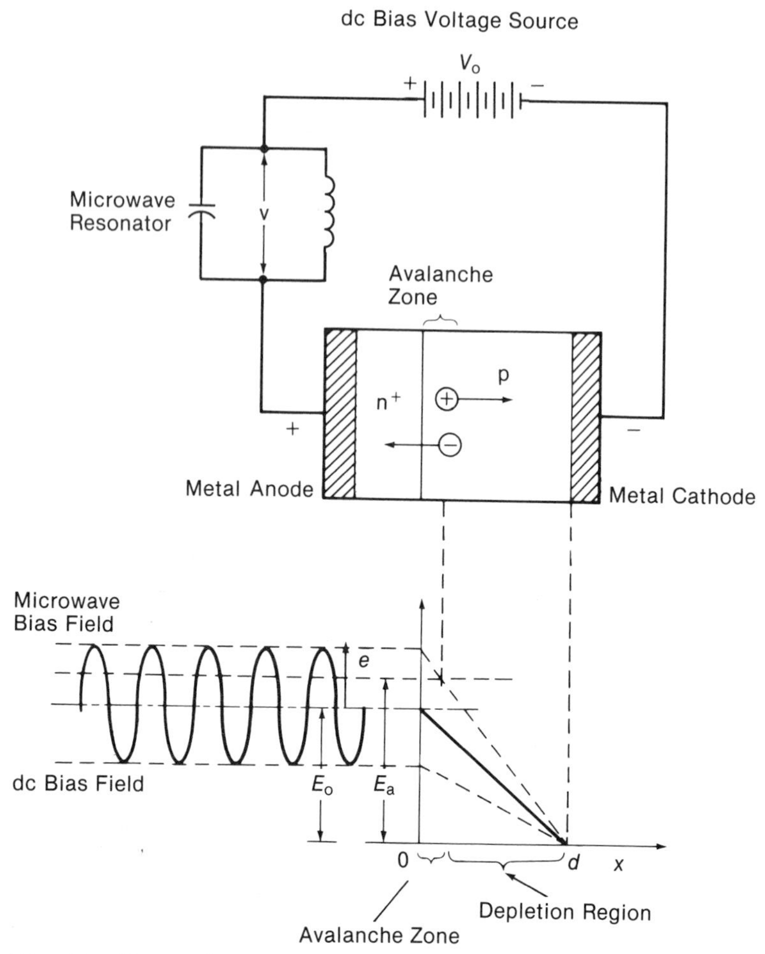

Fig. 6.1 Biasing diagram of an impact avalanche transit time device.

At this moment, the impact avalanche ionization multiplication will occur at the junction. Thus electron–hole pairs are generated and the carrier is injected into the drift region, which is in the p-type semiconductor. The carrier injection is only momentary. As soon as the microwave bias voltage and the dc field $(e + E_o)$ become less than the breakdown voltage E_a,

$$(E_o + e) < E_a \tag{6.1.2}$$

the avalanche breakdown ceases and there will be no more carrier injection. As seen from the illustration in Fig. 6.1, E_a is exceeded by $(E_o + e)$ at least once per period. E_a is known to be 9 kV/cm for Ge and 20 kV/cm for Si in commercially available IMPATT diodes. The generated electron–hole pair is quickly separated. The electrons are immediately discharged at the anode. The holes travel through the p-region to reach the cathode. The drift velocity of the holes is known to be 6×10^6 m/s for Ge and 5×10^6 m/s for Si in commercially available IMPATT diodes. The holes will be discharged at the cathode at $x = d$, thus producing a relatively large current i. If the transit time τ of the holes crossing the p-device coincide with the period T of the small microwave voltage e, or

$$\tau = T \tag{6.1.3}$$

then the large cathode current energizes the microwave resonator when the discharged current i is properly phased with the resonator voltage e. Positive feedback is accomplished and the system goes into oscillation. This type of p–n$^+$ junction is called an *im*pact *a*valanche *t*ransit *t*ime (IMPATT) diode.

The important mechanism here is that the small microwave voltage produces a large microwave current by periodically injecting carriers by means of momentary impact avalanche ionization multiplication.

For example, if an IMPATT diode is operating at $f =$ 26 GHz, then the period $T = 1/f = 1/(26 \times 10^9\ \text{s}^{-1}) = 0.038 \times 10^{-9}$ s. This means that the electron transit time across the drift region is $\tau = T = 0.038$ ns. If the IMPATT diode is made of Si, the drift velocity $u = 5 \times 10^6$ m/s [1]. By taking the average velocity of a triangular velocity distribution throughout the drift region, the thickness of the p-type drift region is therefore approximately $d = u\tau/2 = 0.5(5 \times 10^6\ \text{m/s}) \times 0.038 \times 10^{-9}\ \text{s} = 95 \times 10^{-6}$ m. Now, $E_a = 20$ kV/cm for Si [1]. By taking the average field strength of the E field of a triangular distribution, the dc bias across the drift region must be $V_a = E_a d/2 = 0.5(20 \times 10^5\ \text{V/m})(95 \times 10^{-6}\ \text{m}) = 95$ V. As seen from Fig. 6.1, the voltage drop across thin n$^+$-layer and metallization is negligible compared with the voltage across the p-region. $V_a = 95$ V is considered to be the avalanche breakdown voltage of this 95 μm thick reverse biased IMPATT diode. If the carrier injection angle is θ, then from Fig. 6.1

$$E_o + e\cos\left(\frac{\theta}{2}\right) = E_a \tag{6.1.4}$$

$$\theta = 2\cos^{-1}\left(\frac{E_a - E_o}{e}\right) \tag{6.1.5}$$

For example, if $E_a = E_o$, then $\theta = \pi$. The carrier is injected for the entire half-period of microwave cycle. On the other hand, if θ

is chosen to be $\pi/4$, then

$$\cos\left(\frac{1}{2} \cdot \frac{\pi}{4}\right) = \frac{E_a - E_o}{e}$$

$$\frac{E_a - E_o}{e} = \cos\left(\frac{\pi}{8}\right) = 0.92388$$

Since $E_a = 2 \times 10^6$ V/m, the bias dc field strength E_o must be adjusted so that E_o together with e produces the desired carrier injection angle θ. From Fig. 6.1, it is seen that when the oscillation is weak, E_o must be close to E_a. When oscillation starts, θ is small. As the oscillation gets stronger and approaches steady state, θ gets larger. Ideally, as soon as the bunch of the carriers discharges at the anode, new carriers should be injected and multiplied at the anode. θ should therefore be less than $\pi/4$. In Eq. 6.1.4,

$$1 + \frac{e}{E_o}\cos\left(\frac{\theta}{2}\right) = \frac{E_a}{E_o}$$

$$\frac{e}{E_o} = \left(\frac{E_a}{E_o} - 1\right) \Big/ \cos\left(\frac{\theta}{2}\right) \tag{6.1.6}$$

So if $\theta = \pi/4$, E_o is chosen to be 1.9×10^6 V/m for $E_a = 2 \times 10^6$ V/m, then $e/E_o = [(2 \times 10^6 \text{ V/m}/1.9 \times 10^6 \text{ V/m}) - 1]/\cos(\pi/8) = 0.057$. This means that the amplitude of the microwave field strength $e = 0.057E_o$ V/m $= 0.057 \times 1.9 \times 10^6$ V/m $= 0.1082 \times 10^6$ V/m. E_o is translated to the terminal dc bias voltage using the triangular approximation as shown in Fig. 6.1, $V_o = 0.5E_o d = 0.5(1.9 \times 10^6 \text{ V/m})(95 \times 10^{-6} \text{ m}) = 90.25$ V. The amplitude of the microwave voltage v_o, which is superimposed on the dc bias V_o, is $v_o = 0.5ed = 0.5(0.1082 \times 10^6 \text{ V/m})(95 \times 10^{-6} \text{ m}) = 5.14$ V.

6.2 Oscillation by Impact Avalanche Transit Time Effect

When the p–n$^+$ junction diode is reverse biased at avalanche breakdown voltage, an appreciable amount of shot noise occurs in addition to the thermal agitation noise. If a microwave resonator is placed in the bias circuit, as shown in Fig. 6.1, an appreciable amount of microwave voltage will develop across the resonator due to the filter action and resonance effect. As stated in Section 6.1, this microwave voltage v superimposed on the dc bias V_o will inject carriers into the drift region. The carriers discharge a relatively large amount of current into the cathode and into the resonator, and the oscillation builds up by the positive feedback loop through the discharged current on the cathode.

Suppose that a total hole charge Q (C) is injected at a distance x in Fig. 6.1. This hole charge Q induces the negative charge $-Q_c$ on the cathode at $x = d$ by electrostatic induction. The same hole charge Q induces the negative charge $-Q_a$ on the anode at $x = 0$ by electrostatic induction. The charge neutrality law states that

$$Q = Q_c + Q_a \tag{6.2.1}$$

For simplicity, plane parallel electrodes are assumed at $x = 0$ and $x = d$, and the amount of the charge induced is inversely proportional to the distance. Therefore, at the cathode,

$$Q_c = Q\frac{x}{d} \tag{6.2.2}$$

When the injected hole charge Q approaches the cathode, x increases so Q_c increases. The change in Q_c refers to the induced current flow in the external circuit. This current is due to the electrostatic induction by the moving charge Q. The induction current therefore starts to flow even before the hole charge Q physically reaches the cathode at $x = d$. If the induced external circuit current is I_e,

$$\begin{aligned} I_e &= \frac{\mathrm{d}Q_c}{\mathrm{d}t} = \frac{\mathrm{d}}{\mathrm{d}t}Q\frac{x}{d} = \frac{Q}{d}\frac{\mathrm{d}x}{\mathrm{d}t} \\ &= \frac{Q}{d}u \end{aligned} \tag{6.2.3}$$

where u is the average drift velocity of the injected hole charge in the p-region, then d/u is the hole transit time τ across the p-region.

$$\tau = \frac{d}{u} \tag{6.2.4}$$

Combining Eq. 6.2.3 and Eq. 6.2.4,

$$I_e = \frac{Q}{\tau} \tag{6.2.5}$$

But before the current I_e can flow, the charge Q must be formed at $x = 0$.

If the carrier injection angle is θ_o, then θ_o is the electrical angle of the microwave bias voltage V within *a portion of* the positive half-cycle as illustrated in Fig. 6.1. During this period of θ_o, the hole charge Q is formed. The charge Q is not fully discharged until the injected hole charge reaches the cathode τ seconds later or

$$\theta_o + \theta_t = \omega_o \tau \tag{6.2.6}$$

where θ_t is the hole transit angle across the drift region. Therefore, from the injection of the carrier at the junction to the discharge at the cathode, there is a phase delay angle of

$$\theta = \theta_o + \theta_t \tag{6.2.7}$$

from the peak microwave bias voltage. When the discharge at the cathode, dQ/dt, is at a maximum, it is the peak cathode current. The peak current delays the peak voltage by the angle θ.

$$\frac{\dot{v}}{\dot{i}} = \dot{z} = ze^{j\theta} \tag{6.2.8}$$

where $\dot{v}$ is the microwave bias voltage and $\dot{i}$ is the microwave current at the biasing circuit. It should be noted that there is a small biasing dc current which also flows beside the microwave current $\dot{i}$. Then $\dot{z}$ in Eq. (6.2.8) is the microwave impedance of the diode.

If the delay angle is such that

$$\frac{\pi}{2} < \theta \leq \pi \tag{6.2.9}$$

then $\dot{z}$ contains the negative real part or negative resistance. The negative resistance is

$$r = z \cos \theta = z \cos(\theta_o + \theta_t) \tag{6.2.10}$$

As discussed in Chapters 3, 4, and 5, the negative resistance is responsible for oscillation and amplification. As seen from Eq. 6.2.10, the negative resistance is determined by, among other things, the delay angle θ. The maximum realizable negative resistance occurs at

$$\cos \theta = \cos(\theta_o + \theta_t) = -1 \tag{6.2.11}$$

or

$$\theta = \theta_o + \theta_t = \pi \tag{6.2.12}$$

to meet the condition 6.2.9.

If the circuit is matched to this condition, from Eq. 6.2.8,

$$P_L = ri^2 = R_L i^2 = \frac{v^2}{R_L} \tag{6.2.13}$$

$$X = X_L \tag{6.2.14}$$

where the negative impedance

$$\dot{z} = r(\omega) + jX(\omega) \tag{6.2.15}$$

and the load impedance

$$\dot{Z}_L = R_L(\omega) + jX_L(\omega), \tag{6.2.16}$$

the oscillation frequency is the simultaneous solution of Eqs. 6.2.13 and 6.2.14 with respect to the angular frequency ω. The microwave output power is given by Eq. 6.2.13.

For example, if the IMPATT diode of 26 GHz described in Section 6.1 produces microwave output power $P_L = 500$ mW $= 5 \times 10^{-1}$ W to matched output microwave impedance $R_L = 50\ \Omega$, then the microwave voltage across the output terminal is $v = \sqrt{R_L P_L} = \sqrt{50\ \Omega \times 5 \times 10^{-1}\ \text{W}} = 5$ V. The output microwave current is $i = \sqrt{P_L/R_L} = \sqrt{5 \times 10^{-1}\ \text{W}/50\ \Omega} = 10^{-1}$ A $= 100$ mA. As calculated in Section 6.1, this 5-V, 100-mA, 500-mW microwave power is generated by dc bias $V_o = 90.25$ V $+ 5.19$ V $-$ 5 V $= 90.39$ V. The amount of charge Q generated at the avalanche region can be calculated from Eq. 6.2.3.

$$Q = id/u = (5 \times 10^{-1}\ \text{A})(95 \times 10^{-6}\ \text{m})/5 \times 10^{6}\ \text{m}$$

$$= 9.5 \times 10^{-12}\ \text{C} = 9.5\ \text{pC}.$$

Since the charge of an electron is $q = 1.6 \times 10^{-19}$ C, then the number of holes at the avalanche region is after the avalanche effect, $N_h = Q/q = 9.5 \times 10^{-12}\ \text{C}/1.6 \times 10^{-19}\ \text{C} = 5.94 \times 10^7$. While this does not look like very many, they are confined in a small region of the avalanche region. Now, if the diode has a cross section of $A = 150\ \mu\text{m} \times 150\ \mu\text{m}$ and the avalanche region has a thickness of 9.5 μm, then 5.94×10^7 holes are confined in $150 \times 150 \times 9.5\ \mu\text{m}^3 = 2.14 \times 10^{-13}\ \text{m}^3$. Therefore, the hole density $n_h = 5.94 \times 10^7/2.14 \times 10^{-13}\ \text{m}^3 = 2.78 \times 10^{20}\ \text{m}^{-3} = 2.78 \times 10^{14}\ \text{cm}^{-3}$.

Since Eq. 6.2.5 can be interpreted as an average current during the oscillation period τ, then the dc diode current $I_{dc} = I_e = Q/\tau = Qf = 9.5 \times 10^{-12}\ \text{C} \cdot 26 \times 10^9\ \text{Hz} = 247 \times 10^{-3}$ A. Then the dc power consumption of this diode is $P_o = V_o I_{dc} = 90.39$ V

$\times$ 247 $\times 10^{-3}$ A = 22.3 W. The efficiency is then $(P_L/P_o) \times$ 100% = $(5 \times 10^{-1}$W)/(22.3 W) $\times$ 100% = 2.2% The microwave voltage across the diode is now $v = 5$ V and the microwave current is $i = 0.1$ A. Therefore the diode impedance is, from Eq. 6.2.8, $\dot{z} = \dot{v}/\dot{\imath} = 5$ V/0.1 A$e^{j\pi} = -50\ \Omega$, if $\theta = \pi$. This is the negative resistance. This satisfies $\dot{Z} + \dot{Z}_L = -50\ \Omega + 50\ \Omega = 0$, which is an oscillation condition and the oscillation frequency is $f = u_d/d = 0.5u/d = 0.5\ (5 \times 10^6\ \text{m/s})/95 \times 10^{-6}\ \text{m} = 26 \times 10^9$ Hz = 26 GHz, provided that the resonator is at resonance or, in Eqs. 6.2.15 and 6.2.16, $X(\omega) + X_L(\omega) = 0$. In this specific case, since $X_L(\omega) = 0$, then $X(\omega) = 0$ at resonance, or vice versa.

6.3 Amplification by Impact Avalanche Transit Time Effect

The magnitude of the negative resistance due to the impact avalanche transit time effect can be obtained from Eq. 6.2.8

$$r = z\cos(\theta_o + \theta_t) \tag{6.3.1}$$

under the amplification equilibrium

$$\frac{v^2}{|r|} + \frac{v^2}{AR_L} = \frac{v^2}{R_L} \tag{6.3.2}$$

where v is the microwave voltage across the load after tuning. Then v^2/R_L is the microwave output, v^2/r is the power generated by the negative resistance r, and the output power v^2/R_L divided by the gain A, which is v^2/AR_L, is the input power. Then

$$\frac{1}{AR_L} = \frac{1}{R_L} - \frac{1}{|r|}$$

$$A = \frac{|r|}{|r| - R_L} \tag{6.3.3}$$

Combining Eqs. 6.3.1 and 6.3.3,

$$A = \frac{|z\cos(\theta_o + \theta_t)|}{|z\cos(\theta_o + \theta_t)| - R_L} \tag{6.3.4}$$

In Section 6.2, the noise was the cause of the oscillation, and the circuit was adjusted so that

$$\frac{v^2}{|r|} > \frac{v^2}{R_L} \tag{6.3.5}$$

to start the oscillation. When the oscillation reaches its steady state or equilibrium,

$$\frac{v^2}{|r|} = \frac{v^2}{R_L} \tag{6.3.6}$$

by the nonlinearity of r, which is a function of v. In the case of amplification, instead of the naturally existing noise, there is the input signal voltage to be amplified. In order to start the amplification and yet avoid self-oscillation,

$$\frac{v^2}{|r|} < \frac{v^2}{R_L} \tag{6.3.7}$$

but the condition

$$\left(\frac{v^2}{|r|} + \frac{v^2}{AR_L}\right) > \frac{v^2}{R_L} \tag{6.3.8}$$

is necessary to start the amplification. At the steady state of amplification or under steady conditions, by the nonlinearity of the negative resistance r adjusted by itself,

$$\frac{v^2}{|r|} + \frac{v^2}{AR_L} = \frac{v^2}{R_L} \tag{6.3.9}$$

As seen from Eq. 6.3.1, if

$$\frac{\pi}{2} < (\theta_o + \theta_t) < \frac{3\pi}{2} \tag{6.3.10}$$

or

$$90^\circ < (\theta_o + \theta_t) < 270^\circ$$

then

$$r < 0 \tag{6.3.11}$$

For example, if the load or external circuit impedance $R_L = 50$ Ω then, to meet the condition 6.3.7, the IMPATT diode must be chosen with the impedance $z > R_L$ or adjusted to meet the condition 6.3.7 by adjusting the dc bias voltage V_o or by trimming the external impedance. If $z = 60\ \Omega$ is chosen and $A = 20$ dB =

100 is desired, then from Eq. 6.3.4,

$$100 = \frac{|60\cos(\theta_o + \theta_t)|}{|60\cos(\theta_o + \theta_t)| - 50} \qquad (6.3.12)$$

$$\cos(\theta_o + \theta_t) = 0.84 \qquad (6.3.13)$$

$$\theta_o + \theta_t = 33^\circ + 180^\circ = 213^\circ$$

$$= 3.72 \text{ rad} \qquad (6.3.14)$$

If the operating frequency is 26 GHz then, from Eq. 6.2.6,

$$\omega_o \tau = \theta_p + \theta_t = 3.72 \text{ rad} \qquad (6.3.15)$$

$$\tau = 3.72 \text{ rad}/\omega_o = 3.72 \text{ rad}/2\pi \times 26 \times 10^9 \text{ rad/s}$$

$$= 0.02274 \times 10^{-9} \text{ s} \qquad (6.3.16)$$

This means that the dc bias voltage V_o must be adjusted so that the overall transit time $\tau = 0.02274$ ns to obtain 20 dB power gain using a 60 Ω diode to 50 Ω load impedance. It is interesting to note that the period for 26 GHz microwave is $T = 1/f = 1/26 \times 10^9 = 0.038$ ns.

In general, r and R_L in Eq. 6.3.3 are complicated functions of the operating frequency ω. Then

$$A(\omega) = \frac{|r(\omega)|}{|r(\omega)| - R_L(\omega)} \qquad (6.3.17)$$

At the center frequency of the amplification $\omega = \omega_o$, and

$$A(\omega_o) = \frac{|r(\omega_o)|}{|r(\omega_o)| - R_L(\omega_o)} \qquad (6.3.18)$$

At the frequency band edges, $\omega = \omega^{\pm}$, where ω^+ is the high side of ω_o and ω^- is the low side of ω_o, with the half gain points,

$$\frac{1}{2}A(\omega_o) = \frac{|r(\omega^{\pm})|}{|r(\omega^{\pm})| - R_L(\omega^{\pm})} \qquad (6.3.19)$$

This is usually a complicated transcendental equation of $\omega^{\pm}$. But if the transcendental equation is approximately solved for $\omega^{\pm}$ by graphical or numerical means, then the frequency bandwidth Δf is

$$\Delta f = \frac{\omega^+ - \omega^-}{2\pi} \qquad (6.3.20)$$

Now in the simplified example of 26 GHz IMPATT diode amplifier presented here, at the center frequency of 26 GHz from Eq. 6.3.12,

$$100 = \frac{60 \cos 3.72}{60 \cos 3.72 - 50} \tag{6.3.21}$$

Now if $\dot{z}(\omega) = 60\ \Omega$ varies very slowly with ω when compared with $\theta = \omega\tau$, then at the half gain point,

$$50 = \frac{60 \cos \omega^{\pm}\tau}{60 \cos \omega^{\pm}\tau - 50} \tag{6.3.22}$$

Since V_o is kept constant and τ is constant at $\tau = 0.02274$ ns, then, solving Eq. 6.3.22 for $\omega^{\pm}\tau$,

$$\begin{aligned} \omega^{\pm}\tau &= \cos^{-1} 0.85 = \pm 0.5548 + \pi \text{ rad} \\ &= 3.6964 \text{ rad or } 2.5868 \text{ rad} \end{aligned} \tag{6.3.23}$$

Then $\omega^{+} = 3.6964\ \text{rad}/0.02274 \times 10^{-9}\ \text{s} = 162.6 \times 10^{9}\ \text{rad/s}$ and $\omega^{-} = 2.5868\ \text{rad}/0.02274 \times 10^{-9}\ \text{s} = 113.76 \times 10^{9}\ \text{rad/s}$. The frequency bandwidth is therefore

$$\begin{aligned} \Delta f &= \frac{\omega^{+} - \omega^{-}}{2\pi} = \frac{162.6 \times 10^{9}\ \text{rad/s} - 113.76 \times 10^{9}\ \text{rad/s}}{2\pi} \\ &= 7.77 \times 10^{9}\ \text{Hz} \end{aligned}$$

This means that this 26 GHz amplifier has a half power gain frequency bandwidth of 7.77 GHz.

A larger bandwidth can be obtained by making the IMPATT diode structure in the form of a distributed parameter transmission line. The anode and cathode of the IMPATT diode form a microwave transmission line in this structure [14]. This is therefore an IMPATT traveling wave amplifier.

6.4 Read Structure [4]

The impact avalanche transit time effect can be utilized by making the device structure n^{+}–p–i–p^{+} as originally proposed by W. T. Read. The biasing scheme and the electric field distribution are sketched in Fig. 6.2. To make a good ohmic contact with a metal electrode, the end layers n^{+} and p^{+} are doped heavily. An impact avalanche carrier injection occurs in the avalanche zone around the $n^{+}p$ junction whenever the combination of the dc bias field E_o and the microwave field e exceeds the avalanche breakdown field E_a. The generated electrons are discharged quickly at the anode.

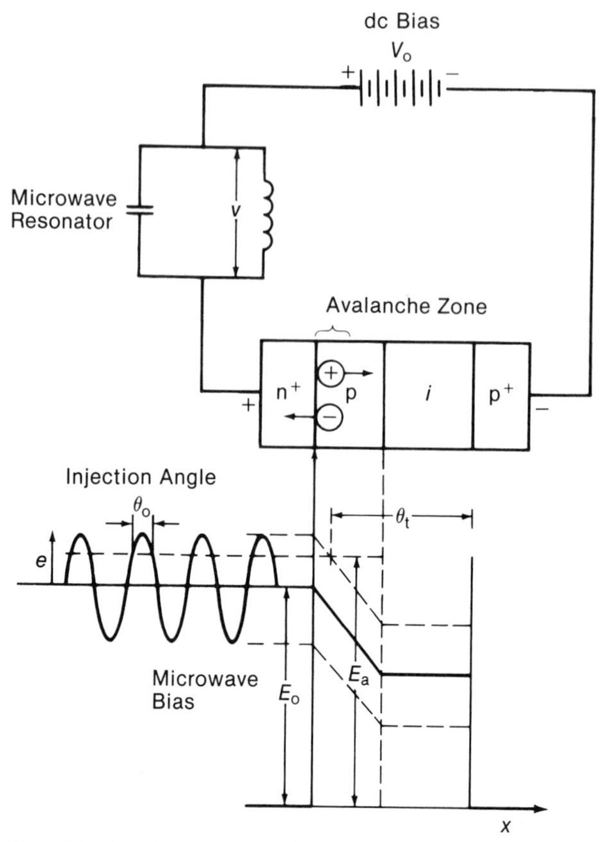

Fig. 6.2 Read structure of an IMPATT device, bias circuit, and the field distribution.

The generated holes must travel through the depletion region, which is created by the reverse bias and the intrinsic semiconductor layer to reverse bias and the intrinsic semiconductor layer to reach the cathode. This n^+–p–i–p^+ structure generally requires a relatively longer electron transit time and a higher dc voltage than the simple n^+–p structure as shown in Section 6.1. This is therefore suitable for use in low-microwave-frequency, high-power IMPATT device. In Si IMPATT devices, the impurity concentration is 10^{20} m^{-3} for N_p or N_n, 10^{24} m^{-3} for N_p+ or N_n+, and even 10^{14} m^{-3} for N_i [9, 10].

6.5 Single Drift Flat Structure

In the structures discussed up to Section 6.4, the impact avalanche transit time effect was generated by holes injected into the drift region by momentary avalanche breakdown. A similar effect can

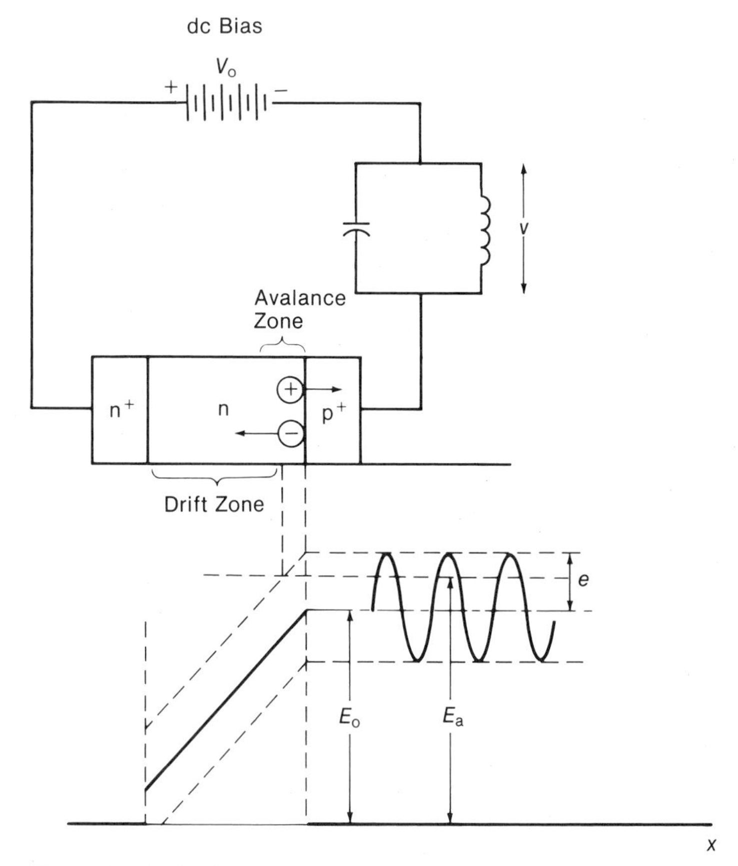

Fig. 6.3 Single drift flat IMPATT device and the electric field distribution.

be generated by electrons if the structure of the device is changed to the one shown in Fig. 6.3.

The structure is n^+–n–p^+. This structure is termed a single drift flat (SDF) structure. When reverse biased, the avalanche zone appears at the n–p^+ junction. When the dc field E_o plus the microwave field e exceeds the avalanche breakdown field E_a, the electron–hole pairs are generated by the impact avalanche effect. The generated holes exit quickly at the cathode, but the generated electrons are injected into the drift zone and drift across the n-layer to reach the anode. In this silicon structure, an efficiency of 5–6.5% has been reported [5].

6.6 Double Drift Flat Structure

In the regular IMPATT structure, generated electrons at the avalanche zone are not utilized. In the SDF structure, the holes

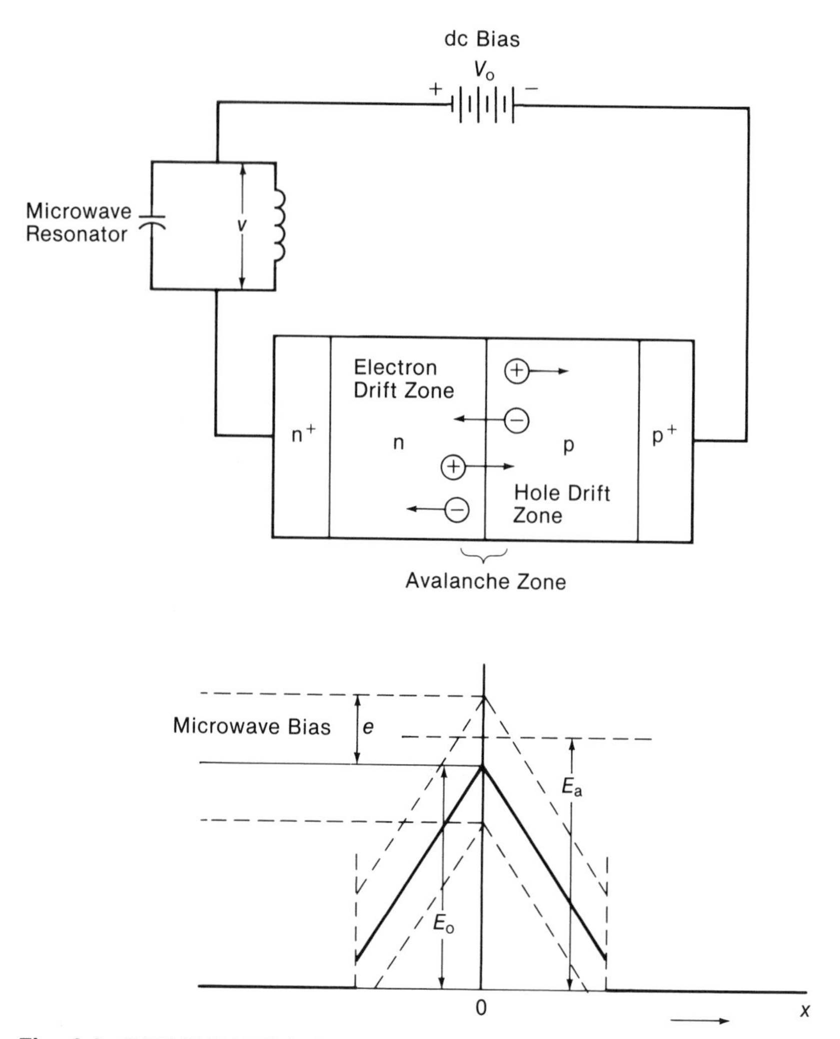

Fig. 6.4 DDF IMPATT device and the electric field distribution.

generated at the avalanche zone are not utilized at all. To utilize both the electrons and the holes generated at the avalanche zone, the double drift flat (DDF) structure has been engineered. A schematic diagram of the DDF structure is shown in Fig. 6.4. When the combined electric field strength of the dc bias and the microwave bias exceeds the avalanche breakdown field strength, there will be a momentary avalanche breakdown at the n–p junction. As illustrated in Fig. 6.4, the generated holes drift across the p-region to reach the cathode, and the generated electrons drift across the n-region to reach the anode. A critical adjustment to be made in fabricating this type of device is good synchronization of drifting electrons and holes. The transit time for the electrons and holes must be statistically equal in order to accomplish this objective. It has been reported that the efficiency of DDF IMPATT devices is 8–11% [5]. This is a significant improvement over the SDF IMPATT.

6.7 Trapped Plasma Avalanche Triggered Transit

In the ordinary IMPATT structure of an n–p device, as shown in Fig. 6.1, if the dc bias electric field E_o is almost equal to the avalanche breakdown electric field E_a by positive feedback, the microwave voltage V grows so large that the microwave field e is almost equal to E_a or

$$E_o + e \approx 2E_a \tag{6.7.1}$$

The avalanche breakdown occurs in a wide region in the p-type semiconductor. In other words, when the IMPATT is driven to a large signal level with a high dc bias, the avalanche zone is extensive, creating a large amount of electron–hole pairs by the impact avalanche ionization multiplication effect. In the positive halfcycle of the microwave field, a large amount of plasma is generated in the diode. The large amount of avalanche breakdown demands a large voltage drop through the microwave resonator impedance, which is connected in series with the dc bias voltage source. This low terminal voltage makes the carrier drift velocity slow. Before all of the electrons and holes drift away from the wide avalanche zone, the microwave field e goes to the negative half-cycle by trapping the plasma in the p-region. In the negative half-cycle, there is a field strength of at least $(E_o - e)$ V. Therefore, the electrons and holes still move slowly toward the anode and the cathode respectively. This makes the current in the negative half-cycle very small. The large amount of plasma in the positive half-cycle makes the current extremely high. In the negative half-cycle, the trapped plasma slowly disintegrates by the electrons and holes drifting away from it or by recombination of them within it. Then the cycle of the process repeats. This mode of operation is termed the trapped plasma avalanche triggered transit or TRAPATT [6,7]. The TRAPATT mode of operation efficiently produces large microwave current and power.

6.8 Barrier Injection Transit Time Devices

Instead of the n–p$^+$ structure in the SDF IMPATT device shown in Fig. 6.3, n–m structure—an n-type semiconductor–metal

Schottky barrier structure—can be utilized. The reverse biased n–m Schottky barrier behaves similarly to the n–p^+ junction in reverse biasec conditions. When the bias field momentarily exceeds the threshold voltage of the avalanche breakdown, electron–hole pairs are generated in the depletion region of the reverse biased Schottky barrier diode. These are initially triggered by tunneling electrons through the barrier from the metal to the semiconductor. The generated holes exist immediately to the metallic cathode. The generated electrons are injected to the drift region of the n-type semiconductor and then drift across the semiconductor and exit through the ohmic contact to the anode. Though the junction structure and mechanism of carrier injection is different from SDF IMPATT, other parts of their actions are similar to the one with the SDF IMPATT. This device is termed the barrier injection transit time device or the BARITT device. The actual BARITT diode is quite complicated [8], but the essential elements of the theory are as stated in this section.

6.9 Mounting Configuration of IMPATT Devices

The biasing structure of IMPATT devices is quite different from the biasing schemes of tunnel diodes, Gunn diodes or avalanche diodes. As shown throughout Chapter 6, the microwave resonator must be connected in series with the dc bias circuit of an IMPATT device. In other microwave active diodes, the microwave circuit can be either in series or parallel to the bias circuit. The microwave resonator circuit discussed in Chapter 3 is connected in parallel with the active diodes.

A typical waveguide mounting scheme for an IMPATT device is shown in Fig. 6.5. As seen from this diagram, the IMPATT diode is connected in series with a disk resonator, also called a hat resonator. The diameter of the disk is approximately one half-wavelength.

If a pill type IMPATT diode is used for an MMIC or microstrip circuit, the circuit configuration as shown in Fig. 3.5 is applicable. If a planer IMPATT diode is used, a different circuit configuration, as shown in Fig. 6.6, can be used to accomodate different diode biasing requirements. An important design concept is that the diode must be properly biased and the microwave voltage must be superimposed on the dc bias voltage [18, 19]. The configuration shown in Fig. 6.6 uses two half-wavelength resonators, one spreading over both sides of the diode and the other one at the right-hand side of the diode. The diode is biased through the microstripline resonator and $\lambda_l/4$ microwave choke

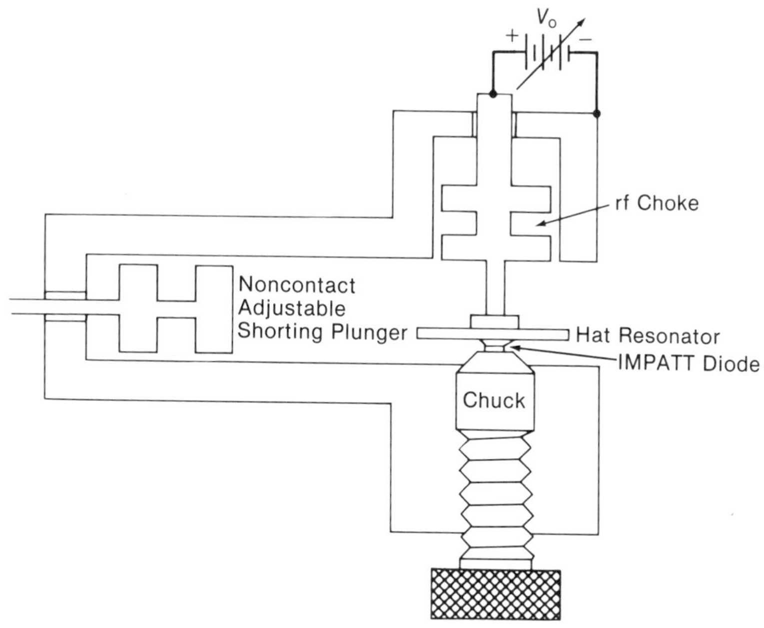

Fig. 6.5 Mounting and biasing configurations of an IMPATT diode.

lines which include the capacitor pads. An extremely thin microstrip allows for high impedance. Input impedance of the $\lambda_l/4$ thin bias line is extremely high because the line is thin and capacitively short circuited at a point $\lambda_l/4$ away from the feeding point at the microstripline resonator. At this point, microwave impedance is almost zero because it is $\lambda_l/4$ away from the open

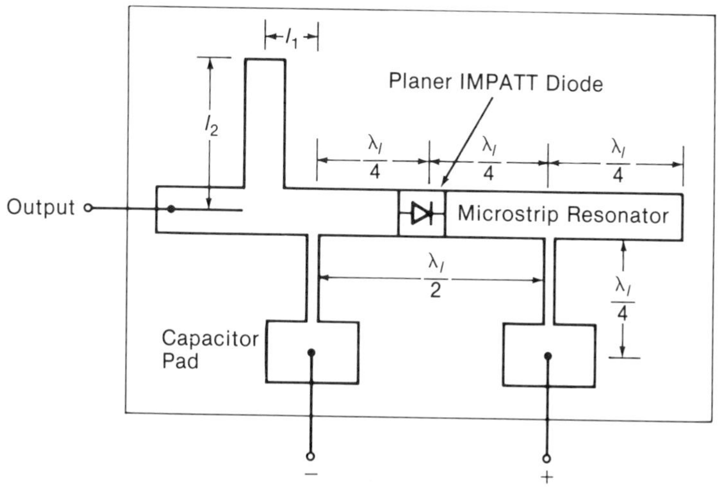

Fig. 6.6 Schematic diagram of a planer IMPATT diode microstripline oscillator circuit board.

circuit end of the microstrip resonator line. These points are therefore points of minimum disturbance to microwave oscillation for feed points of dc bias voltage. The impedance of oscillators is matched by single-stub matching. By adjusting l_1 and l_2 as shown in Fig. 6.6, desired impedance-matching or coupling to the output circuit is accomplished.

The microstrip IMPATT oscillator or amplifier shown in the figure uses microstripline resonators. Instead of microstripline resonators, dielectric resonators can be used. If so desired, by simply placing a dielectric resonator close to the microstripline, the oscillator can be fabricated [11].

For example, if the IMPATT diode operates at 26 GHz, the free space wavelength is $\lambda_o = c/f = 3 \times 10^8$ m/s/26×10^9 s^{-1} $= 1.15 \times 10^{-2}$ m = 1.15 cm. If the microstripline board is made of an alumina substrate of $\varepsilon_r = 10$, then the transmission line wavelength $\lambda_l \approx \lambda_o/\sqrt{\varepsilon_r} = 1.15$ cm$/\sqrt{10} = 0.364$ cm. The entire circuit board of Fig. 6.6 can be built on a $\lambda_l/4 \times \lambda l/4 = 0.45$ cm $\times$ 0.45 cm chip.

6.10 Impact Avalanche Transit Time Effect Devices

Several different types of IMPATT devices are studied in this chapter. In spite of some variations, the following points are common.

1. The junction is reverse biased.
2. The bias voltage is on the verge of the threshold of avalanche breakdown.
3. Superimposed microwave fields cause momentary avalanche breakdown and, by the avalanche impact ionization multiplication, a number of electron–hole pairs are generated in the avalanche zone of the junction.
4. During the momentary avalanche breakdown, and after the breakdown ceases, the carriers are injected into the drift zone and the bunch of carriers drifts across the drift zone producing a half-cycle of microwave current in the outside circuit.
5. The process repeats.
6. Good synchronization of the tuning of the resonance circuit and the electron or hole transit time across the device is needed.

The IMPATT devices are relatively high power devices when compared with other active diodes such as tunnel diodes or Gunn diodes. However, because of the nature of the avalanche breakdown, the device is noisy. Therefore, it is wise to use IMPATT devices for transmitting rather than for receiving in an rf amplifier or a local oscillator of a microwave receiver [12].

Like any other semiconductor device, the electrical characteristics of an IMPATT device varies with environmental temperature. For example, the avalanche breakdown voltage changes with temperature. Even the thermal resistivity of the semiconductor itself is a function of temperature [15]. Therefore, cooling mechanisms and heat sinking devices must be carefully designed.

There are some other versions of IMPATT devices. In addition to TRAPATT and BARITT devices, there are TUNNETT and QWITT devices. All are distinguishable by how carriers are injected into the drift region. Both the TUNNETT (tunnel injection transit time) devices and the QWITT (quantum well injection transit time) devices employ the tunnel effect to inject electrons into the drift region [16], but the TUNNETT has a homojunction structure at the injection junction and the QWITT has a heterojunction structure at the injection junction. A QWITT device is generally considered to have a better injection efficiency than a TUNNETT device [16].

Problems

6.1 Define an IMPATT device.

6.2 Explain the IMPATT action in an n^+–p structure.

6.3 In an n^+–p IMPATT structure, doping in the p-type semiconductor is graded toward the cathode. State the reason the electric field distribution gets smaller toward the cathode.

6.4 Estimate the thickness of an n^+–p IMPATT diode operating at 10.525 GHz. Compute for both Si and Ge devices.

6.5 Point out what starts IMPATT oscillation.

6.6 Sketch the waveform of the induced current in the cathode during the IMPATT action of n^+–p junction devices.

6.7 Comparatively discuss the magnitude of induced current and discharged current at the cathode.

6.8 Explain, step by step, the formation of negative impedance in an n^+–p IMPATT device.

6.9 Explain how the oscillation frequency is determined from the concept of negative impedance for an IMPATT device.

6.10 Explain what determines the oscillation power level in IMPATT diodes.

6.11 Explain, step by step, the mechanism of microwave amplification by IMPATT diodes.
6.12 Comparatively discuss the advantages and disadvantages of the n^+–p structure and the n^+–p–i–p^+ structure for IMPATT action.
6.13 Comparatively discuss the advantages and disadvantages of the n^+–p IMPATT and SDF IMPATT.
6.14 Point out the advantages of the DDF structure over the SDF structure in an IMPATT oscillator.
6.15 Explain, step by step, the principle of operation for TRAPATT devices.
6.16 State the difference, between SDF IMPATT and BARRITT devices in principles and the structure.
6.17 Sketch the proper method to bias an IMPATT device in a microwave circuit.
6.18 If a microwave resonator is to be connected in parallel to an IMPATT device, is there any way to effectively produce the microwave bias voltage in series with the dc bias voltage?
6.19 Point out the common characteristics among BARRITT, TUNNETT, and QWITT devices.

References

1 B. G. Streetman, "Solid-State Electronic Devices." Prentice-Hall, Englewood Cliffs, New Jersey, 1972.
2 J. E. Carrol, "Hot Electron Microwave Generators." Elsevier, New York, 1970.
3 R. J. Chaffin, "Microwave Semiconductor Devices." Wiley, New York, 1973.
4 W. T. Read, A proposed high frequency negative resistance diode. *Bell Syst. Tech. J.* **37**, 401–446 (March 1958).
5 W. J. Ress, IMATT sources: Striving for higher power and efficiency. *Microwave Syst. News* **8**(4), 76–82 (April 1978).
6 B. G. Streetman, "Solid-State Electronic Devices," 2nd ed. Prentice-Hall, Englewood Cliffs, New Jersey, 1980.
7 A. van der Ziel, "Solid-State Physical Electronics," 2nd ed. Prentice-Hall, Englewood Cliffs, New Jersey, 1976.
8 S. Ahmad and J. Freyer, Design and development of high power microwave silicon BARITT diodes. *IEEE Trans. Electron Devices* **ED-26**(9), 1370–1373 (1979).
9 S. Y. Liao, "Microwave Solid-State Devices." Prentice-Hall, Englewood Cliffs, New Jersey, 1985.
10 R. L. Johnson, B. C. DeLoach, and G. B. Cohen, A silicon diode microwave oscillator. *Bell Syst. Tech. J.* **44**, 369–372 (February 1965).
11 M. Dydyk and H. Iwer, Planar IMPATT diode oscillator using dielectric resonator. *Microwaves RF* **23**(10), 145–151 (1984).

12 D. Roddy, "Microwave Technology." Prentice-Hall, Englweood Cliffs, New Jersey, 1986.
13 S. Y. Liao, "Microwave Devices and Circuits," 2nd ed. Prentice-Hall, Englewood Cliffs, New Jersey, 1985.
14 M. Matsumoto, M. Tsutsumi, and N. Kumagai, A simplified field analysis of a distributed IMPATT diode using multiple uniform layer approximation. *IEEE Trans. Microwave Theory Tech.* **MTT-36**(8), 1283–1285 (1988).
15 M. J. Bailey, Anamolous temperature variation of the thermal resistivity of GaAs IMPATT diodes. *IEEE Trans. Electron Devices* **ED-35**(9), 1565–1567 (1988).
16 V. P. Kesan, D. P. Neikirk, P. A. Blakey, B. G. Streetman, and T. D. Linton, The influence of transit time effects on the optimum design and maximum oscillation frequency of quantum well oscillators. *IEEE Trans. Electron Devices*, **ED-35**(4), 405–413 (1988).
17 J. T. Coleman, "Microwave Devices." Reston Publ., Reston, Virginia, 1982.
18 B. Bayraktaroglu, Monolithic 60GHz GaAs CW IMPATT oscillators. *IEEE Trans. Microwave Theory Tech.* **MTT-36**(12), 1925–1929 (1988).
19 N.-L. Wang, W. Stacey, R. C. Brooks, K. Donegan, and W. E. Hoke, Millimeter-wave monolithic GaAs IMPATT VCO. *IEEE Trans. Microwave Theory Tech.* **MTT-36**(12), 1942–1947 (1988).

7

Microwave Autodyne and Homodyne Detectors

7.1 Autodyne and Homodyne Detection by Nonlinear Diodes

7.1.1 Autodyne Detection

In microwave electronics, autodyne detection usually means converting carrier microwaves to lower frequency signals, including the direct current or zero frequency signals. Carrier microwaves are simply rectified or detected. When the carrier signal is amplitude modulated, the demodulation takes place after detection.

Normally, an impedance $\dot{Z}$ is defined as the ratio of the voltage across the impedance $\dot{v}$ to the current $\dot{i}$:

$$\dot{Z} = \frac{\dot{v}}{\dot{i}} \tag{7.1.1}$$

or

$$\dot{v} = \dot{Z}\dot{i}. \tag{7.1.2}$$

This means $\dot{Z}$ is the dynamic impedance or the differential impedance of the microwave frequency, and is not the static impedance. If the voltage–current relationship is linear, the impedance is a constant, independent of the voltage or the current. However, if the impedance $\dot{Z}$ is dependent on $\dot{v}$ or $\dot{i}$, the voltage–current relationship is no longer linear. The impedance is then said to be nonlinear. If the microwave current $\dot{i}$ flows through the nonlinear impedance $\dot{Z}(v)$ when the microwave voltage $\dot{v}$ is applied across the nonlinear impedance $\dot{Z}(v)$, then

$$\dot{i} = \frac{\dot{v}}{\dot{Z}(v)} = \dot{Y}(v)\dot{v} \tag{7.1.3}$$

where $\dot{Y}(v)$ is the nonlinear admittance. If $\dot{v}$ is the input microwave voltage to be detected, then $\dot{i}$ contains the detected current. If $\dot{v}$ is a periodic function with respect to time, in a form of ωt, where ω is the angular frequency and t is time, then the Fourier expansion of Eq. 7.1.3 gives

$$\dot{i}(t) = \dot{Y}[v(t)]\dot{v}(t);$$

after tuning,

$$i(t) = G[v(t)]v(t) \tag{7.1.4}$$

where $G[v(t)]$ is the conductance, and

$$i(\omega t) = \frac{a_o}{2} + \sum_{n=1}^{\infty} (a_n \cos n\omega t + b_n \sin n\omega t) \tag{7.1.5}$$

where a_o, a_n, and b_n are Fourier coefficients. The dc term

$$a_o = \frac{1}{\pi}\int_{-\pi}^{\pi} i(\omega t)\mathrm{d}(\omega t) \tag{7.1.6}$$

where ω is the fundamental angular frequency. For the dc component I_o,

$$\mathrm{n} = 0 \tag{7.1.7}$$

and

$$I_o = \frac{1}{2}a_o = \frac{1}{2\pi}\int_{-\pi}^{\pi} i(\omega t)\mathrm{d}(\omega t) = \frac{1}{2\pi}\int_{-\pi}^{\pi} G[v(\omega t)]v(\omega t)\mathrm{d}(\omega t). \tag{7.1.8}$$

This is the detected current by the nonlinear admittance $\dot{Y}(v)$ or the nonlinear impedance $\dot{Z}(v)$ after tuning out the susceptance $B(v)$ or reactance $X(v)$. In practice, most of the time I_o is the self-bias current. This must be taken into consideration for integration of Eq. 7.1.8.

If $v_1(t)$ microwave voltage is sufficiently small in comparison with the dc bias V_o then

$$\frac{v_1(t)}{V_o} \ll 1 \quad \text{and} \quad v(t) = V_o + v_1(t) \tag{7.1.9}$$

by

$$G[v(t)] = G(V_o) + G(V_o)v_1 + \frac{G(V_o)}{2!}v_1^2 + \cdots + \frac{G^{(n)}(V_o)}{n!}v_1^n + \cdots \approx G(V_o) + G'(V_o)v_1(t) \tag{7.1.10}$$

and substituting Eq. 7.1.10 in Eq. 7.1.8,

$$I_o = \frac{1}{2\pi}\int_{-\pi}^{\pi}\left[G(V_o)v_1(\omega t) + G'(V_o)v_1^2(\omega t)\right]\mathrm{d}(\omega t) \tag{7.1.11}$$

If $v_1(\omega t)$ is a small microwave input signal of amplitude v_i, then

$$v_1(\omega t) = v_i\cos(\omega t + \phi) \tag{7.1.12}$$

and

$$\begin{aligned} I_o &= \frac{1}{2\pi}\left[\int_{-\pi}^{+\pi}\acute{G}(V_o)v_i\cos(\omega t + \phi)\mathrm{d}(\omega t) + \int_{-\pi}^{+\pi}G'(V_o)v_1^2\cos^2(\omega t + \phi)\mathrm{d}(\omega t)\right] \\ &= \frac{v_1^2}{2\pi}G'(V_o)\int_{-\pi}^{\pi}\cos^2(\omega t + \phi)\mathrm{d}(\omega t) \\ &= \frac{v_1^2}{2\pi}G'(V_o)\pi = \frac{G'(V_o)}{2}v_1^2. \end{aligned} \tag{7.1.13}$$

This is called the square law detection. The detected current is proportional to the square of the input voltage.

This rule holds to the self or externally biased detector. $G'(V_o)$ indicates that the derivative of the differential conductance at the microwave input is zero with the dc bias voltage at V_o. Biasing a crystal detector to find a maximum $G'(V_o)$ is a useful technique [1] because the detected current is proportional to $G'(V_o)$ as seen from Eq. 7.1.13.

For example, in the case of a 5082–2750 microwave Schottky barrier diode which is a metal–semiconductor contact diode, the measured current–voltage characteristic is as shown in Fig. 7.1 [7].

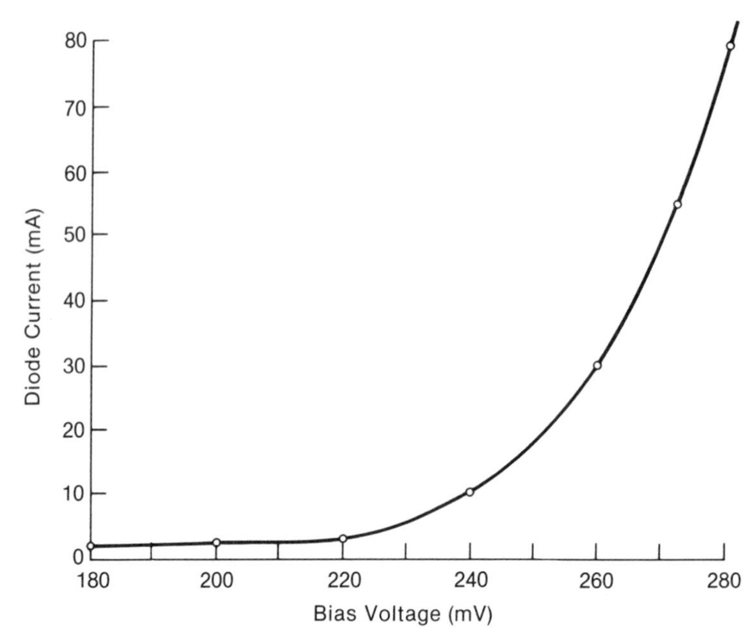

Fig. 7.1 Current–voltage characteristic curve of a 5082–2750 microwave Schottky barrier detector diode.

The diode current I is plotted against the diode bias voltage V_o. From this data, the value of dynamic conductance $G(V_o) \approx |\Delta I/\Delta V_o|_{V_o}$ at various bias voltage points V_o can be calculated and plotted. The results are shown in Fig. 7.2. In this figure the calculated dynamic conductance $G(V_o)$ is plotted against the bias voltage V_o. As seen in Fig. 7.2, the differential conductance is nonlinear. The differential conductance starts low and then increases rapidly after the bias voltage exceeds 180 mV. From this data, the derivative of the differential conductance $G'(V_o) \equiv |dG/dV_o|_{V_o}$ can be calculated and plotted. $G'(V_o)$ is plotted against the bias voltage V_o in Fig. 7.3. As seen in Fig. 7.3, $G'(V_o)$ is also nonlinear. At a bias voltage greater than 180 mV, $G'(V_o)$ increases rapidly but reaches a maximum at $V_o = 240$ mV. According to Eq. 7.1.13, the detected current I_o is proportional to $G'(V_o)$. If I_o is to be maximized for the given microwave amplitude v_i then, in the case of the 5082–2750 diode, the bias voltage should be approximately 240 mV. If this is compared with the zero bias case, estimating from Fig. 7.3, $G'(0 \text{ mV}) = 1.0$ S/V, then $G'(240 \text{ mV})/G'(0 \text{ mV}) = 30 \text{ S/V}/1.0 \text{ S/V} = 30$. This means that the detected current ratio $I_o(240 \text{ mV})/I_o(0 \text{ mV}) = 30 = 20\log_{10} 30$ dB = 20.5 dB. By simply dc biasing the diode at 240 mV, the detected current increases 30 times or 29.5 dB.

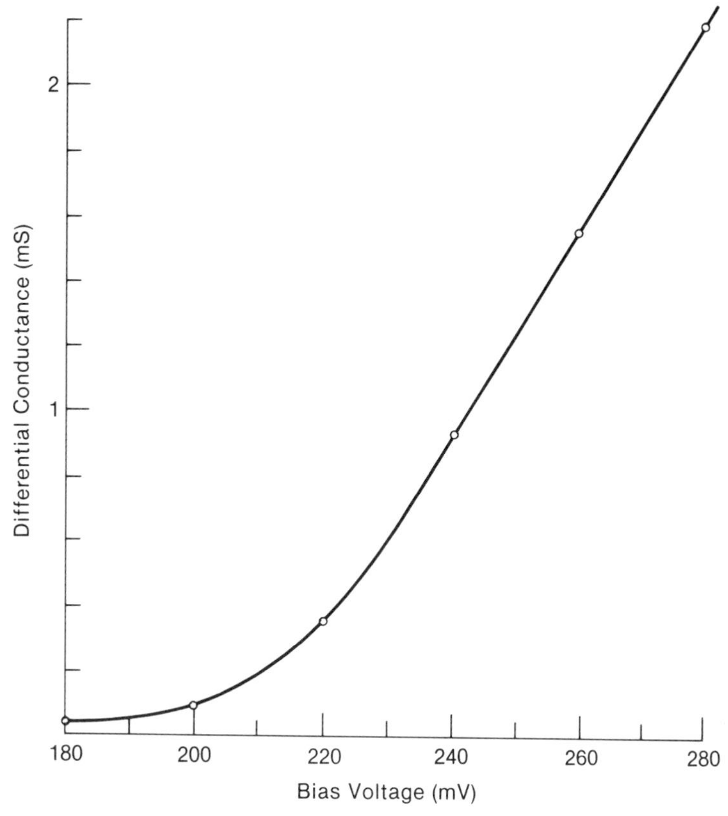

Fig. 7.2 Differential conductance of a 5802–2750 microwave Schottky diode as a function of bias voltage. $G(V_o) = \mathrm{d}I/\mathrm{d}V_o$.

It should not be confused with I_o from the dc bias current of the diode. I_o is the detected current of microwave voltage $\dot{v}_i$. For the same amount of microwave signal strength, the 5082–2750 diode has a maximum detected current at the bias voltage 240 mV. Fig. 7.1 shows that the bias voltage 240 mV would produce a dc bias current of 10 mA. According to the specifications of the diode [7], the tangential signal sensitivity of this diode is at a maximum in the neighborhood of the bias current 10 mA. The maximum tangential signal sensitivity is listed as -57.5 dBm [7]. The tangential signal sensitivity is, by definition, the signal input power level which produces a unity signal-to-noise power ratio at the output. In industrial practice it is not easy to measure the signal-to-noise ratio at the input directly because of the low noise level involved. Instead, the detected signals are amplified and the input microwave signal level which makes the amplified output signal-to-noise ratio 0 dB is observed to be the tangential signal

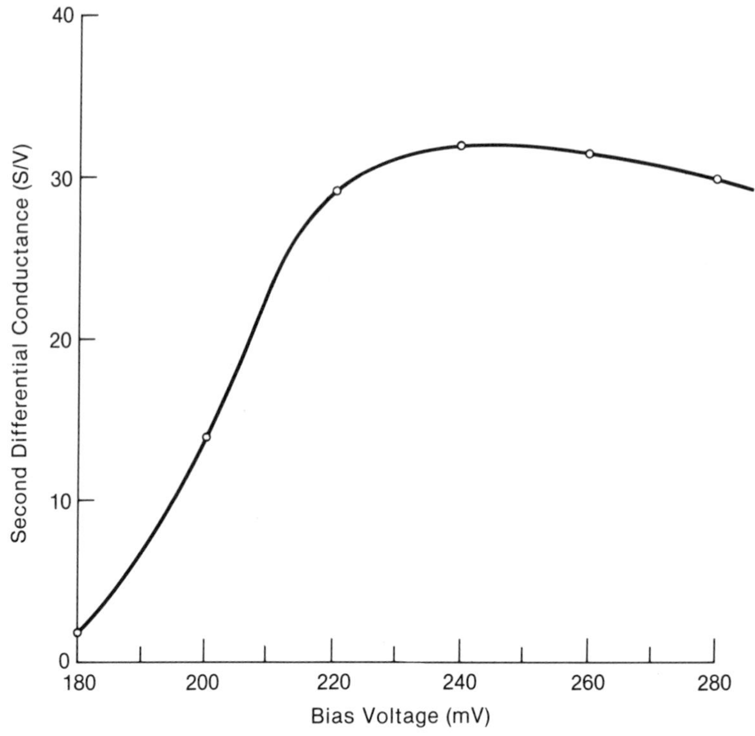

Fig. 7.3 Second differential conductance of a 5082–2750 Schottky barrier diode as a function of bias voltage. $G'(V_o) = dG/dV_o$.

sensitivity. This means that $(S/N)_{output} = 1 = 0$ dB. The $(S/N)_{input}$ can be obtained from the definition of the noise figure of the measuring system which is by definition

$$\text{N.F.} = (S/N)_{input} \tag{7.1.14}$$

Then

$$(S/N)_{input} = (\text{N.F.})(S/N)_{output} \tag{7.1.15}$$

and

$$(S/N)_{input} = 8 \text{ dB} \tag{7.1.16}$$

is very well accepted in industry [7]. This means that $(S/N)_{output} = 1 = 0$ dB and N.F. = 8 dB is implied. The actual measurement of N.F. may not be 8 dB, but when $(S/N)_{output} = 1$, then $(S/N)_{input}$ = N.F. of the measuring system.

The tangential signal sensitivity of this particular diode is specified as -57.5 dBm. This is usually for $R = 50\ \Omega$ resistance. $P_i = -57.5 \text{ dBm} = [\log_{10}^{-1}(57.5/10)]^{-1}$ (mw) $= 1.78 \times 10^{-6}$ mW $= 1.78$ nW. If the microwave input signal voltage at the peak value is v_i volts, then $P_i = 0.5(v_i^2/R)$. Then $v_i = \sqrt{2RP_i} = \sqrt{2 \times 50 \times 1.78 \times 10^{-9}} = 4.2 \times 10^{-4}$ V = 0.42 mV.

The microwave power of −57.5 dBm means it is 1.78 nW and it will develop 0.42 mV or 420 μV across the 50 Ω impedance. This voltage is indeed small when compared with the bias voltage of 240 mV.

As far as autodyne detection by a nonlinear diode is concerned, beside the tangential signal sensitivity (TSS), the current sensitivity β, the voltage sensitivity γ and the video resistance R_V are considered as specifying or characterizing parameters. If the microwave carrier frequency CW voltage in peak value is v_i V and the detected current is I_o A, then the ratio

$$\beta = \frac{I_o}{v_i/\sqrt{2}} \quad \text{(A/V)} \tag{7.1.17}$$

is called the current sensitivity. The dimensions of β are the same as for the admittance. To avoid confusion, A/V (amperes/volts) is used as the unit of the current sensitivity instead of S (Siemens). If Eqs. 7.1.13 and 7.1.17 are combined, then

$$\beta = \frac{G'(V_o)}{\sqrt{2}} v_i \quad \text{(A/V)} \tag{7.1.18}$$

For example, in the case of the 5082–2750 diode, at bias voltage 240 mV, G'(240 mV) = 30 S/V, and for microwave input power of −57.5 dBm, v_i = 0.42 mV. Therefore the current sensitivity for −57.5 dBm CW input is

$$\beta = \frac{30 \text{ S/V}}{\sqrt{2}} \times 0.42 \text{ mV} = 8.91 \times 10^{-3} \text{ A/V}$$

$$= 89.1 \ \mu\text{A/mV}.$$

In the microwave detector diode industry, the frequency of the detected current or voltage including the detected dc component is commonly called the video frequency. The resistance of the diode at this video frequency is called the video resistance. For example, the video resistance of the 5082–2750 diode is specified as R_V = 1.2–1.6 Ω [7]. If R_V = 1.6 Ω is chosen for small input power, then the detected voltage V_D across the diode is, by the definition of R_V in combination with Eq. 7.1.13,

$$V_D = I_o R_V = \frac{G'(V_o)}{2} v_i^2 R_V \tag{7.1.19}$$

At $V_o = 240$ mV, $v_i = 0.42$ mV, $G'(240 \text{ mV}) = 30$ S/V, the detected current I_o μA is then

$$I_o = \frac{G'(V_o)}{2} v_i^2 = \frac{30 \text{ S/V}}{2} (0.42 \times 10^{-3} \text{ V})^2 = 6.3 \, \mu\text{A}$$

Then from Eq. 7.1.19, the detected output voltage is $V_D = I_o R_V = 6.3 \, \mu\text{A} \times 1.6 \, \Omega = 10 \, \mu\text{V}$. The detected output current I_o is Thevenin's short circuit current. The detected output voltage is Thevenin's open circuit voltage. The detected output current I_o is in the short circuited "video" circuit. The detected output voltage is with the open circuited video circuit. V_D is therefore considered as the emf in the "video" circuit. The ratio V_D/P_i is called the voltage sensitivity.

$$\gamma \equiv V_D/P_i \tag{7.1.20}$$

In this example, the 5082–2750 diode produced $V_D = 10 \, \mu$V for microwave input power $P_i = 1.78$ nW. The voltage sensitivity of this diode therefore is $\gamma = 10 \, \mu\text{V}/1.78 \text{ nW} = 5.62 \, \mu\text{V/nW} = 5.62 \text{ mV}/\mu\text{W}$.

7.1.2 Power Rectification

In large microwave power detection for utility purposes, nonlinear diodes are used and modeled as an on–off switch. [8]. When a microwave emf applied to a utility circuit is in the positive half-cycle, the diode is turned on. When the microwave emf is in the negative half-cycle, the diode is turned off. This is a straightforward microwave rectification. If the microwave emf to be rectified is

$$E = E_o \sin(\omega t) \tag{7.1.21}$$

then the rectified output emf is

$$E_R = \frac{1}{2\pi\sqrt{2}} \int_0^{\pi} E_o \sin(\omega t) d(\omega t) = \frac{E_o}{\pi\sqrt{2}} \tag{7.1.22}$$

where E_o is the peak value of the applied microwave emf in the utility circuit.

For example, in a waveguide of characteristic impedance $Z_o = 450 \, \Omega$, if a microwave of $P = 1$ kW is applied to a rectifying diode, the microwave emf is approximately

$$E_o = \sqrt{2 Z_o P} = \sqrt{2 \times 450 \times 10^3}$$
$$= 9.49 \times 10^2 \text{ V} \tag{7.1.23}$$

then

$$E_R = \frac{E_o}{\pi\sqrt{2}} = \frac{949 \text{ V}}{\pi\sqrt{2}} = 212 \text{ V} \tag{7.1.24}$$

If the rectifying diode's output current is driving a loaded dc motor which requires $P_M = 0.5$ kW motor power and $V_M = 200$ V dc then the motor current is

$$I_M = \frac{P_M}{V_M} = \frac{500 \text{ W}}{200 \text{ V}} = 2.5 \text{ A} \tag{7.1.25}$$

The dynamic resistance of the motor is then

$$R_M = \frac{V_M}{I_M} = \frac{200 \text{ V}}{2.5 \text{ A}} = 80 \ \Omega \tag{7.1.26}$$

If the video resistance of this rectifying diode is R_V Ω, then

$$I_M = \frac{E_R}{R_V + R_M} \tag{7.1.27}$$

$$R_V = \frac{E_R}{I_M} - R_M = \frac{212 \text{ V}}{2.5 \text{ A}} - 80 \ \Omega = 4.8 \ \Omega \tag{7.1.28}$$

To date, to the author's knowledge, there is no single microwave rectifying diode of $P_M = 500$ W, $V_M = 200$ V, and $I_M = 2.5$ A with $R_V = 4.8$ Ω. A number of experiments have been done by synthesizing a diode array utilizing a large number of microwave detector diodes in the past [9,10]. Therefore, by using MMIC technology, there will be an MMIC power rectifier diode package in the near future.

7.1.3 Envelope Detection

In microwave communications, if v_i is amplitude modulated to

$$v_i = v_{io}\left[1 + m \sin(pt + \phi_p)\right] \tag{7.1.29}$$

where m is the depth of modulation, p is the modulation angular frequency, v_{io} is the amplitude of microwave carrier, and ϕ_p is the initial phase angle of the modulation signal, then the detected current is, by combining Eqs. 7.1.13 and 7.1.29,

$$I_o(pt) = \frac{G'(V_o)}{2} v_{io}^2 \left[1 + m \sin(pt + \phi_p)\right]^2$$

$$= \frac{G'(V_o)}{2} v_{io}^2 \left[1 + 2m \sin(pt + \phi_p) + m^2 \sin^2(pt + \phi_p)\right]$$

$$= \frac{G'(V_o)}{2} v_{io}^2 \left\{1 + 2m \sin(pt + \phi_p) + \frac{m^2}{2}\left[1 - \cos 2(pt + \phi_p)\right]\right\}$$

$$= \frac{G'(V_o)}{2} v_{io}^2 \left[\left(1 + \frac{m^2}{2}\right) + 2m \sin(pt + \phi_p) - \frac{m^2}{2} \cos 2(pt + \phi_p)\right] \quad (7.1.30)$$

In this equation, the first term is a dc component, the second term is the fundamental frequency of the modulation signal, and the last term is the second harmonic of the modulation signal. Thus, the square law detector output current contains the dc component and the second harmonic. If the dc and the second harmonic outputs are filtered out, the rest is the fundamental frequency current

$$I_1 = mG'(V_o) v_{io}^2 \sin(pt + \phi_p) \quad (7.1.31)$$

Thus the detected and filtered current preserves the waveform of modulation signals shown in Eq. 7.1.29, or the modulated microwave signals are demodulated or detected. The envelope of the microwave carrier is $mv_{io} \sin(pt + \phi_p)$. Therefore Eq. 7.1.31 shows that the envelope is detected.

For example, for the case of the 5082–2750 diode, if $v_{io} = 0.42$ mV and $m = 0.5$, then the microwave input signal is

$$v_1 = v_{io} \left[1 + m \sin(pt + \phi_p)\right] \cos(\omega t + \phi)$$

$$= 0.42 \times 10^{-3} \text{ V} \left[1 + 0.5 \sin(pt + \phi_p)\right] \cos(\omega t + \phi) \quad (7.1.32)$$

The detected current is, from Eq. 7.1.31,

$$I_1 = (0.5 \times 30 \text{ S/V})(0.42 \times 10^{-3} \text{ V})^2 \sin(pt + \phi_p)$$
$$= 2.646 \sin(pt + \phi_p) \ \mu\text{A} \tag{7.1.33}$$

Therefore, for the microwave autodyne detection by nonlinear conductance, $G'(V_o)$ is important as well as $G(V_o)$. The nonlinear conductance is created by various solid-state devices in microwave electronics. One of the most common structures is a Schottky barrier diode.

7.1.4 Homodyne Detection

In the case of homodyne detection, the detector diode is biased with microwaves from a local oscillator whose frequency is equal to the incident microwave signals to be detected. If the local oscillator signal is represented by

$$v_{lo}(\omega t) = v_{loo} \cos(\omega t + \psi) \tag{7.1.34}$$

Then the total voltage impressed on the detector diode v_t is, with Eq. 7.1.12,

$$v_t = v_1 + v_{lo} = v_i \cos(\omega t + \phi) + v_{loo} \cos(\omega t + \psi) \tag{7.1.35}$$

or in the phaser form

$$\dot{v}_t = \dot{v}_1 + \dot{v}_{lo} = v_1 e^{j\phi} + v_{lo} e^{j\psi} \tag{7.1.36}$$

In a triangle formed by $\dot{v}_1 + \dot{v}_{lo}$ and $\dot{v}_t$, the angle against $\dot{v}_t$, θ, is by inspection of the triangle

$$\theta = \phi + \pi - \psi \tag{7.1.37}$$

Then, by the use of a trigonometric identity,

$$v_t^2 = v_1^2 + v_{lo}^2 - 2v_1 v_{lo} \cos\theta$$
$$= v_1^2 + v_{lo}^2 + 2v_1 v_{lo} \cos(\phi - \psi) \tag{7.1.38}$$

Combining Eqs. 7.1.13 and 7.1.38, the detected current is

$$I_o = \frac{G'(V_o)}{2}\left[v_1^2 + v_{lo}^2 + 2v_1 v_{lo} \cos(\phi - \psi)\right]$$
$$= \frac{G'(V_o)}{2}\left[v_i^2 + v_{loo}^2 + 2v_i v_{loo} \cos(\phi - \psi)\right] \tag{7.1.39}$$

Since the information is contained in v_i, the amount of information containing detected current I_d is

$$I_d = \frac{G'(V_o)}{2}\left[v_i^2 + 2v_i v_{loo}\cos(\phi - \psi)\right] \qquad (7.1.40)$$

The local oscillator voltage v_{loo} does not contain any information at all, but it produces a self dc bias V_o and strengthens information containing the second term of Eq. 7.1.40. The detected current is optimized by choosing

$$\phi - \psi = 0 \qquad (7.1.41)$$

This is achieved by synchronizing the local oscillator signal to incident microwave signals to be detected.

The relationship between v_i and V_D in Eq. 7.1.24 in autodyne power rectifier detectors

$$V_o \approx \frac{V_{loo}}{\pi\sqrt{2}} \qquad (7.1.42)$$

For example, if the 5082–2750 Schottky diode is used as a homodyne detector, and the local oscillator provides $P_{lo} =$ 10 mW of power to $Z = 50\ \Omega$ load, then $v_{loo} = \sqrt{2P_{lo}Z} = \sqrt{2 \times 10 \times 10^{-3}\ \text{W} \times 50\ \Omega} = 1.0$ V. If this voltage is applied across the diode, then the dc self bias voltage V_o is developed and, from Eq. 7.1.42, $V_o \approx 1.0\ \text{V}/\pi\sqrt{2} = 0.225\ \text{V} = 225$ mV. If the incoming microwave signals to be detected and the local oscillator signals are synchronized to each other, then $\phi = \psi$ and $\cos(\phi - \psi) = 1$. By using Eq. 7.1.40, the detected current is $I_d = (30\ \text{S/V}/2)[(4.2 \times 10^{-4}\ \text{V})^2 + 2(4.2 \times 10^{-4}\ \text{V}) \times 1\ \text{V} \times 1] = 12.6 \times 10^{-3}\ \text{A} = 12.6$ mA. In this calculation, the value of $G'(225\ \text{mV}) = 30$ S/V is found in Fig. 7.3. This means that the homodyne produces 12.6 mA of detected current for 0.42 mV of microwave input signals. For the same input signal voltage, if this is an autodyne detection, it would produce only 6.3 μA at the optimum. The current gain increase of homdyne detection over autodyne detection is $20\log[12.6 \times 10^{-3}\ \text{A}/6.3 \times 10^{-6}\ \text{A}] =$ 66 dB. This is a significantly larger gain over the autodyne detection.

7.2 Detection by Schottky Barrier

At the contact of the metal and the n-type semiconductor, the electrons in the metal repel electrons in the semiconductor and create the depletion layer at the contact. This carrier depletion layer is also created by natural electron diffusion from the metal to the semiconductor neutralizing holes. This layer is known as the Schottky depletion layer. It creates a potential barrier off ϕ_b at the contact with respect to the metal's Fermi level where q is the amount of electronic charge. For example, $q\phi_b = 0.6$ eV for AlSi, 0.8 eV for AuSi, 0.8 eV for AlGaAs and 0.48 eV for AlGe diodes [2]. If the semiconductor is negatively biased by the amount of $-V$ volts the effective potential barrier is $q(\phi_b - V)$ electron-volts and, with the electrons in the semiconductor, the potential barrier is easy to overcome when the bias V is increased. The bias which makes the semiconductor side negative is termed the forward bias. Electrons in the semiconductor gain kinetic energy and overcome the effective potential barrier $(\phi_b - V)$ electron-volts and drift into the conduction band of the metal. The total energy of the electrons overcoming the barrier potential $q(\phi_b - V)$ is generally greater than that of the electrons originally in the metal. The electrons drifting out from the semiconductor to the metal are called hot electrons or hot carriers. Therefore the diodes utilizing these hot electrons are called hot carrier diodes.

Adapting the Richardson–Dushman equation of thermionic emission [2] to the semiconductor, the forward current due to electron emission from the semiconductor to the metal is given by

$$I_f = SRT^2 e\left(\frac{-q(\phi_b - V)}{kT}\right) \quad \text{(A)} \tag{7.2.1}$$

where S is the contact cross section m^2, R is the effective Richardson constant $4\pi q m^* k^2/h^3$, k is the Boltzmann constant 1.38×10^{-23} J/K $= 8.62 \times 10^{-5}$ eV/K, and q is the electronic charge 1.6×10^{-19} C; m^* is the effective mass of the electron, h is Planck's constant, 6.625×10^{-34} J-s, and $q\phi_b$ is the Schottky barrier potential created by the Schottky depletion layer.

For example, in Eq. 7.2.1, in a room temperature $T = 300$ K, $q = 1.6 \times 10^{-19}$ C, and if m^* is assumed to be equal to $m = 9.1 \times 10^{-31}$ kg, then the effective Richardson constant

$$\begin{aligned}
R &= 4\pi(1.6 \times 10^{-19}\ \text{C})(9.1 \times 10^{-31}\ \text{kg}) \\
&\quad \times (1.38 \times 10^{-23}\ \text{J/K})^2/(6.625 \times 10^{-34}\ \text{J-s})^3 \\
&= 1.20 \times 10^6 \left(\frac{\text{C-kg}}{\text{K}^2\text{-J-s}^2}\right) = 1.20 \\
&\quad \times 10^6 \frac{\text{C}}{\text{s}} \frac{1}{\text{K}^2} \frac{\text{kg}}{\text{kg-(m/s)}^2\text{-s}^2} \\
&= 1.20 \times 10^6\ \text{A/m}^2\text{-K}^2
\end{aligned} \tag{7.2.2}$$

The cross sectional area of a detector depletion layer S can be calculated from Eq. 7.2.1 as follows:

$$S = \frac{I_\mathrm{f}}{\mathrm{R}T^2} e^{(q(\phi_\mathrm{b} - V)/\mathrm{k}T)} \tag{7.2.3}$$

From Fig. 7.1, for the case of the 5082–2750 diode as an example, $I_\mathrm{f} = 80$ mA at the forward bias $V = 280$ mV, and if $T = 300$ K, Then

$$S = \frac{80 \times 10^{-3}\ \mathrm{A}}{(1.20 \times 10^{6}\ \mathrm{A/m^2\text{-}K^2})(300\ \mathrm{K})^2}$$

$$\cdot e^{(1.6 \times 10^{-19}\ \mathrm{C})(0.6 - 0.28\ \mathrm{V})/(1.38 \times 10^{-23}\ \mathrm{J/K})(300\ \mathrm{K})}$$

$$= 7.05 \times 10^{-13}\ \mathrm{m^2} e^{(0.124 \times 10^2)} = 1.712 \times 10^{-7}\ \mathrm{m^2}$$

$$= 0.414\ \mathrm{mm} \times 0.414\ \mathrm{mm} = 414\ \mu\mathrm{m} \times 414\ \mu\mathrm{m}$$

where AlSi contact is assumed. This is a relatively large contact area. If the equivalent depletion layer is $d = 10\ \mu$m thick, then the depletion layer capacitance is approximately

$$C_\mathrm{d} = \varepsilon_\mathrm{o}\varepsilon_\mathrm{r}\frac{S}{d} = 8.854 \times 10^{-12}\ \mathrm{F/m} \times 11.8 \times \frac{1.712 \times 10^{-7}\ \mathrm{m^2}}{10 \times 10^{-6}\ \mathrm{m}}$$

$$= 1.78 \times 10^{-12}\ \mathrm{F}.$$

The energy diagram of the metal n-type semiconductor Schottky barrier diode is shown in Fig. 7.4. As seen from Fig. 7.4 and Eq. 7.2.1, when the forward bias voltage V is increased, $(\phi_\mathrm{b} - \mathrm{V})$ becomes negatively large. And as seen from Eq. 7.2.1, the forward current increases exponentially. At

$$V = \phi_\mathrm{b}, \tag{7.2.4}$$

from Eq. 7.2.1

$$I_\mathrm{f} = S\mathrm{R}T^2 \tag{7.2.5}$$

at the zero bias, or

$$V = 0. \tag{7.2.6}$$

$$I_\mathrm{f} = S\mathrm{R}T^2 e^{(-q\phi_\mathrm{b}/\mathrm{k}T)} \tag{7.2.7}$$

but the net current at zero bias must be zero. So at the zero bias, the same amount of the current must flow in the reverse direction.

$$I_\mathrm{b} = S\mathrm{R}T^2 e^{(-q\phi_\mathrm{b}/\mathrm{k}T)} \tag{7.2.8}$$

therefore the net current at zero bias is

$$I = I_\mathrm{f} - I_\mathrm{b} = 0 \tag{7.2.9}$$

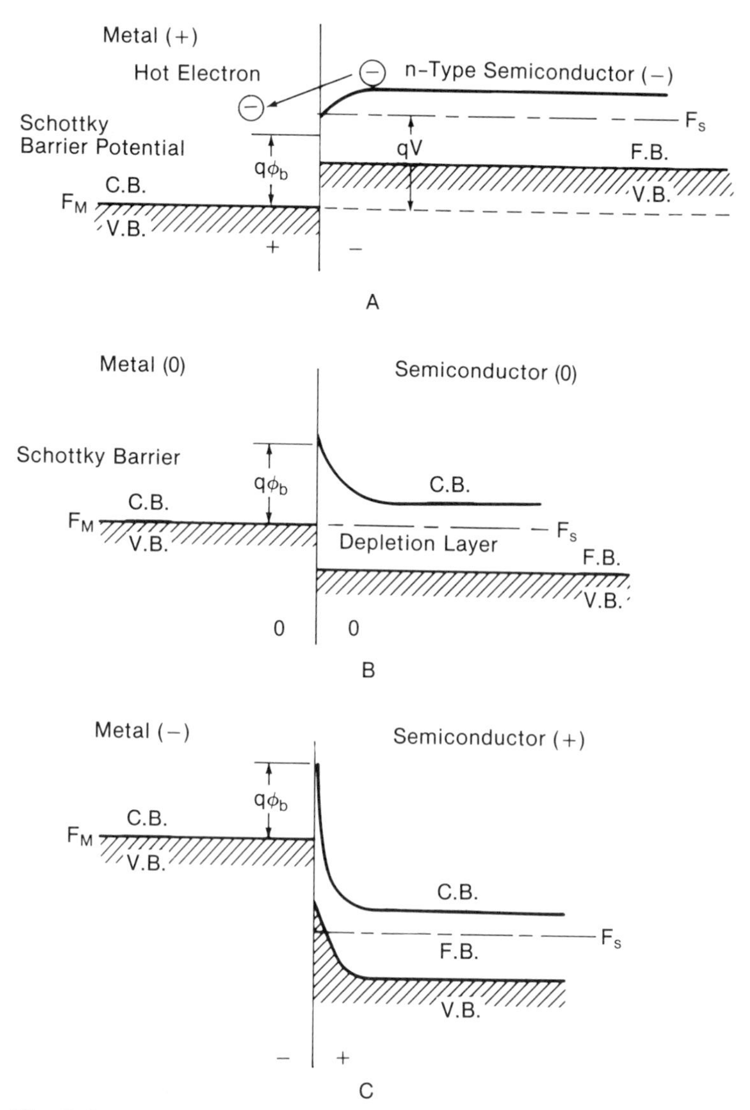

Fig. 7.4 Energy diagram of a Schottky barrier diode. A. Forward bias. B. Zero bias. C. Reverse bias. C.B., conductance band; V.B., variance band; F.B., forbidden band.

Then the net current at any bias voltage either forward or reverse should be, by interpreting V algebraically,

$$\begin{aligned} I &= I_{\mathrm{f}} - I_{\mathrm{b}} \\ &= SRT^2 e^{(-q(\phi_{\mathrm{b}} - V)/\mathrm{k}T)} - SRT^2 e^{(-q\phi_{\mathrm{b}}/\mathrm{k}T)} \\ &= SRT^2 e^{(-q\phi_{\mathrm{b}}/\mathrm{k}T)} \left(e^{(qV/\mathrm{k}T)} - 1 \right). \end{aligned} \tag{7.2.10}$$

For the reverse direction, $V < 0$ and I approaches

$$I \rightarrow -S\mathrm{R}T^2 e^{(-q\phi_b/\mathrm{k}T)} \equiv I_s \qquad (7.2.11)$$

This is the reverse saturation current then combining Eqs. 7.2.10 and 7.2.11

$$I = I_s\left(e^{(qV/\mathrm{k}T)} - 1\right) \qquad (7.2.12)$$

The current–voltage curve of a Schottky barrier diode or a hot carrier diode is sketched in Fig. 7.5. For example, if the contact is Al–Si, then $\phi_b = 0.6$ V. Using the contact area $S = 1.712 \times 10^{-7}\,\mathrm{m}^2$, the effective Richardson constant is $\mathrm{R} = 1.2 \times 10^6\ \mathrm{A/m^2\text{-}\ K^2}$, when the forward bias $V = \phi_b = 0.6$ V, from Eq. 7.2.3 at room temperature $T = 300$ K, $I_f = S\mathrm{R}T^2 = (1.712 \times 10^{-7}\ \mathrm{m^2})(1.2 \times 10^6\ \mathrm{A/m^2\text{-}\ K^2})(300\ \mathrm{K})^2 = 18.6 \times 10^3$ A. Evidently, the diode is destroyed before reaching the forward bias $V = 0.6$ V or 600 mV. When the forward bias $V = 0$ V, according to Eq. 7.2.7 the forward current of

$$\begin{aligned} I_f &= S\mathrm{R}T^2 e^{(-q\phi_b/\mathrm{k}T)} \\ &= \left(18.6 \times 10^3\ \mathrm{A}\right) e^{(-(1.6\times10^{-19}\ \mathrm{C})(0.6\ \mathrm{V})/(1.38\times10^{-23}\ \mathrm{J/^\circ K})(300\ \mathrm{K}))} \\ &= 1.58 \times 10^{-6}\ \mathrm{A} \end{aligned}$$

is supposed to be flowing even though it is small. In reality, there is no forward current. The electrons trying to diffuse into the metal from the semiconductor form the forward current I_f, which is counterbalanced by the reverse current I_b of the same amount which is formed by electrons diffusing from the metal to the

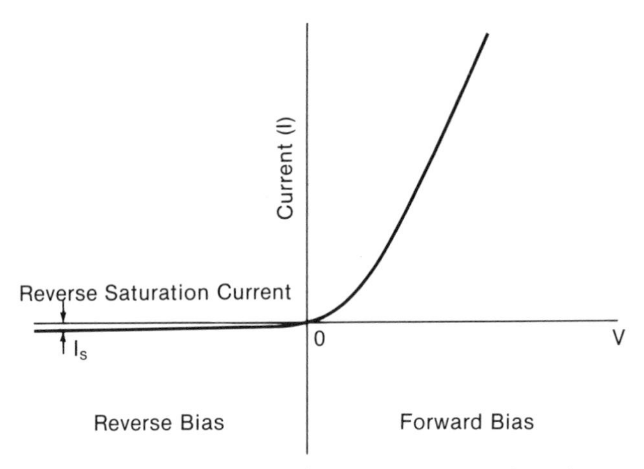

Fig. 7.5 Current–voltage characteristic curve of a Schottky barrier diode.

semiconductor. So in this case, $I_b = 1.58 \times 10^{-6}$ A, and the total current $I_o = I_f - I_b = 0$ A.

According to Eq. 7.2.11, the reverse saturation current in this case is $I_s = SRT^2 e^{(-q\phi_b/kT)} = 1.58 \times 10^{-6}$ A since $q/kT = 1.6 \times 10^{-19}$ C/(1.38 $\times 10^{-23}$ J/K)(300 K) = 38.6, the current–voltage curve of this diode must be $I = 1.58 \times 10^{-6}(e^{38.6V} - 1)$ A. If $V = 280$ mV = 0.28 V, then $I = (1.58 \times 10^{-6}$ A$(e^{(38.6 \times 0.28)} - 1)$ ≈ 0.08 A. If $V = 240$ mV = 0.24 V, then $I = 1.58 \times 10^{-6}$ A $(e^{(38.6 \times 0.24)} - 1) = 0.0167$ A = 16.7 mA. According to the measured value shown in Fig. 7.1, this is supposed to be 10 mA. At $V = 180$ mV = 0.18 V, $I = (1.58 \times 10^{-6}$ A$)(e^{(38.6 \times 0.18)} - 1) =$ 0.0016 A = 1.6 mA. According to the measured value from Fig. 7.1, this is supposed to be 1.5 mA.

In spite of the many rough assumptions made, Eq. 7.2.12 is a workable equation with reasonably acceptable errors.

The differential admittance G', which is useful for detection as seen from Eqs. 7.1.13 and 7.1.31, is obtained by differentiating Eq. 7.2.10 with

$$G(V) = \frac{dI}{dV} = \frac{qI_s}{kT} e^{(qV/kT)} \tag{7.2.13}$$

$$G'(V) = \frac{dG(V)}{dV} = \left(\frac{q}{kT}\right)^2 I_s e^{(qV/kT)} \tag{7.2.14}$$

Substituting Eq. 7.2.14 in Eq. 7.1.13, the detected dc current is

$$I_o = \frac{1}{2}\left(\frac{q}{kT}\right)^2 I_s e^{(qV/kT)} v_i^2 \tag{7.2.15}$$

where V is the dc bias voltage either by self-biasing or by being applied externally. If there is zero bias,

$$I_o = \frac{1}{2}\left(\frac{q}{kT}\right)^2 I_s v_i^2 \tag{7.2.16}$$

Therefore the current gain of dc biased autodyne detection is

$$\begin{aligned} G_I &= 20 \log \frac{I_o(\mathrm{V})}{I_o(0)} \\ &= 20 \log e^{(q\mathrm{V}/kT)} \\ &= 20 \frac{q}{kT} V \log e \\ &= 20(38.6)V \times 0.4343 = 335V \,\mathrm{dB} \end{aligned} \tag{7.2.17}$$

This formula is useful only in the square law region.

For the amplitude modulated microwaves, if the diode is dc biased V volts, then from Eq. 7.1.31 and 7.2.14 the detected modulation frequency current is

$$I_1(pt) = m\left(\frac{q}{\mathrm{k}T}\right)^2 I_s e^{(qV/\mathrm{k}T)} v_{io}^2 \sin(pt + \phi_p) \quad (7.2.18)$$

The envelope detection is accomplished. If the diode is not biased at all, the demodulation current is

$$I_1(pt) = m\left(\frac{q}{\mathrm{kT}}\right)^2 I_s v_{io}^2 \sin(pt + \phi_p) \quad (7.2.19)$$

Comparing Eqs. 7.2.18 and 7.2.19, it is known that the current gain due to dc biasing is obtained by Eq. 7.2.17.

7.3 Detection by Degenerate p – n Junction

For tunnel diodes discussed in Chapter 3, if the doping is light in the degenerate semiconductors, the Fermi level of the p-type semiconductor approaches the surface of the valence band, and the Fermi level of the n-type semiconductor approaches the lower edge of the conduction band in Fig. 3.1. Light doping means that there are not many holes available in the p-type side to accept the tunneling electrons, if they are tunneling from the n-type side. This means that the forward tunnel current is small in comparison with the heavily doped regular tunnel diode. In Eq. 3.3.11, lightly doped degenerate p–n junction diodes or tunnel diodes have smaller k_c, k_v, k_{cv}, ΔE_n, ΔE_p, $E - E_{F_p}$, $E - E_{F_n}$, $E_v - E$, and $E - E_c$ values. Therefore the forward tunnel current is small until the bias voltage reaches the valley voltage V_v. After that, the drift current begins to flow. The forward characteristic of the lightly doped degenerate p–n junction is shown in Fig. 7.6. Under reverse biased conditions, the area facing the conduction band of the n-type semiconductor against the valence band of the p-type semiconductor through the thin forbidden gap increases rapidly. This means that in Eq. 3.3.11, large k_c, k_v, k_{cv}, $E - E_{F_p}$, $E_v - E$, and $E - E_c$ values are expected. Therefore, a rapid increase in the reverse tunnel current results from the increasing reverse voltage, as shown in Fig. 7.6. The current–voltage curve of this lightly doped tunnel diode resembles an upside-down and backward regular p–n junction diode. This is the reason that this diode is called a back diode or backward diode.

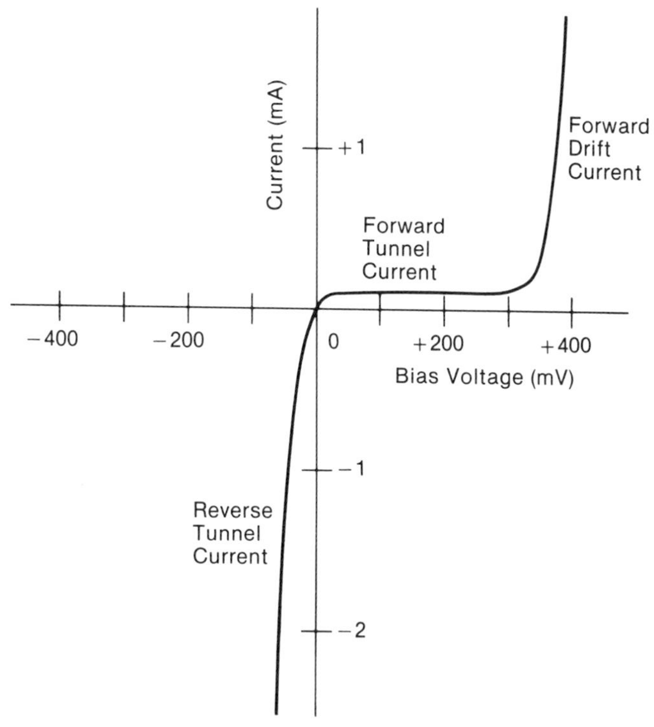

Fig. 7.6 Current–voltage characteristic curve of a lightly doped degenerate p–n junction (back diode).

The back diode has a large second differential conductance at $V = 0$ and $V = V_c$ where conduction band-to-conduction band drifting starts to occur.

$$G'(V) = \frac{d^2 I}{dV^2} \tag{7.3.1}$$

This value is obtained graphically from Fig. 7.6 or numerically from the numerical integration of Eq. 3.3.11. Once the value of $G'(V)$ is obtained, the detected dc current can be obtained from Eq. 7.1.13, and the demodulated signal from Eq. 7.1.31.

Usually, practical back diodes appear to have large differential conductances $G(V) = dI/dV$ at $V = 0$ or $V = V_c$ where hot electrons or conduction band-to-conduction band drifting begins to occur, when compared with Schottky barrier diodes. In contrast, the second differential conductance $G'(V) = d^2I/dV^2$ is not as large. For example, in a commercial back diode at $V = 0$ V, $G'(0) \leq 1.0$ S/V, and at $V_c = 400$ mV $G'(0.4\text{ V}) \leq 8$ S/V [4].

In microwave detection, one of the most important factors is $G'(V) = d^2I/dV^2$. As a result, heavily doped degenerate p–n homojunction diodes, which are common tunnel diodes [5], and heavily doped degenerate p–n heterojunction diodes [6] can be used as microwave detectors. For that matter, a common nondegenerate p–n junction or BJT or FET can also be used for microwave detection as long as the junction capacitance, lead inductance, and spreading resistance or bulk resistance are small enough.

7.4 Detector Circuits

Schottky barrier diodes, back diodes, or any other detector diodes can be mounted on a microstrip circuit or a waveguide circuit as shown in Fig. 3.5 and Fig. 3.6. It is important to tune out the diode susceptance, and this can be done by using tuning stubs or tuners. The diode susceptance is formed by the junction capacitance, package capacitance, lead inductance, and lead and bulk resistances. When the detector susceptance is tuned out, detection by nonlinear conductance takes place. Higher sensitivity is obtained under this condition. As seen in Figs. 3.5 and 3.6, it is important to have the rf choke and the dc return through a dc load or video load. In Figs. 3.5 and 3.6, for detector use, the dc bias terminals are replaced by the detected output. The dc load or the video load is connected to this detected output connector. If an external dc bias is desired for the detector diode, the bias source is added in series with the dc load or the video frequency load [3].

7.5 Autodyne Detection and Homodyne Detection

Microwave autodyne detection entails the extraction of video frequency signals from microwave signals. The extraction is done by applying microwave signals to a nonlinear impedance. The nonlinear impedance is made of various kinds of semiconductor diodes such as Schottky barrier diodes, back diodes, and tunnel diodes. One of the most important parameters is the second differential conductance of the diode $G'(V) = \mathrm{d}^2I/\mathrm{d}V^2$. This is a key parameter in determining the conversion efficiency from microwave signals to video signals. The conversion efficiency can be optimized by choosing a proper dc bias voltage V for optimizing $G'(V)$. In the case of homodyne detection, instead of a dc bias, microwave signals of the same frequency as those to be detected will be used for the bias from a local oscillator. The strength of the bias microwave must be adjusted to optimize $G'(V)$ for conversion efficiency optimization. The synchronization of the local oscillator signal to the incident microwave signal is also required for optimization. Both autodyne and homodyne detection are used for the envelope detection in microwave communication utilizing AM, ASK, and OOK type modulation. In homodyne detection, by adding a frequency tracking circuit to the local oscillator, FM, PM, FSK, or PSK type modulation can be handled for demodulation.

In microwave utility power transmission, autodyne detection is used for rectification. For better efficiency, the video resistance of the rectifier diode must be as small as possible.

In addition to seeking larger $G'(V)$ values, the lead inductance, lead resistance, junction capacitance, and package capacitance must be made small enough to operate at microwave frequencies. For optimum operation, the diode must be impedance matched for both the microwave frequency circuit and the video frequency circuit. That is, the diode must be impedance matched at the video frequency of the video circuit, and it must be impedance matched at the microwave frequency of the microwave circuit.

Problems

7.1 Discuss the possibility of microwave detection by nonlinear susceptance.

7.2 Discuss the effect of junction capacitance on detection efficiency.

7.3 Discuss an advantage of Schottky barrier detection over p–n junction detection in microwave frequencies.

7.4 Define microwave detection.

7.5 Derive an equation for square law detected current.

7.6 Point out a characteristic necessary to obtain sensitive square law detection.

7.7 For a large signal continuous wave input, devise a graphical method to obtain a dc detected current.

7.8 Discuss the role of $G(V_o)$ in amplitude modulated wave detection.

7.9 Is it possible to tune out any size junction capacitance? What are the problems of a tuned out detector?

7.10 Is it possible to perfectly and completely tune out a nonlinear admittance by a linear circuit?
7.11 Define a Schottky barrier.
7.12 Define forward bias in a Schottky barrier diode.
7.13 Define hot electrons in a Schottky barrier diode.
7.14 What causes the reverse current at zero bias to make the net current zero?
7.15 Define reverse saturation current.
7.16 Find the ratio of the detected currents of a Schottky barrier diode detector operating at 40 °C and at −30 °C.
7.17 Explain why the back diode's current–voltage characteristics are so.
7.18 Draw an equivalent circuit of a packaged detector diode mounted on a waveguide circuit.
7.19 Draw an equivalent circuit of a packaged detector diode mounted in a microstrip circuit.

References

1 T. K. Ishii and A. L. Brault, Noise output and noise figure of biased millimeter wave detector diodes. *IRE Trans. Microwave Theory Tech.* **MTT-10**, 258–262 (July 1962).
2 E. S. Yang, "Fundamentals of Semiconductor Devices." McGraw-Hill, New York, 1978.
3 T. K. Ishii and A. L. Brault, Noise figure reduction of millimeter wave receiver by crystal biasing. *Electronics* **34**, 65 (June 1961).
4 H. E. Thomas, "Handbook of Microwave Techniques and Equipment." Prentice-Hall, Englewood Cliffs, New Jersey, 1972.
5 FEI Microwave, Planar tunnel diode detectors with high power handling capability. *Microwaves RF* **27**(9), 13 (September 1988).
6 J. M. Gering, T. J. Rudnick, and P. D. Coleman, Microwave detection using the resonant tunneling diode. *IEEE Trans. Microwave Theory Tech.* **MTT-36**(7), 1145–1150 (1988).
7 Hewlett Packard, "Diode and Transistor Designer's Catalog." Hewlett Packard, San Jose, California, 1980.
8 T. K. Ishii, Theory of microwave distributed parameter power rectifiers. *J. Microwave Power* **4**(2), 70–74 (1969).
9 W. C. Brown, The receiving antenna and microwave power rectification. *J. Microwave Power* **5**(4), 235–260 (1970).
10 W. C. Brown, Progress in the efficiency of free space microwave power transmission. *J. Microwave Power* **7**(3), 223–230 (1972).

8

Microwave Superheterodyne Detectors

8.1 Microwave Superheterodyne [1]

In Eq. 7.1.4, the detected current is

$$i(t) = G[v(t)]v(t)$$

This is for autodyne detection [1]. Therefore, the only microwave signal being fed is at signal frequency ω_s and microwave signal voltage $v_s(\omega_s t)$. Equation 7.1.4 can be rewritten as

$$i_s(\omega_s t) = G[v_s(\omega_s t)]v_s(\omega_s t) \tag{8.1.1}$$

If the differential nonlinear conductance G is controlled by a local oscillator voltage $v_l(\omega_l t)$, then

$$i(\omega_s t, \omega_l t) = G[v_l(\omega_l t)]v_s(\omega_s t) \tag{8.1.2}$$

This is for heterodyne detection [5]. In heterodyne detection, the diode differential conductance is controlled by the local oscillator. If the local oscillator voltage impressed across the heterodyne detector diode is

$$v_l = v_{lo}\cos(\omega_l t + \psi) \tag{8.1.3}$$

and the incident microwave signal voltage or the microwave rf (radio frequency) voltage to be detected and impressed across the heterodyne detector diode is

$$v_s = v_{so}\cos(\omega_s t + \phi) \tag{8.1.4}$$

then similar processes occur in the case of homodyne detection. Equations 7.1.35–7.1.40 will apply when both v_l and v_s are impressed across the heterodyne detector diode simultaneously. In Eq. 7.1.40, for practical cases, $v_i = v_s \ll v_{lo} = v_{loo}$. The heterodyne detected current is therefore

$$\begin{aligned} I_d &\approx G'(V_o)v_s v_l \cos[(\omega_s t + \phi) - (\omega_l t + \psi)] \\ &= G'(V_o)v_s v_l \cos[(\omega_s - \omega_l)t + (\phi - \psi)] \end{aligned} \tag{8.1.5}$$

I_d is called the heterodyne detected current [9]. If $\omega_l = \omega_s$, it degenerates into homodyne detection, Eq. 7.1.40. The self-biased voltage V_o is, as in the case of homodyne Eq. 7.1.42,

$$V_o \approx \frac{V_{lo}}{\pi\sqrt{2}} \tag{8.1.6}$$

The difference frequency between the microwave signal frequency $f_s = \omega_s/2\pi$ and the local oscillator frequency $f_l = \omega_l/2\pi$, which is $f_{if} \equiv (\omega_s - \omega_l)/2\pi$, is called the intermediate frequency (IF). In this detection technique, the incident microwave signals are detected by observing the intermediate beat frequency with the local oscillator. This detection, by taking the beat, is called heterodyne detection. If the beat frequency or the intermediate frequency is beyond audio frequencies, then the heterodyne detection is called superheterodyne detection.

At the detector diode, two frequency signals, one from the incident rf microwaves f_s and the other from the local oscillator (LO), are mixed. The diode is therefore called the mixer diode. In practice, the mixer diode contains other frequency components and harmonics other than the intermediate frequency. Usually a filter is attached to filter out the intermediate frequency for use.

As seen from Eq. 8.1.5, the mixer, with the IF filter, is the subtractor for the frequencies. This is also called the down converter, because microwave frequencies are converted down to IF. In addition, Eq. 8.1.5 shows that it is a multiplier for the amplitude, and produces a high gain in detection. In homodyne detection, I_d in Eq. 7.1.40 is dc. A dc current is generally difficult to amplify with high gain, low noise, and without interferences. In superheterodyne detection, I_d in Eq. 8.1.5 is an intermediate frequency. The IF signals are easy to amplify using a high gain IF amplifier.

For example, $f_{if} = 30$ MHz is one of many common frequencies used as the IF signal in a microwave mixer. For example, if the incident microwave rf is $f_s = 9.376$ GHz, then the local oscillator frequency $f_l = f_s - f_{if} = 9376 \times 10^6$ Hz $- 30 \times 10^6$ Hz $= 9346 \times 10^6$ Hz. If the incident microwave rf signal power $P_s = -95$ dBm $= [\log_{10}^{-1}(95/10)]^{-1}$ mW $= 3.16 \times 10^{-10}$ mW $= 3.16 \times 10^{-13}$ W $= 0.316$ pW, then the input voltage across the $Z_o = 50\ \Omega$ line is

$$v_s = \sqrt{2P_sZ_o} = \sqrt{2 \times 3.16 \times 10^{-13}\ \text{W} \times 50\ \Omega}$$

$$= 5.62 \times 10^{-6}\ \text{V}$$

A 5082–2285 mixer Schottky barrier diode has low noise with the local oscillator power of 0 dBm [6]. So for $P_l = 0$ dBm $=$ 1 mW $= 10^{-3}$ W, and $Z_o = 50\ \Omega$,

$$v_l = \sqrt{2P_lZ_o} = \sqrt{2 \times 10^{-3} \times 50} = 0.316\ \text{V}.$$

This will develop a self-bias for the mixer diode. The self-bias voltage is obtained by the use of Eq. 8.1.6.

$$V_o = \frac{V_{lo}}{\pi\sqrt{2}} = \frac{0.316\ \text{V}}{\pi\sqrt{2}} = 0.071\ \text{V} = 71\ \text{mV}$$

From the data sheet of the current–voltage curve for the 5082–2200 series, $G'(71\ \text{mV}) \approx 7.5$ mS/V [6]. Then from Eq. 8.1.5,

$$|I_d| = (7.5 \times 10^{-3}\ \text{S/V})(5.62 \times 10^{-6}\ \text{V})(0.316\ \text{V})$$

$$= 15.174 \times 10^{-9}\ \text{A}$$

According to the specification sheet, the IF impedance at 0 dBm local oscillator power injection at 9.376 GHz for the 5082–2285 diode is $Z_{if} \approx 100\ \Omega$. This means that the IF output voltage is $v_{if} = Z_{if}I_d = 100\ \Omega \times 15.174 \times 10^{-9}\ \text{A} = 1.5174 \times 10^{-6}$ V. This means that at $v_s = 5.62\ \mu$V, a 9.376 GHz microwave signal is converted down to $v_{if} = 1.52\ \mu$V 30 MHz IF. The conversion loss $L_c = 20 \log v_s/v_{if} = 20 \log(5.62\ \mu\text{V}/1.52\ \mu\text{V}) = 11.35$ dB. The conversion loss can be decreased by increasing v_l. For example, if $P_l = 3$ dBm instead of 0 dBm as it is in this sample calculation, then the conversion loss $L_c \approx 8.35$ dB. If $P_l = 6$ dBm, then $L_c \approx 5.35$ dB. However, according to the specification sheet, the noise figure is higher for this particular diode if the local oscillator power exceeds 8 dBm [6]. For most commercial mixer diodes in the 9–12 GHz range, the conversion loss is about 7 dB, and the noise figure is about 6–7 dB.

As seen from Eq. 8.1.5, the important parameters to increase the detected output are $G'(V_o)$ and v_l. If the diode is chosen, then the local oscillator power injection P_l controls both $G'(V_o)$ and v_l.

8.2 Microwave Superheterodyne through a Schottky Barrier Diode

The current–voltage characteristic of a Schottky barrier diode is expressed by Eq. 7.2.10

$$I = S\mathrm{R}T^2 e^{(-q\phi_b/\mathrm{k}T)}\left(e^{(qV/\mathrm{k}T)} - 1\right) \tag{8.2.1}$$

thus the conductance at any bias voltage V is, from Eq. 7.2.14,

$$G'(V) = \frac{\mathrm{d}^2 I}{\mathrm{d}V^2} = \left(\frac{q}{\mathrm{k}T}\right)^2 I_s e^{(qV/\mathrm{k}T)} \tag{8.2.2}$$

where

$$I_s = S\mathrm{R}T^2 e^{(-q\phi_b/\mathrm{k}T)} \tag{8.2.3}$$

as defined in Eq. 7.2.11.

As seen from Eq. 8.1.6,

$$V \approx \frac{V_{lo}}{\pi\sqrt{2}} \tag{8.2.4}$$

The detected IF current is from Eq. 8.1.5

$$I_d = G'(V)v_s v_l \cos[(\omega_s - \omega_l)t + (\phi - \psi)]$$

If the IF impedance is Z_{if} Ω, then the IF output voltage v_{if} is

$$v_{if} = I_d Z_{if} \tag{8.2.5}$$

An example of a 5082–2285 Schottky mixer diode is given in Section 8.1. For this diode, as calculated in Section 7.2, using $\phi_b = 0.6$ V, $q/\mathrm{k}T = 38.6\ \mathrm{V}^{-1}$ at $T = 300$ K, and $\mathrm{R} = 1.2 \times 10^6$ A/m^2-K^2, and from Section 8.1, it was found that $G'(V) = G'(71\ \mathrm{mV}) = 7.5$ mS/V. From Eq. 8.2.2,

$$7.5 \times 10^{-3}\ \mathrm{S/V} = (38.6^2\ \mathrm{V}^{-2}) I_s e^{(38.6\ \mathrm{V}^{-1})(71\times10^{-3}\ \mathrm{V})}$$

$$I_s = 7.5 \times 10^{-3}\ \mathrm{S/V}/38.6^2\ (\mathrm{V}^{-1}) \times 15.5$$

$$= 3.25 \times 10^{-7}\ \mathrm{A}$$

From Eq. 8.2.3,

$$S = I_s / RT^2 e^{(-q\phi_b/kT)}$$

$$= 3.25 \times 10^{-7}\ \text{A}/1.2 \times 10^6\ (\text{A/m}^2\text{-}^\circ\text{K}^2)\, e^{(-38.6\ \text{V}^{-1})(0.6\ \text{V})}$$

$$= 3.1 \times 10^{-9}\ \text{m}^2$$

$$= (0.557 \times 10^{-4}\ \text{m})^2$$

$$= (55.7 \times 10^{-6}\ \text{m})^2$$

This is the cross-sectional area of the Schottky junction. Assuming the equivalent depletion layer is $d = 2\ \mu\text{m}$ thick, the junction capacitance of this mixer diode is

$$C_j = \varepsilon_o \varepsilon_r \frac{S}{d} = (8.854 \times 10^{-12}\ \text{F/m})(11.8)\frac{3.1 \times 10^{-9}\ \text{m}^2}{2 \times 10^{-6}\ \text{m}}$$

$$= 0.162\ \text{pF}$$

This mixer diode has less contact area than the detector diode of the example presented in Chapter 7.

8.3 Microwave Superheterodyne by Back Diode

As seen from Fig. 7.6, in the neighborhood of zero bias, a back diode presents a large second differential conductance $G'(V_o)$. As seen from Eq. 8.1.5, this is an advantage for larger IF output current. As was explained in Chapter 7, the back diode is a lightly doped tunnel diode. The tunnel diode current is given by Eq. 8.3.1 as

$$I_d = qA \int_{E_c}^{E_v} Z_{cv} [1 - f_v(E)] f_c(E) \rho_v(E) \rho_c(E)\, dE \quad (8.3.1)$$

As seen in Fig. 3.1, under biased condition, $E_v - E_c$ depends on bias. So the bias current of the back diode under the bias voltage V_o is, from Eq. 3.3.11,

$$I_d = \frac{aAk_c k_v k_{cv} V_o^{1/2}}{4} \int_{E_c}^{E_c + \Delta E_n + \Delta E_p + qV_o}$$

$$\times \left(1 - \frac{E - E_{F_p}}{2kT}\right)\left(1 - \frac{E - E_{F_n}}{2kT}\right)(E_v - E)^{1/2}(E - E_c)^{1/2}\, dE$$

$$(8.3.2)$$

then the second differential conductance

$$G'(V_o) = \frac{d^2 I_d}{dV_o^2}$$

$$= \frac{d^2}{dV_o^2} \frac{qAk_c k_v k_{cv} V_o}{4} \int_{E_c}^{E_c + \Delta E_n + \Delta E_p + qV_o}$$

$$\times \left(1 - \frac{E - E_{F_p}}{2kT}\right)\left(1 - \frac{E - E_{F_n}}{2kT}\right)$$

$$\times (E_v - E)^{1/2} (E - E_c)^{1/2} \, dE \qquad (8.3.3)$$

The IF output current is, from Eq. 8.1.5,

$$I_{if} = G'(V_o) v_s v_l \cos[(\omega_s - \omega_l)t + (\phi - \psi)] \qquad (8.3.4)$$

where V_o is determined by the local oscillator voltage v_l as

$$V_o \approx \frac{v_{lo}}{\pi\sqrt{2}} \qquad (8.3.5)$$

For example, if the back diode presented in Section 7.3 is used as a mixer diode, then with the local oscillator power injection of 0 dBm (10^{-3} W), 9.346 GHz produces 316 mV for v_{lo} and 71 mV for bias voltage V_o across the back diode. From Fig. 7.6, at 71 mV bias, G'(71 mV) ≈ 0. With additional dc bias, the bias point should therefore be corrected to be 10 mV to maximize the value of $G'(V)$. Referring to Fig. 7.6, G'(10 mV) ≈ 0.33 S/V. If the incident microwave rf signal is −95 dBm (3.16×10^{-13} W) at 9.375 GHz with $v_s = 5.62 \times 10^{-6}$ V, then the IF output current at 30 MHz is, from Eq. 8.3.4, I_{if} = (0.33 S/V(5.62×10^{-6} V)(0.316 V) = 0.586×10^{-6} A. If the IF impedance of this back diode is Z_{if} = 50 Ω, then $v_{if} = Z_{if} I_{if}$ = (50 Ω (0.586×10^{-6} A = 29.3×10^{-6} V. The conversion loss is $L_c = 20 \log(v_s/v_{if})$ = 20 log(5.62 μV/29.3 μV) = −14.3 dB. In this case it is the conversion gain, rather than the loss. If the bias point is not optimized with any dc bias correction, that is if V_o = 71 mV, then the conversion loss is a loss and it is higher. At any rate, this example shows that, if properly biased, the back diode can provide less conversion loss than a comparable Schottky diode mixer.

8.4 Microwave Mixer Circuits [2,3,7–11]

The microstrip circuit shown in Fig. 3.5 and the waveguide configuration shown in Fig. 3.6 are applicable to the mixer circuit for microwave superheterodyning with a slight modification. The input will be no longer a single frequency ω_s only. Both the incident signals of frequency ω_s and the signal from the local oscillator ω_l are fed through the same input to the mixer diode. The generated IF or beat frequency signals are taken out from the port which is used as a dc bias supply port. At this point the RFC must pass at least the IF frequency. An external dc bias may be fed through the same terminal. The first stage IF transformer will follow the IF output terminal. Though it is not commonly used, a regular tunnel diode is also usable as a mixer diode. In Figs. 3.5 and 3.6, when a Schottky barrier diode or a back diode is used as a mixer diode, the tunnel diode must be replaced by those diodes mentioned. The stud length or screw tuner adjustment must be modified to tune out the susceptance of the mixer diode. Various mixer diodes have different admittances which are unique to every device.

The mixer diode mounted as shown in Fig. 3.5 or Fig. 3.6 can be arranged in microwave circuits as shown in Fig. 8.1 A, B, and C. In these figures, transmission lines such as coaxial lines, waveguides, and microstriplines are represented by line drawings. Transmission lines in 8.1 A and C can be coaxial lines, waveguides, or microstriplines. Transmission lines in Fig. 8.1 B are waveguides only.

In Fig. 8.1 A, the microwave signals of ω_s are directly fed to the mixer diode and the local oscillator output of ω_l is simultaneously fed to the mixer diode through a directional coupler. Both signals beat with each other and the IF current of $\omega_{if} = \omega_s - \omega_l$ will flow through the mixer diode. The intermediate frequency transformer (IFT) is connected to the diode as shown in Fig. 8.1 A. Both the primary and the secondary circuits of the IFT are tuned to ω_{if}. The heterodyned frequency is therefore emphasized and filtered out to the IF output terminal. This is a single ended mixer. No provision is made to suppress the electrical noise from the local oscillator.

In the balanced mixers as shown in Figs. 8.1 B and C, the local oscillator noise is suppressed. For example, in the hybrid T-balanced mixer shown in Fig. 8.1 B, the microwave input signal of frequency ω_s is fed through the E-arm of a waveguide hybrid-T. By the very nature of the hybrid-T, the signals of ω_s split into the collinear arm with opposite phase of each other and excite mixer diodes D_1 and D_2 which are mounted at equal distances from the junction [3]. Mixer diodes D_1 and D_2 are excited by the signals of ω_s at opposite phase to each other. The two mixer diodes are connected to the primary of an IFT. The signals of ω_s do not couple to the H-arm of the hybrid-T due to its operational

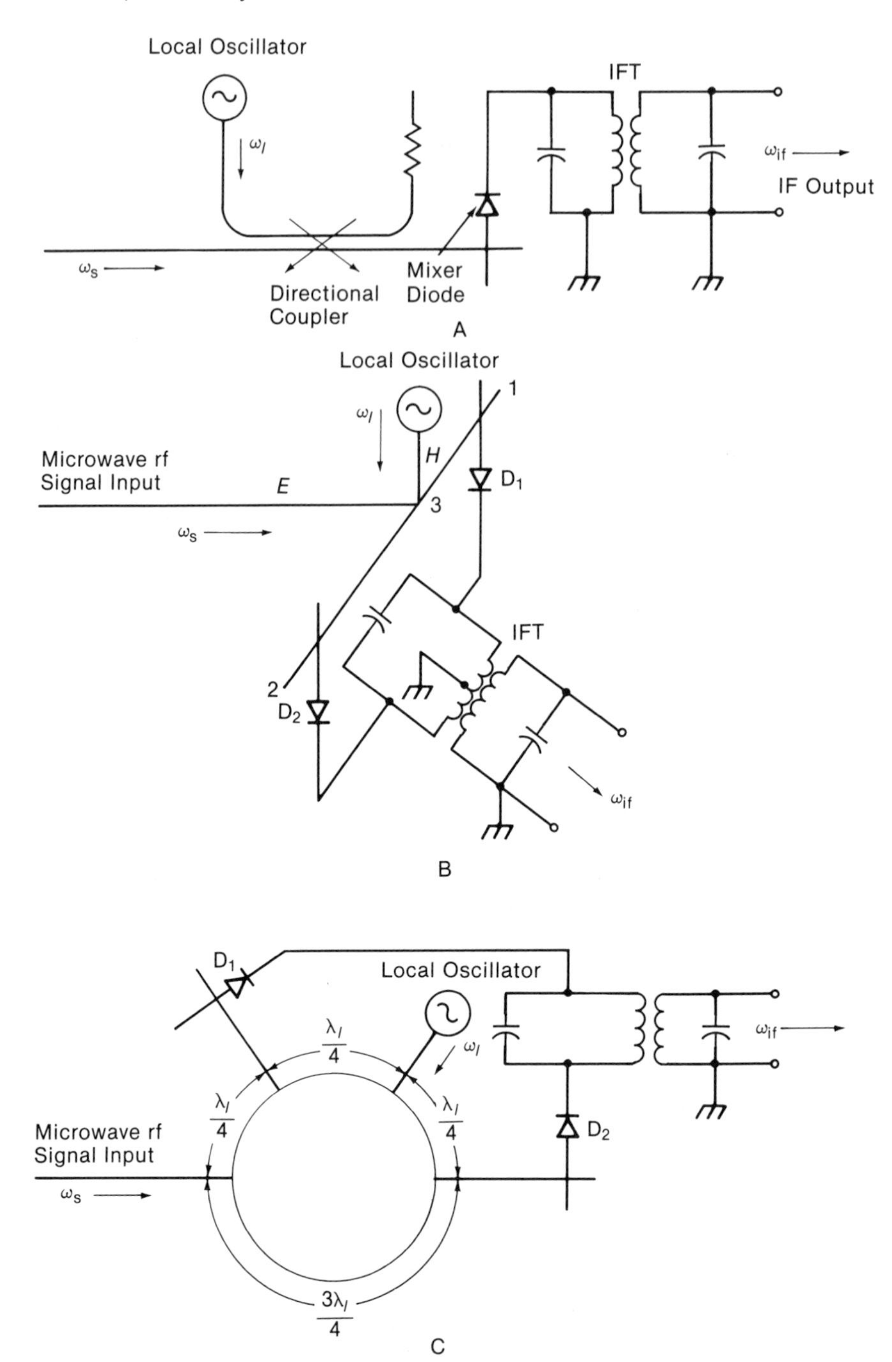

Fig. 8.1 Schematic diagrams of mixer circuits. A. Single ended mixer. B. Hybrid-T balanced mixer. C. Hybrid ring balanced mixer.

principle [3]. The output from a local oscillator of ω_l is fed through the H-arm of the hybrid-T. The local oscillator power of frequency ω_l is split into the collinear arm in phase by the very nature of the hybrid-T and excites mixer diodes D_1 and D_2 simultaneously [3]. Therefore the diodes are excited by the local oscillator in the same phase. By the design of the hybrid-T, the local oscillator power does not couple out of the E-arm [3]. Now both signals ω_s and the local oscillator power ω_l coexist in diodes D_1 and D_2. The IF signals $\omega_{if} = \omega_s - \omega_l$ are therefore generated in the diodes and flow into the primary winding of the tuned IFT. In addition, the diodes are connected in a push–pull configuration for microwave signals and a push–push configuration for the local oscillator against the primary winding of the IFT. Therefore, at IF frequency ω_{if}, the signal is strengthened but the local oscillator noise is suppressed at the primary winding because the diodes are connected in the push–push configuration to the IFT and the local oscillator noise. The IF signals of ω_{if} are coupled out through the tuned secondary winding of the IFT. Thus, the balanced mixer has the advantage of local oscillator noise suppression.

Since the hybrid-T is usually made of three dimensional waveguides, it is not adaptable to a microwave microstripline circuit. In such cases, the configuration shown in Fig. 8.1 C is used. This is a hybrid ring balanced mixer. By the very nature of the hybrid ring, the microwave signals fed into junction 1 appear at junctions 2 and 4 due to equal phase path lengths, where mixer diodes D_1 and D_2 are mounted [3]. The input signals excite diodes D_1 and D_2 in opposite phase to each other. Path length 1–2 is $\lambda_l/4$ and path length 1–4 is $3/4\lambda_l$. This means that the waves traveling from 1 to 2 and from 1 to 4 have a phase difference of 180°.

The local oscillator power of frequency ω_l is fed at the junction 3 as shown in Fig. 8.1 C. By the very nature of the hybrid ring, the local oscillator power appears at junctions 2 and 4 only, where D_1 and D_2 are mounted. Since junctions 2 and 4 are mounted at equal distances, from junction 3 respectively, both diodes D_1 and D_2 are excited by the local oscillator in the same phase. Both the incident microwave signals and the local oscillator power coexist in diodes D_1 and D_2. The IF current ω_{if} is therefore generated and fed into the primary winding of the tuned IFT which is connected to the two diodes D_1 and D_2. The generated IF current is coupled out of the secondary of the IFT. Again the diodes are connected in push–push from for local oscillator noise and push–pull form against the input signals. Thus the local oscillator noise is suppressed and the signals are intensified.

If the operating microwave frequency is in UHF or low L-band, the distribution parameter transmission line circuits as shown in Fig. 8.1 get inconveniently bulky and large.

For example, at $f_s = 30$ GHz, the free space wavelength is $\lambda_o = 10$ mm. If an alumina substrate of $\varepsilon_r = 10$ is used as a microstripline circuit board, the transmission line wavelength $\lambda_l = \lambda_o/\sqrt{\varepsilon_r} = 10\ \text{mm}/\sqrt{10} = 3.16$ mm. Therefore, the hybrid ring size of Fig. 8.1 C can be developed on an area of less than 3 mm × 3 mm. On the other hand, in UHF of 300 MHz, $\lambda_o =$ 100 cm. Therefore with the same alumina substrate, $\lambda_l = 31$ cm. This means that approximately 31 cm × 31 cm area of circuit board is needed to make the hybrid ring. This is too bulky and inconvenient. In UHF and L-band frequency applications, a lumped parameter approach is shown in Fig. 8.2 is therefore preferred. Fig. 8.2 A shows a schematic diagram of a balanced mixer and Fig. 8.2 B shows a schematic diagram of a double balanced mixer. In Fig. 8.2 A, both mixer diodes are excited with both the local oscillator (LO) power of frequency ω_l and the incident microwave rf signal of frequency ω_s. The beat frequency current of frequency ω_{if} is generated in the diodes. The IF voltage then appears across terminals A and C and it is coupled out through an IFT. Since the bridge A–B–C–D–A is balanced at ω_{if}, there is no IF voltage across terminals B and D. Across terminals B and D, only the microwave rf voltage of ω_s exists. Since the local oscillator voltage is fed through the neutral point of the windings of the transformer, as shown in Fig. 8.2 A, as far as the local oscillator frequency ω_l is concerned, the local oscillator frequency potentials at B and D are equal to each other and the local oscillator potentials at A and C are equal to each other. This means that B and D are electrically tied together and A and C are electrically tied together even though these points are physically separated from each other. This means that branches A–B and A–D are connected in parallel to each other, and branches C–B and C–D are connected in parallel to each other, and also combined branches B–A and B–D and other combined branches C–B and C–D are connected in parallel to each other at the local oscillator frequency. In short, D_1 and D_2 are connected in a push–push configuration against the local oscillator power. Thus the local oscillator noise is suppressed. Since the bridge is balanced at the local oscillator frequency, the local oscillator power does not leak out through the transformers shown above. The local oscillator voltages across terminals B and D and A and

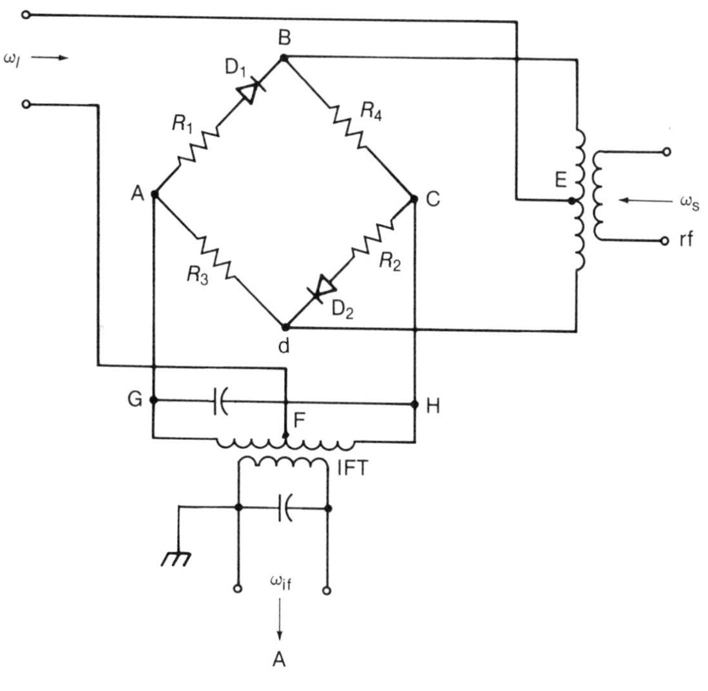

Fig. 8.2 Lumped parameter mixers. A. Balanced mixer. B. Double balanced mixer.

C are zero. At the microwave rf input signal frequency ω_s, the diodes D_1 and D_2 are connected in a push–pull configuration for the impressed voltage across terminals B and D, or alternatively activated for each successive and alternate half-cycle of ω_s. The same applies to these diodes at ω_{if} across terminals A and C also. The bridge is also balanced at ω_s, so there is no voltage across terminals A and C at ω_s. There cannot be any voltage across E and F at ω_s either. The conversion loss from microwave rf to IF can be further reduced by adding two more mixer diodes to the bridge circuit as shown in Fig. 8.2 B. This is a double balanced mixer. The principle of the double balanced mixer is almost identical to that of the balanced mixer. All four diodes carry current at ω_s, the beat frequency current at ω_{if} is generated, and the IF is coupled out of the IFT. All four diodes are equivalently connected back to back and are parallel to each other at ω_l. These diodes operate in a push–pull configuration at ω_s and ω_{if}.

For example, an RMS 2 lumped parameter microwave UHF double balanced mixer is designed for f_s or $f_l = 5$ to 1000 MHz and IF = 0 to 500 MHz [7]. This is packaged in a size 0.25 in × 0.30 in × 0.2 in or 6.35 mm × 7.62 mm × 4.9 mm. If f_s = 500 MHz microwave UHF, the free space wavelength is $\lambda_o = 60$ cm. The longest dimension of the package is 7.62 mm/600 mm = 0.013 of a wavelength. A properly packaged lumped parameter mixer saves space. Typical conversion loss of the RMS 2 is specified as 6.5 dB. At midband, the isolation between the local oscillator terminal to microwave rf input terminal is rated at 35 dB and the isolation between the local oscillator terminal and the IF terminal is specified at 30 dB. This tiny surface-mount technology (SMT) package, contains everything shown in a schematic diagram of Fig. 8.2 B.

8.5 Microwave Mixing

The process of producing IF by beating incident microwave rf with the LO output is called microwave mixing. The beating is done through a nonlinear impedance. Semiconductor diodes such as the Schottky barrier diodes and back diodes are easily used for the mixers nonlinear impedance component. The nonlinear characteristics of these diodes are considered to be a logic multiplier; thus beating occurs. Conversion from microwave rf to IF is not always positive. But this conversion loss can be easily compensated by the usual practice of a high gain IF amplifier following the mixer. One of the most important mixer diode parameters is the second differential conductance $G'(V) = d^2I/dV^2$. The larger the $G'(V)$, the larger the IF output. Mixers can take the form of single ended, balanced, or double balanced mixers, depending on desired mixer performance. Conversion output and noise characteristics are usually better in balanced mixers than single ended mixers. Balanced mixers are more complicated in design, fabrication, and tuning than single ended mixers. The mixers can be fabricated in a transmission line circuit using coaxial lines, waveguides, or microstriplines. These mixers can also be fabricated in compact lumped parameter circuits in the microwave UHF range.

Problems

8.1 Differentiate autodyne, homodyne, heterodyne, and superheterodyne techniques from each other.

8.2 Show graphically that G_l' is controlled by the local oscillator signal.

8.3 Draw a current–voltage curve for a commercial Schottky barrier microwave diode and check the validity of $G_l' \approx G'(V) \sin \omega_l t$.

8.4 Superheterodyning produces a conversion loss from the incident signal frequency power to the intermediate frequency output power. Show the conversion loss analytically. Is the loss positive or negative?

8.5 In a Schottky barrier diode mixer with zero bias, find $G'(0)$.

8.6 Is it possible to mix or superheterodyne through a perfectly linear conductance? Justify the conclusion.

8.7 In back diode mixing, the diode current may be a runaway condition in the reverse direction. Is $G'(0)$ also in a runaway condition?

8.8 Explain how balanced mixers work.

8.9 Explain how double balanced mixers work.

References

1 A. B. Carlson, "Communication Systems." McGraw-Hill, New York, 1968.

2 H. E. Thomas, "Handbook of Microwave Techniques and Equipment." Prentice-Hall, Englewood Cliffs, New Jersey, 1972.

3 T. K. Ishii, "Microwave Engineering." Ronald Press, New York, 1966.

4 A. Papoulis, "Signal Analysis." McGraw-Hill, New York, 1977.

5 C. A. Holt, "Electronic Circuits." Wiley, New York, 1978.

6 Hewlett Packard, "Diode and Transistor Designer's Catalog. Hewlett Packard, San Jose, California, 1980.
7 Mini-Circuits, The world's smallest surface mount mixers. *Microwave J.*, **30**(1), 204 (1987).
8 T. S. Laverghetta, "Practical Microwaves." Sams & Co., Indianapolis, Indiana, 1984.
9 A. van der Ziel, "Solid-State Physical Electronics," 2nd ed. Prentice-Hall, Englewood Cliffs, New Jersey, 1976.
10 H. H. G. Zirath, R. Irwin, R. Curby, R. Forse, and K. VanBuren , Temperature-variable noise and electrical characteristics of Au-GaAs Schottky millimeter wave mixer diodes. *IEEE Trans. Microwave Theory Tech.* **MTT-36**(11), 1469–1475 (1988).
11 J. K. Izadian, S. M. Nilsen, H. Hjelmgren, L. P. Ramberg, and E. L. Kohlberg, Uniplanar microwave double balanced mixer using miniature beam-lead crossover Schottky diode quad. *IEEE MTT-S Int. Microwave Symp. Dig.*, **2**, 691–694 (May 1988).

9

Parametric Amplification

9.1 Junction Capacitance [1,2,4]

In a depletion layer of a p–n junction as illustrated in Fig. 9.1, the one-dimensional Poisson's equation is

$$\frac{d^2V}{dx^2} = -\frac{qN_d}{\varepsilon} \tag{9.1.1}$$

where q is the electronic charge, N_d is the donor density in the n-type semiconductor, and ε is the permittivity of the depletion layer in the junction of the semiconductor. Integrating Eq. 9.1.1,

$$\frac{dV}{dx} = -\frac{qN_d}{\varepsilon}x + A \tag{9.1.2}$$

where A is an integrating constant to be determined by the following boundary condition

$$\left.\frac{dV}{dx}\right|_{x=w} \approx 0 \tag{9.1.3}$$

where $x = w$ is the end of the depletion layer and $x = 0$ is the beginning of the depletion layer. Then from Eq. 9.1.2,

$$A = \frac{qN_d w}{\varepsilon} \tag{9.1.4}$$

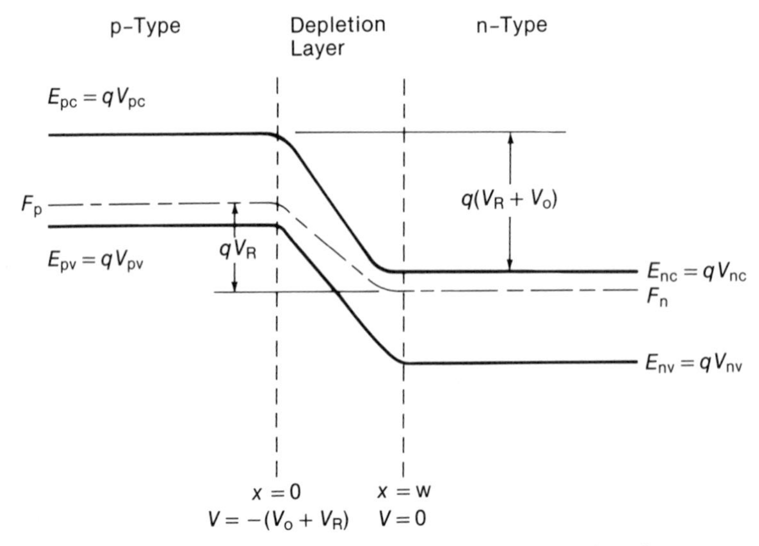

Fig. 9.1 Energy band diagram of a reverse biased p–n junction.

Substituting Eq. 9.1.4 in Eq. 9.1.2,

$$\frac{dV}{dx} = -\frac{qN_d}{\varepsilon}(x - w)$$

$$= \frac{qN_d}{\varepsilon}(w - x) \tag{9.1.5}$$

Integrating once more,

$$V = -\frac{qN_d}{2\varepsilon}x^2 + \frac{qN_d w}{\varepsilon}x + B \tag{9.1.6}$$

B is an integrating constant to be determined by the following boundary condition

$$V|_{x=w} \approx 0 \tag{9.1.7}$$

Then

$$0 = \frac{qN_d}{2\varepsilon}w^2 + B \tag{9.1.8}$$

$$B = -\frac{qN_d}{2\varepsilon}w^2 \tag{9.1.9}$$

substituting 9.1.9 in 9.1.6,

$$V = \frac{qN_d}{2\varepsilon}x^2 + \frac{qN_d w}{\varepsilon}x - \frac{qN_d}{2\varepsilon}w^2 \tag{9.1.10}$$

As seen from Fig. 9.1,

$$V|_{x=0} = -(V_o + V_R) \tag{9.1.11 A}$$

where V_R is the reverse bias voltage and V_o is the magnitude of the contact potential barrier. Applying condition 9.1.11 A to Eq. 9.1.10,

$$-(V_o + V_R) = -\frac{qN_d}{2\varepsilon}w^2 \tag{9.1.11 B}$$

$$w = \sqrt{\frac{2\varepsilon}{qN_d}(V_o + V_R)} \tag{9.1.12}$$

If the cross-sectional area of the diode is S (m^2), then the charge stored in the depletion layer is

$$Q = qN_d S w = S\sqrt{2qN_d\varepsilon(V_o + V_R)} \tag{9.1.13}$$

The differential capacitance or the dynamic capacitance C is defined as

$$C \equiv \frac{dQ}{dV_R} \tag{9.1.14}$$

substituting Eq. 9.1.13 in Eq. 9.1.14,

$$C = S\sqrt{2qN_d\varepsilon} \cdot \tfrac{1}{2} \cdot (V_o + V_R)^{-1/2}$$

$$= S\sqrt{\frac{qN_d\varepsilon}{2(V_o + V_R)}} \tag{9.1.15}$$

This dynamic capacitance at the junction, called the differential junction capacitance, as seen in Eq. 9.1.15, is nonlinear with respect to the reverse bias voltage V_R. This diode, made intentionally sensitive to the bias voltage, is called the variable capacitance diode or the varactor.

For example, in a typical microwave silicon varactor, $S = 50\ \mu m \times 50\ \mu m = 2500\ \mu m^2 = 2.5 \times 10^{-9}\ m^2$, $N_d = 2 \times 10^{15}\ cm^{-3} = 2 \times 10^{21}\ m^{-3}$, $q = 1.6 \times 10^{-19}$ C, $\varepsilon = \varepsilon_r\varepsilon_o = 11.8 \times 8.854 \times 10^{-12}$ F/m $= 1.04 \times 10^{-10}$ F/m and $V_o = 0.8$ V at $V_R = 2$ V.

Then, from Eq. 9.1.15,

$$C = 2.5 \times 10^{-9}\ \mathrm{m}^2 \times \sqrt{\frac{(1.6 \times 10^{-19}\ \mathrm{C})(2 \times 10^{21}\ \mathrm{m}^{-3})(1.04 \times 10^{-10}\ \mathrm{F/m})}{2(0.8\ \mathrm{V} + 2.0\ \mathrm{V})}}$$

$$= 1.9 \times 10^{-13}\ \mathrm{F} = 0.19\ \mathrm{pF}\ [1].$$

In most parametric amplifiers, the varactor is biased by a local oscillator power, which is called the pump power, in addition to dc bias voltage V_R. For example, if the local oscillator power is $P_{Lo} = 30$ mW to $Z_o = 50\ \Omega$, then the voltage across the 50 Ω load is $v_l = \sqrt{2Z_oP_{Lo}} = \sqrt{2 \times 50\ \Omega \times 30 \times 10^{-3}\ \mathrm{W}} = 1.73$ V. In this case, the minimum V_R is $V_{R_{min}} = 2\ \mathrm{V} - 1.73\ \mathrm{V} = 0.27$ V and the maximum V_R is $V_{R_{max}} = 2\ \mathrm{V} + 1.73\ \mathrm{V} = 3.73$ V. For $V_{R_{min}}$, from Eq. 9.1.15, $C_{max} = 0.31$ pF. For $V_{R_{max}}$, $C_{min} = 0.167$ pF. This means that when the local oscillator voltage swings between 1.73 V and −1.73 V, the junction capacitance of the varactor diode changes from 0.31 pF to 0.167 pF with a Q-point dynamic capacitance of 0.19 pF.

Equation 9.1.15 gives several suggestions for a varactor design. The value of the dynamic capacitance C can be adjusted by proper selection of S, N_d and ε. A large value of C lowers the impedance of the diode and lessens the efficiency. A small value of C limits power handling capability. S can be as small as 10 μm × 10 μm and as large as 250 μm × 250 μm. ε_r can be 11.8 for Si, 16 for Ge, and 13.2 for GaAs [5].

The difference $(C_{max} - C_{min})$ is maximized also by maximizing S and ε within reason. However, the ratio C_{max}/C_{min} is a function of $(V_o + V_R + V_l)$ and not a function of S, N_d, and ε.

9.2 Parametric Mixing

From Eq. 9.1.15, the dynamic junction capacitance of a reverse biased p–n junction is given by

$$C = S\sqrt{\frac{qN_d\varepsilon}{2}}\,(V_o + V_R)^{-1/2} \tag{9.2.1}$$

If the reverse bias voltage V_R is modulated by a local oscillator or a pump oscillator in the following form

$$V_R = V_{Ro} + v_p \sin(\omega_p t + \phi_p) \tag{9.2.2}$$

where

V_{Ro} is the dc reverse bias voltage,
v_p is the amplitude of the local oscillator voltage, which is called the pump oscillator of a parametric amplifier,
ω_p is the angular frequency of the pump oscillator, and
ϕ_p is the initial phase angle of the pump oscillator output.

Then the dynamic junction capacitance is

$$C = S\sqrt{\frac{qN_d\varepsilon}{2}}\left[V_o + V_{Ro} + v_p \sin(\omega_p t + \phi_p)\right]^{-1/2} \tag{9.2.3}$$

or

$$C = S\sqrt{\frac{qN_d\varepsilon}{2(V_o + V_{Ro})}}\left[1 + \frac{v_p}{V_o + V_{Ro}}\sin(\omega_p t + \phi_p)\right]^{-1/2} \tag{9.2.4}$$

For the capacitance C to be real, it is desirable that

$$\frac{v_p}{V_o + V_{Ro}} \leq 1 \tag{9.2.5}$$

If Eq. 9.2.4 is expanded by a binomial expansion and the higher order terms are dropped because they produce frequencies other than ω_p, then

$$\begin{aligned} C &= S\sqrt{\frac{qN_d\varepsilon}{2(V_o + V_{Ro})}}\left[1 - \frac{v_p}{2(V_o + V_{Ro})}\sin(\omega_p t + \phi_p)\right] \\ &= C_o\left[1 - \frac{v_p}{2(V_o + V_{Ro})}\sin(\omega_p t + \phi_p)\right] \end{aligned} \tag{9.2.6}$$

where

$$C_o = S\sqrt{\frac{qN_d\varepsilon}{2(V_o + V_{Ro})}} \tag{9.2.7}$$

C_o is the junction capacitance without the pump oscillator output applied. If the input voltage applied across the capacitance is

$$v_s = v_{so} \sin(\omega_s t + \phi_s) \tag{9.2.8}$$

where v_{so} is the amplitude, ω_s is the angular frequency, and ϕ_s is the initial phase angle of the incident microwave signals to be amplified, respectively, then the capacitor current is

$$\begin{aligned}
i_c &= \frac{d}{dt}(Cv_s) = C\frac{dv_s}{dt} + v_s\frac{dC}{dt} \\
&= C_o\left[1 - \frac{v_p}{2(V_o + V_{Ro})}\sin(\omega_p t + \phi_p)\right]\omega_s v_{so}\cos(\omega_s t + \phi_s) \\
&\quad - v_{so}\sin(\omega_s t + \phi_s)\frac{v_p C_o \omega_p}{2(V_o + V_{Ro})}\cos(\omega_p t + \phi_p) \\
&= C_o\omega_s v_{so}\cos(\omega_s t + \phi_s) \\
&\quad - \frac{\omega_s v_{so} v_p C_o}{2(V_o + V_{Ro})}\sin(\omega_p t + \phi_p)\cos(\omega_s t + \phi_s) \\
&\quad - \frac{\omega_p v_{so} v_p C_o}{2(V_o + V_{Ro})}\cos(\omega_p t + \phi_p)\sin(\omega_s t + \phi_s) \\
&= \omega_s C_o v_{so}\cos(\omega_s t + \phi_s) \\
&\quad - \frac{v_{so} v_p C_o}{2(V_o + V_{Ro})}\left[\omega_p\cos(\omega_p t + \phi_p)\sin(\omega_s t + \phi_s)\right. \\
&\quad \left. + \omega_s\sin(\omega_p t + \phi_p)\cos(\omega_s t + \phi_s)\right] \\
&= \omega_s C_o v_{so}\cos(\omega_s t + \phi_s) \\
&\quad - \frac{v_{so} v_p C_o}{2(V_o + V_{Ro})}\left\{\frac{\omega_p}{2}\sin\left[(\omega_s - \omega_p)t + (\phi_s - \phi_p)\right]\right. \\
&\qquad + \frac{\omega_p}{2}\sin\left[(\omega_s + \omega_p)t + (\phi_s + \phi_p)\right] \\
&\qquad + \frac{\omega_s}{2}\sin\left[(\omega_p - \omega_s)t + (\phi_p - \phi_s)\right] \\
&\qquad \left. + \frac{\omega_s}{2}\sin\left[(\omega_p + \omega_s)t + (\phi_p + \phi_s)\right]\right\} \\
&= \omega_s C_o v_{so}\cos(\omega_s t + \phi_s) \\
&\quad - \frac{v_{so} v_p C_o}{4(V_o + V_{Ro})}\left\{(\omega_s - \omega_p)\sin\left[(\omega_p - \omega_s)t + (\phi_p - \phi_s)\right]\right. \\
&\qquad \left. + (\omega_s + \omega_p)\sin\left[(\omega_p + \omega_s)t + (\phi_p + \phi_s)\right]\right\} \\
&= \omega_s C_o v_{so}\cos(\omega_s t + \phi_s) \\
&\quad + \frac{v_{so} v_p C_o}{4(V_o + V_{Ro})}\left\{(\omega_p - \omega_s)\sin\left[(\omega_p - \omega_s)t + (\phi_p - \phi_s)\right]\right. \\
&\qquad \left. - (\omega_p + \omega_s)\sin\left[(\omega_p + \omega_s)t + (\phi_p + \phi_s)\right]\right\} \qquad (9.2.9)
\end{aligned}$$

In parametric amplifier technology, the difference frequency $(\omega_p - \omega_s)$ is called the idler frequency.

$$\omega_i \equiv \omega_p - \omega_s \tag{9.2.10}$$

The magnitude of the idler frequency current is

$$|i_I| = \frac{(\omega_p - \omega_s) v_{so} v_p C_o}{4(V_o + V_{Ro})} \tag{9.2.11}$$

If the pump frequency ω_p is adjusted to

$$\omega_p = 2\omega_s \tag{9.2.12}$$

then, from Eq. 9.2.10,

$$\omega_i = \omega_s \tag{9.2.13}$$

Substituting Eq. 9.2.12 in Eq. 9.2.11,

$$|i_I| = \omega_s C_o v_{so} \frac{v_p}{4(V_o + V_{Ro})} \tag{9.2.14}$$

In addition to this current in the diode, the original signal circuit current

$$|i_s| = \omega_s C_o v_{so} \tag{9.2.15}$$

is flowing as seen from Eq. 9.2.9. After adjusting

$$\phi_p = 2n\pi \tag{9.2.16}$$

where n is 0 or an integer, the total current $|i_T|$ at ω_s is then

$$|i_T| = |i_I| + |i_s| \tag{9.2.17}$$

In a practical parametric amplifier circuit, the varactor is loaded parallel to both the microwave signal circuit ω_s and the idler circuit ω_i. To increase the excitation efficiency, both circuits are usually tuned to their individual resonance frequencies so they do not interfere with each other except through the diode. If the quality factors of the signal circuit resonator and the idler circuit resonator are Q_s and Q_i, respectively, then the total current in the diode under resonance is

$$|i_T|_R = |i_I|\sqrt{Q_i} + |i_s|\sqrt{Q_s} \tag{9.2.18}$$

Substituting Eqs. 9.2.14 and 9.2.15 in Eq. 9.2.18,

$$|i_T|_R = \omega_s C_o v_{so} \left(\frac{v_p \sqrt{Q_i}}{4(V_o + V_{Ro})} + \sqrt{Q_s} \right) \tag{9.2.19}$$

Under this condition, the voltage across the capacitance C_o is

$$v_o = v_{so}\left(\frac{v_p\sqrt{Q_i}}{4(V_o + V_{Ro})} + \sqrt{Q_s}\right) \tag{9.2.20}$$

Then the voltage gain of this amplifier is

$$A = \frac{v_o}{v_{so}} = \frac{v_p\sqrt{Q_i}}{4(V_o + V_{Ro})} + \sqrt{Q_s} \tag{9.2.21}$$

For example, if the same pump oscillator as shown in the previous example of 30 mW is used, the oscillator produces $v_l = 1.73$ V across the 50 Ω line. If the same voltage is applied to a pump cavity of $Q_p = 200$ to which the varactor is loaded, then the pump voltage across the varactor is $v_{po} = v_l\sqrt{Q_p} = (1.73\ \text{V})\sqrt{200} = 24.5$ V.

From Eq. 9.2.5, it is desirable that $v_{po} \approx V_o + V_{Ro}$ or $V_{Ro} \approx v_{po} - V_o = 24.5\ \text{V} - 0.8\ \text{V} = 23.7\ \text{V}$. The diode must be reverse biased 23.7 V for $P_{lo} = 30$ mW operation. If the incident microwave signal is $f_s = 0.375$ GHz, $P_s = -90$ dBm $= [\log_{10}^{-1}(90/10)]^{-1} = 10^{-9}$ mW $= 10^{-12}$ W, then the voltage across $Z_o = 50$ Ω line is $v'_{so} = \sqrt{2Z_oP_s} = \sqrt{2 \times 50\ \Omega \times 10^{-12}\ \text{W}} = 10 \times 10^{-6}$ V.

If the input circuit resonator's loaded Q_s value is 200, then the voltage is stepped up to $v_{so} = v'_{so}\sqrt{Q_s} = 10 \times 10^{-6}\ \text{V} \times \sqrt{200} = 141.4 \times 10^{-6}$ V. Now from Eq. 9.2.21, choosing $Q_i = 200$, $A = 3 \times 24.5\ \text{V}\ \sqrt{200}/4(0.8\ \text{V} + 23.7\ \text{V}) + \sqrt{200} = 17.6 = (20 \log 17.6)\ \text{dB} = 24.9$ dB. To obtain this gain, condition 9.2.12 must be fulfilled. The local oscillator frequency must be adjusted to $f_p = 2f_s = 2 \times 9.375$ GHz $= 18.75$ GHz. This type of amplification, or the amplification through a variable junction capacitance, is called parametric amplification.

In Eq. 9.2.9, the frequency $(\omega_p + \omega_s)$ is called the up-conversion frequency. The voltage gain of the up-convertor is then, through a procedure similar to the one used to obtain Eq. 9.2.21,

$$A_{up} \equiv \frac{v_{up}}{v_{so}} = \frac{v_p\sqrt{Q_{up}}}{4(V_o + V_{Ro})} \tag{9.2.22}$$

where v_{up} is the magnitude of the voltage across the junction capacitor at the up-converted frequency $(\omega_p + \omega_s)$, and Q_{up} is the quality factor of the resonance circuit which is tuned at the up-converted frequency $(\omega_p + \omega_s)$.

For example, if $f_s = 600$ MHz, $f_p = 1.2$ GHz, and $P_p = 10$ mW $= 10^{-2}$ W, then $v_p = \sqrt{2P_pZ_o} = \sqrt{2 \times 10^{-2} \times 50} = 1.0$ V for a 50 Ω line. If the varactor is mounted in a resonator of 1.2 GHz with $Q_p = 200$, then the voltage across the varactor is $v_p = \sqrt{Q}\,v_p' = \sqrt{200} \times 1.0$ V $= 14.14$ V. The dc reverse bias required is then $V_{Ro} = 14.14$ V $- 0.8$ V $= 13.34$ V. If $Q_{up} = 200$, the conversion gain is, from Eq. 9.2.22,

$$A_{up} = \frac{(14.14\ \text{V})\sqrt{200}}{4(0.8\ \text{V} + 13.34\ \text{V})} = 3.53$$

$$= (20 \log 3.53)\ \text{dB} = 10.96\ \text{dB}$$

and the frequency is converted from $f_s = 600$ MHz to $f_s + f_p = 600$ MHz $+ 1200$ MHz $= 1800$ MHz.

In Eq. 9.2.9, the frequency $(\omega_p - \omega_s)$ is called the down-conversion frequency. The voltage gain at the down-conversion frequency is

$$A_{down} = \frac{v_{out,\,down}}{v_{so}} = \frac{v_p\sqrt{Q_d}}{4(V_o + V_{Ro})} \tag{9.2.23}$$

where Q_d is the quality factor of the output resonator circuit which resonates at $(\omega_p - \omega_s)$.

For example, if $f_s = 9.2$ GHz and $f_p = 12.4$ GHz, the down-converted frequency should be $f_p - f_s = 12.4$ GHz $- 9.2$ GHz $= 3.2$ GHz. For the pump oscillator in the previous example, $P_p = 10$ mW, $Q_p = 200$ and $v_p = 14.14$ V. If $Q_d = 200$ as in Eq. 9.2.23, the conversion gain is

$$A_{down} = \frac{14.14\ \text{V}\sqrt{200}}{4(0.8\ \text{V} + 13.34\ \text{V})}$$

$$= 3.53 = (20 \log 3.53)\ \text{dB} = 10.96\ \text{dB}$$

The pump power $P_p = 10$ mW necessitates the dc reverse bias voltage of 13.34 V through the pump frequency circuit of $Q_p = 200$.

In this parametric mixing, the signal frequency ω_s and the pump frequency ω_p are mixed with each other through a nonlinear junction capacitance C. The output is at the signal frequency ω_s through parametric amplification, at the sum frequency $\omega_p + \omega_s$ through the up-conversion, and at the difference frequency $(\omega_p - \omega_s)$ through the down-conversion [2]. This parametric mixing has been viewed classically by the Manley and Rowe relationships [6,7]. One of the latest alternative processes among many other possible versions is presented in this section [4,5,9–11].

9.3 Noise in Parametric Amplification

Noise in parametric amplifiers is considered smaller than other types of microwave amplifiers except for masers. This is the reason that parametric amplifiers are used when noise is critical as exemplified by preamplifiers for earth stations of a satellite communication system or a radio telescope. In many cases, these amplifiers are mounted near the focus of large parabolic reflector antennas.

The voltage gain of a parametric amplifier across the junction capacitance is given by Eq. 9.2.21 and

$$A = \frac{v_p\sqrt{Q_i}}{4(V_o + V_{Ro})} + \sqrt{Q_s} \qquad (9.3.1)$$

The available noise input power to the amplifier is

$$N_i = kT\Delta f \qquad (W) \qquad (9.3.2)$$

where T is the temperature of the input circuit, k is the Boltzmann constant 1.28×10^{-23} J/K, and Δf is the frequency bandwidth of this amplifier [3]. Then the available noise output power due to the available noise input power is

$$N_{oi} = N_i A^2 = \left[\frac{v_p\sqrt{Q_i}}{4(V_o + V_{Ro})} + \sqrt{Q_s}\right]^2 kT\Delta f \qquad (W) \quad (9.3.3)$$

This is because A is the voltage gain. The power gain $A_p = A^2$. In addition, the amplifier itself generates noise and delivers the noise to the output. The available noise output caused by internally generated noise is thermal agitation noise and shot noise [4].

The parametric variable capacitance p–n junction has a leakage resistance R to which the junction capacitance C is considered to be connected in parallel. The thermal agitation noise

(Johnson–Nyquist theorem) emf across the junction resistance R is

$$v_{\mathrm{RN}} = \sqrt{4\mathrm{k}TR\,\Delta f} \qquad (\mathrm{V}) \tag{9.3.4}$$

The available noise power to a matched external load resistance, which includes the input resistance and the output resistance, is therefore

$$R_{\mathrm{L}} = R \qquad (\Omega) \tag{9.3.5}$$

$$N_{\mathrm{oR}} = \frac{v_{\mathrm{RN}}^2}{4R} = \mathrm{k}T\Delta f \qquad (\mathrm{W}) \tag{9.3.6}$$

This noise power will be split into two parts, one going back to the input and the other going on to the load resistance. The noise power going to the load is

$$N_{\mathrm{oRL}} = \tfrac{1}{2}\mathrm{k}T\Delta f \tag{9.3.7}$$

The diode is considered to be a noise current source of shot noise. The shot noise current is given by

$$i_{\mathrm{N}} = \sqrt{2qI_{\mathrm{d}}\,\Delta f} \qquad (\mathrm{A}) \tag{9.3.8}$$

where q is the charge of an electron and I_{d} is the diode current [4]. This noise current flows back into the input circuit, out of the diode itself and into the output load.

In this equivalent circuit of noise current, to calculate the *available* noise under resonance, the condition 9.3.5 is met by

$$R_{\mathrm{out}} = R_{\mathrm{in}} = 2R_{\mathrm{L}} = 2R \tag{9.3.9}$$

where R_{out} is the resistance of the output circuit, and R_{in} is the resistance of the input circuit, and both R_{out} and R_{in} are connected in parallel with R. Then the *available* noise output is

$$N_{\mathrm{s}} = \frac{(i_{\mathrm{N}}/4)^2}{2} R_{\mathrm{out}} = \frac{1}{32} R_{\mathrm{out}} i_{\mathrm{N}}^2$$

$$= \frac{1}{16} R_{\mathrm{out}} q I_{\mathrm{d}}\,\Delta f \qquad (\mathrm{W}) \tag{9.3.10}$$

The total noise output power of this parametric amplifier is then

$$P_N = N_{oi} + N_{oRL} + N_s$$
$$= \left[\frac{v_p}{4(V_o + V_{Ro})}\sqrt{Q_i} + \sqrt{Q_s}\right]^2 kT\Delta f + \frac{1}{2}kT\Delta f + \frac{1}{16}R_{out}qI_d\,\Delta f \quad (9.3.11)$$

The noise figure F is defined as

$$F \equiv \frac{P_N}{N_{oi}} \quad (9.3.12)$$

Substituting Eqs. 9.3.3 and 9.3.11 into Eq. 9.3.12,

$$F = 1 + \frac{1}{2\left[\dfrac{v_p\sqrt{Q_i}}{4(V_o + V_{Ro})} + \sqrt{Q_s}\right]^2} + \frac{R_{out}qI_d}{16\left[\dfrac{v_p\sqrt{Q_i}}{4(V_o + V_{Ro})} + \sqrt{Q_s}\right]^2 kT} \quad (9.3.13)$$

In most parametric amplifiers

$$\frac{2(V_o + V_{Ro})}{v_p} \approx 2 \quad (9.3.14)$$

and

$$I_d \approx 0 \quad (9.3.15)$$

Therefore, F in Eq. 9.3.13 is small. Usually F is about 2 or 3 dB or less. As seen from Eq. 9.3.11, if the amplifier is refrigerated, the noise will be decreased. Refrigerated parametric amplifiers are common in satellite communication systems and in radio telescopes with the parametric amplifier at the focal point of the parabolic reflector.

In the example of the parametric amplifier presented in Section 9.2, the amplifier has a gain of $A = 24.9$ dB at $f_s = 9.375$ GHz with the pump power injection $P_{Ro} = 30$ mW of reverse bias voltage of $V_{Ro} = 23.7$ V and $Q_s = 200$. $Q_s = 200$ means that $Q_s = f_s/\Delta f_s = 200$ or $\Delta f_s = f_s/200 = 9375\text{ MHz}/200 = 46.9$ MHz. Therefore, according to Eq. 9.3.2, at room temperature $T = 300$ K, the available noise input power to the amplifier is $N_i = kT\Delta f_s = (1.38 \times 10^{-23}\text{ J/K})(300\text{ K})(46.9 \times 10^6\text{ s}^{-1}) = 1.94 \times 10^{-13}\text{ W} =$

0.194 pW. If this noise is amplified by the amplifier is $A =$ 24.9 dB, since this is in terms of power gain, $A_p = \log_{10}^{-1}(A/10) = \log_{10}^{-1}(2.49) = 309.0$ times. So the noise output is, according to Eq. 9.3.3, $N_{oi} = N_i A_p = 0.194$ pW $\times$ 309.0 = 59.95 pW. This is the noise output of the parametric amplifier if the amplifier itself does not generate noise.

If the differential resistance of the reverse biased varactor diode is $R = 10$ kΩ, then the noise voltage across the diode due only to the thermal agitation noise in the signal channel is, from Eq. 9.3.4,

$$v_{RN} = \sqrt{4kTR\,\Delta f}$$

$$= (4 \times 1.38 \times 10^{-23}\ \mathrm{J/K} \cdot 300\ \mathrm{K} \cdot 10^4\ \Omega \cdot 46.9 \times 10^6\ \mathrm{s}^{-1})^{\frac{1}{2}}$$

$$= 88.1 \times 10^{-6}\ \mathrm{V}$$

The corresponding thermal noise power is, from Eq. 9.3.6,

$$N_{oR} = kT\Delta f = 1.38 \times 10^{-23}\ \mathrm{J/K} \cdot 300\ \mathrm{K} \cdot 46.9 \times 10^6\ \mathrm{s}^{-1}$$

$$= 0.194\ \mathrm{pW}.$$

Only half of this power goes to the load. From Eq. 9.3.7, $N_{oRL} = \frac{1}{2} N_{oR} = 0.097$ pW.

If the reverse leakage current is $I_d = (V_o + V_{Ro})/R_R = (0.8\ \mathrm{V} + 23.7\ \mathrm{V})/10^4\ \Omega = 2.45$ mA, then dc dissipation power of 23.7 V $\times$ 2.45 mA = 58.1 mW. Here it is assumed that $R \approx R_R$ which is the dc reverse resistance of the diode at that bias point. From Eq. 9.3.8, the shot noise current is $i_N = \sqrt{2qI_d\,\Delta f} = (2 \times 1.6 \times 10^{-19}\ \mathrm{C} \times 2.45 \times 10^{-3}\ \mathrm{A} \times 46.9 \times 10^6\ \mathrm{s}^{-1})^{1/2} = 0.19 \times 10^{-6}$ A. A part of this shot noise current flows out the output circuit. From Eq. 9.3.10, the shot noise power that appears at the output circuit is, together with Eq. 9.3.9,

$$N_s = \tfrac{1}{16} R_{out} q I_d\,\Delta f$$

$$= \tfrac{1}{8} R q I_d\,\Delta f$$

$$= \tfrac{1}{8} \times 10^4\ \Omega \times 1.6 \times 10^{-19}\ \mathrm{C} \times 2.45 \times 10^{-3}\ \mathrm{A} \times 46.9 \times 10^6\ \mathrm{s}^{-1}$$

$$= 22.98 \times 10^{-12}\ \mathrm{W} = 22.98\ \mathrm{pW}.$$

From Eq. 9.3.11, the total noise power in the output circuit at resonance is therefore

$$P_N = N_{oi} + N_{oRL} + N_s$$

$$= 59.95\ \mathrm{pW} + 0.097\ \mathrm{pW} + 22.98\ \mathrm{pW} = 83\ \mathrm{pW}$$

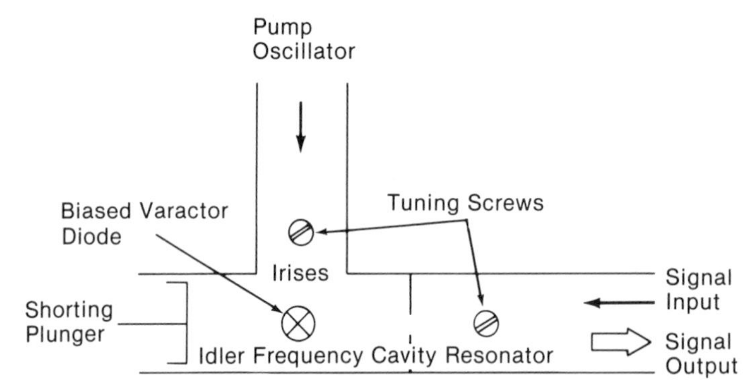

Fig. 9.2 Simplified diagram of a waveguide parametric amplifier.

From Eq. 9.3.12, the noise figure of this parametric amplifier is

$$F = \frac{P_N}{N_{oi}} = \frac{83 \text{ pW}}{59.95 \text{ pW}} = 1.38 = (10 \log 1.38) \text{ dB} = 1.4 \text{ dB}$$

This is considered to be a low noise amplifier at $f_s = 9.375$ GHz, $\Delta f_s = 46.9$ MHz with the power gain of $A_p = 24.9$ dB at room temperature.

9.4 Parametric Amplifier Circuit

A simplified schematic diagram of a variable capacitance diode parametric amplifier is shown in Fig. 9.2. A variable capacitance diode or varactor is mounted in a rectangular cavity resonator as shown in Fig. 9.2. This cavity resonator is adjusted by a shorting plunger to resonate at the idler frequency. This is the reason that this cavity resonator is called the idler cavity. The idler frequency is the difference between the pump oscillator frequency and the incident signal frequency.

The pump frequency power is fed through a waveguide iris to the idler cavity. Impedance matching is accomplished by adjusting a tuning screw at the pump frequency. A similar arrangement is made for the signal frequency. If the pump oscillator is on and the signals are fed into the idler cavity, there will be parametric mixing of the incident signal and the pump frequency. Parametric mixing generates a beat frequency signal at the idler frequency which is stronger than the incident signal. The idler frequency power couples out of the iris and comes out to the waveguide where the incident waves are fed. In other words, if a small signal is fed at the input signal frequency to the idler cavity, amplified large waves at idler frequency come out of the idler cavity. If the pump oscillator frequency is adjusted to exactly twice the input signal frequency, then the idler frequency is exactly equal to the incident signal frequency. The amplified signal is therefore at the signal frequency. The idler cavity must be tuned to the signal frequency. If the idler cavity is tuned at the difference frequency between the pump frequency and the incident signal frequency, the system is a down-converter. If the amplified output is at the difference frequency between the pump frequency and the incident signal frequency, the outpout screw tuner must be tuned to the difference frequency. If the idler cavity's shorting plunger is adjusted so that the cavity resonates at the sum frequency of the pump and the incident frequencies, then it is an up-converter. The output screw tuner must then be tuned to the sum frequency. It is possible that the pump power may leak through the idler cavity to the signal frequency waveguide. Good waveguide filters are therefore needed to operate a parametric amplifier properly.

9.5 Parametric Diodes

Reverse biased p–n junction diodes present nonlinear capacitance. The diode is therefore a capacitive reactor. When the incident microwave signal is mixed with pump power at the nonlinear capacitor diode, depending on the tuning of the output circuit, the system becomes

1. a parametric amplifier of the incident microwave signals,
2. an up-converter for the intensified signals of the sum frequency of the incident microwave signal frequency and the pump oscillator frequency, or
3. a down-converter for the intensified signals of the difference frequency between the incident microwave signal frequency and the pump oscillator frequency.

The nonlinear capacitance is created by the widening of the depletion region of a p–n junction due to the reverse bias.

It should be noted that the junction capacitance considered in this chapter is the differential capacitance or the dynamic capacitance. It should not be confused with the static capacitance Q/V. The dynamic capacitance is $\mathrm{d}Q/\mathrm{d}V$.

It should also be noted that the reverse diode resistance discussed in this chapter is not the static resistance V_R/I_R. It is the dynamic resistance or the differential resistance $|\mathrm{d}V_R/$

$dI_R|_{V_R=V_{Ro}}$. If a p–n junction diode is reverse biased, the static resistance could be as high as 10^3 Ω. The dynamic resistance under proper bias can be higher, as high as 10^4 Ω.

The intrinsic resistivity of Si is 2.5×10^5 Ω-cm. For Ge it is only 43 Ω-cm and for GaAs it is 4×10^8 Ω-cm [8].

One feature of varactor parametric amplifiers is the low noise. The shot noise contribution over the amplified and transferred noise from the input is very small. This makes the noise figure of the parametric amplifier small.

Problems

9.1 Write down the necessary boundary conditions for a depletion layer of a p–n junction to integrate Poisson's equation.

9.2 Integrate Poisson's equation in the depletion layer while grounding the p-type semiconductor and ungrounding the n-type semiconductor.

9.3 Show that the junction capacitance of a p–n junction is inversely proportional to the square root of the reverse voltage.

9.4 Obtain the static junction capacitance of a reverse biased p–n junction.

9.5 Design a varactor diode with the dynamic maximum capacitance of 10 pF and minimum capacitance of 6 pF.

9.6 Obtain both the static and dynamic capacitance of a forward biased varactor diode.

9.7 Plot the family of curves for the dynamic junction capacitance with the reverse bias voltage for various parameters.

9.8 Show that the dynamic capacitance of a varactor is a nonlinear capacitance with respect to the reverse bias voltage.

9.9 Try to sketch a frequency spectrum created by parametric mixing.

9.10 Identify the output frequency of a parametric amplifier. Is it at the idler frequency or at the incident input signal frequency?

9.11 Design a parametric amplifier with a power gain of 10 dB. Specify the pump oscillator power output.

9.12 Design an up-converter with a power gain of 10 dB.

9.13 Design a down-converter with a power gain of 10 dB.

9.14 State the reason that the noise of parametric amplifiers is considered to be smaller than the noise of other types of microwave amplifiers, except for masers.

9.15 Compute the input thermal noise for an amplifier of 1 MHz bandwidth.

9.16 Design a parametric amplifier where the transferred noise, or amplified noise, is 3 dB above the input noise.

9.17 Identify the noise generated in a varactor diode and compute the comparative noise power of each identified noise.

9.18 Sketch a schematic diagram of a parametric amplifier circuit.

9.19 Sketch a schematic diagram of a down-converter circuit.

9.20 Sketch a schematic diagram of an up-converter.

References

1 E. S. Yang, "Fundamentals of Semiconductor Devices." McGraw-Hill, New York, 1978.

2 K. K. N. Chang, "Parametric and Tunnel Diodes." Prentice-Hall, Englewood Cliffs, New Jersey, 1964.

3 T. K. Ishii, "Microwave Engineering." Ronald Press, New York, 1966.

4 A. van der Ziel, "Solid-State Physical Electronics," 2nd ed. Prentice-Hall, Englewood Cliffs, New Jersey, 1976.

5 B. G. Streetman, "Solid-State Electronic Devices," 2nd ed. Prentice-Hall, Englweood Cliffs, New Jersey, 1980.

6 J. M. Manley and E. Peterson, Negative resistance effect in saturable reactor circuits. *Trans. AIEE* **65**, 870 (1946).

7 J. M. Manley and H. E. Rowe, Some general properties of non-linear elements. *Proc. IRE* **44**, 904 (1956).

8 S. Y. Liao, "Microwave Devices and Circuits," 2nd ed. Prentice-Hall, Englewood Cliffs, New Jersey, 1985.

9 S. Y. Liao, "Microwave Solid-State Devices." Prentice-Hall, Englewood Cliffs, New Jersey, 1985.

10 J. T. Coleman, "Microwave Devices." Reston Publ., Reston, Virginia, 1982.

11 D. Roddy, "Microwave Technology." Prentice-Hall, Englewood Cliffs, New Jersey, 1986.

10

Microwave Harmonic Generator Diodes

10.1 Harmonic Generation by Nonlinear Junctions

A p–n junction is basically considered to be an "on–off" switch. When it is forward biased, it is "on" and when it is reversed biased it is "off". Under this simplified consideration, when a sinusoidal voltage is applied across the diode as shown in Fig. 10.1, the voltage is rectified. The rectified voltage is expanded by Fourier series. Fourier's theory states that for a periodic function $f(\omega t)$ [1, 11],

$$f(\omega t) = \frac{a_o}{2} + \sum_{n=1}^{\infty} \left(a_n \cos(n\omega t) + b_n \sin(n\omega t) \right) \quad (10.1.1)$$

and

$$a_n = \frac{1}{\pi} \int_{-\pi}^{\pi} f(\omega t) \cos(n\omega t)\, d(\omega t)$$

$$n = 0, 1, 2, 3, \ldots \quad (10.1.2)$$

$$b_n = \frac{1}{\pi} \int_{-\pi}^{\pi} f(\omega t) \sin(n\omega t)\, d(\omega t)$$

$$n = 1, 2, 3, \ldots \quad (10.1.3)$$

For the voltage waveform across the diode shown in Fig. 10.1,

$$f(\omega t) = 0 \quad \text{for } -\pi < \omega t < 0 \tag{10.1.4}$$

$$f(\omega t) = V_o \sin(\omega t) \quad \text{for } 0 < \omega t < \pi \tag{10.1.5}$$

Substituting Eqs. 10.1.4 and 10.1.5 in Eqs. 10.1.2 and 10.1.3, it can be shown that [1]

$$\left.\begin{aligned} \frac{a_o}{2} &= \frac{V_o}{\pi} \\ b_1 &= \frac{V_o}{2} \end{aligned}\right\} \tag{10.1.6}$$

$$a_n = -\frac{2V_o}{(n-1)(n+1)\pi}$$

$$n = 2, 4, 6, \ldots \tag{10.1.7}$$

$$b_n = 0$$

$$n = 2, 3, 4, 5, 6, \ldots \tag{10.1.8}$$

Inspection of Eqs. 10.1.6, 10.1.7, and 10.1.8 shows that the voltage waveform as shown in Fig. 10.1 B contains a dc component, a fundamental frequency, and even harmonics rapidly decreasing in magnitude with the order of harmonics. A frequency spectrum is sketched in Fig. 10.2. Usually the harmonic generating diode is mounted inside a cavity resonator which resonates at a desired harmonic to enhance the desired harmonic frequency. Band rejection filters are attached to suppress the undesired harmonics.

In reality, practical diodes are quite different from the ideal diode just discussed. The difference between the ideal diode and a practical diode is illustrated in Fig. 10.3. Due to the differences in the current–voltage characteristic curve, the practical diode current takes a complicated waveform as shown in Fig. 10.3 C. These waveforms are usually not describable by analytic functions. As a result, analytical integration of the Fourier coefficients 10.1.2 and 10.1.3 is difficult, if not impossible. It has to be done numerically or graphically. If this is done, it will usually be found that nonideal diodes produce stronger harmonics than ideal diodes.

For example, in the extreme case of a nonideal diode in Fig. 10.3, the current would approach a rectangular pulse train as illustrated in Fig. 10.4 A. The Fourier coefficients of this type of

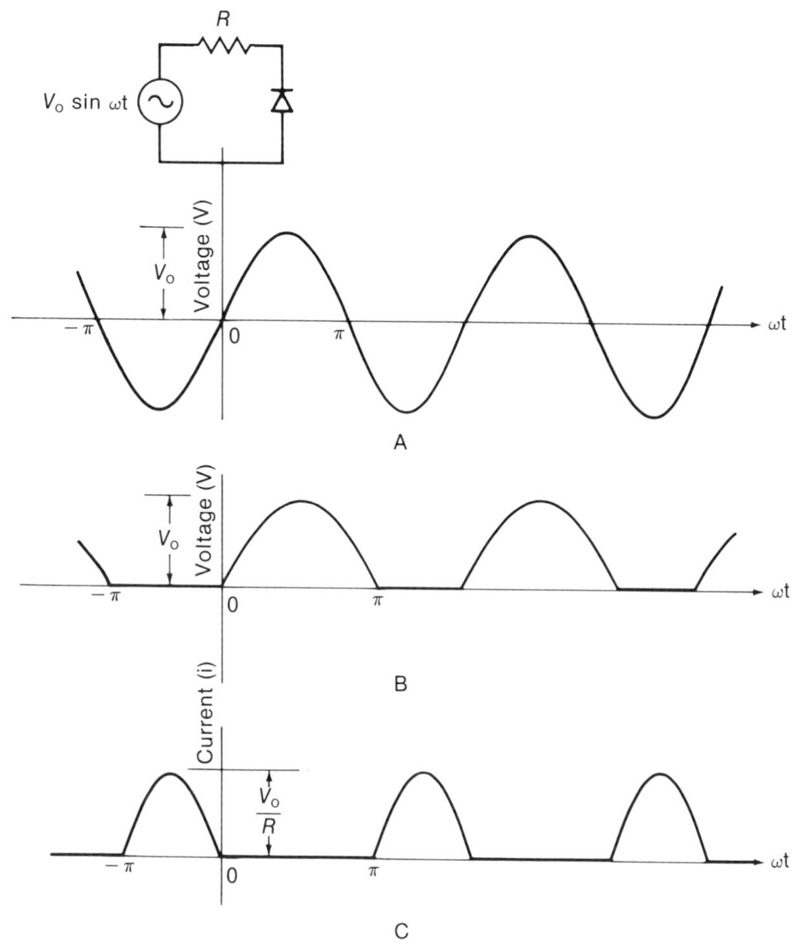

Fig. 10.1 Ideal harmonic generating diode. A. Source voltage. B. Voltage across diode. C. Current through diode.

waveform have been calculated and found to be [1]

$$\frac{a_o}{2} = \frac{i_o \omega \tau}{2\pi} \tag{10.1.9}$$

$$a_n = \frac{2i_o}{n\pi} \sin \frac{n\omega\tau}{2} \tag{10.1.10}$$

$$b_n = 0 \tag{10.1.11}$$

Inspection of Eqs. 10.1.9 and 10.1.10 shows that the amplitudes of the harmonics decrease more slowly than in the case of the ideal diode as the order of the harmonic n is increased. Furthermore, odd harmonics are generated. In other words, the practical diode generates richer harmonics than the ideal diode. This sharp and

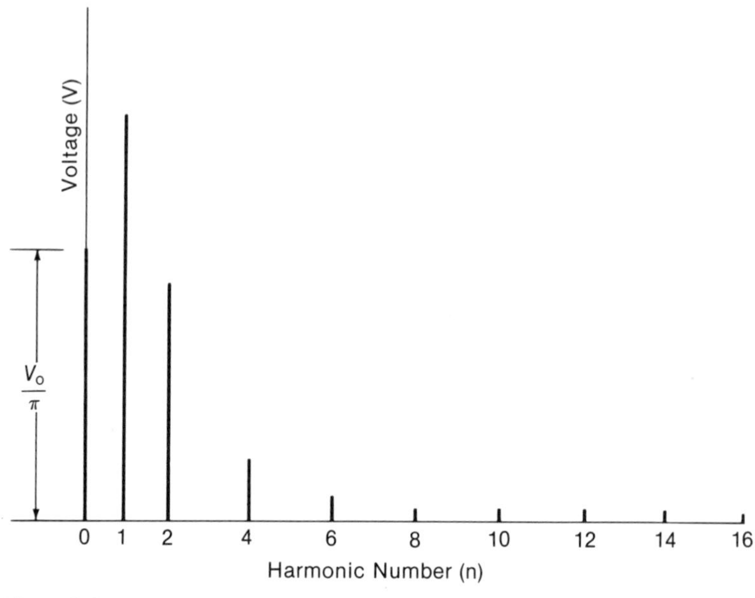

Fig. 10.2 Frequency spectrum of an ideal diode.

narrow rectangular pulse current train can be generated by a deep cut-in voltage and a sharp rise in the current–voltage characteristic curve in the forward direction of the diodes as shown in Fig. 10.3 B. If the diode has a deep cut-in voltage but the current rises gradually with the forward bias voltage as shown in Fig. 10.3 B, the diode current approaches the triangular pulse train as illustrated in Fig. 10.4 B. The Fourier coefficients of this triangular pulse train are found to be [1]

$$\frac{a_o}{2} = \frac{i_o \omega \tau}{4\pi} \tag{10.1.12}$$

$$a_n = \frac{4 i_o \left(1 - \cos \dfrac{n \omega \tau}{2}\right)}{n^3 \omega \tau} \qquad n = 1, 2, 3, 4, \ldots \tag{10.1.13}$$

$$b_n = 0 \tag{10.1.14}$$

Inspection of Eq. 10.1.13 shows that the harmonics of higher orders of this triangular pulse train are not as strong as the higher harmonics of the rectangular pulse train. Ordinary p–n junction diodes, Schottky junction diodes, PIN diodes, varactor diodes, avalanche diodes, Zener diodes, back diodes, and step recovery diodes are alternatives available for microwave harmonic generation.

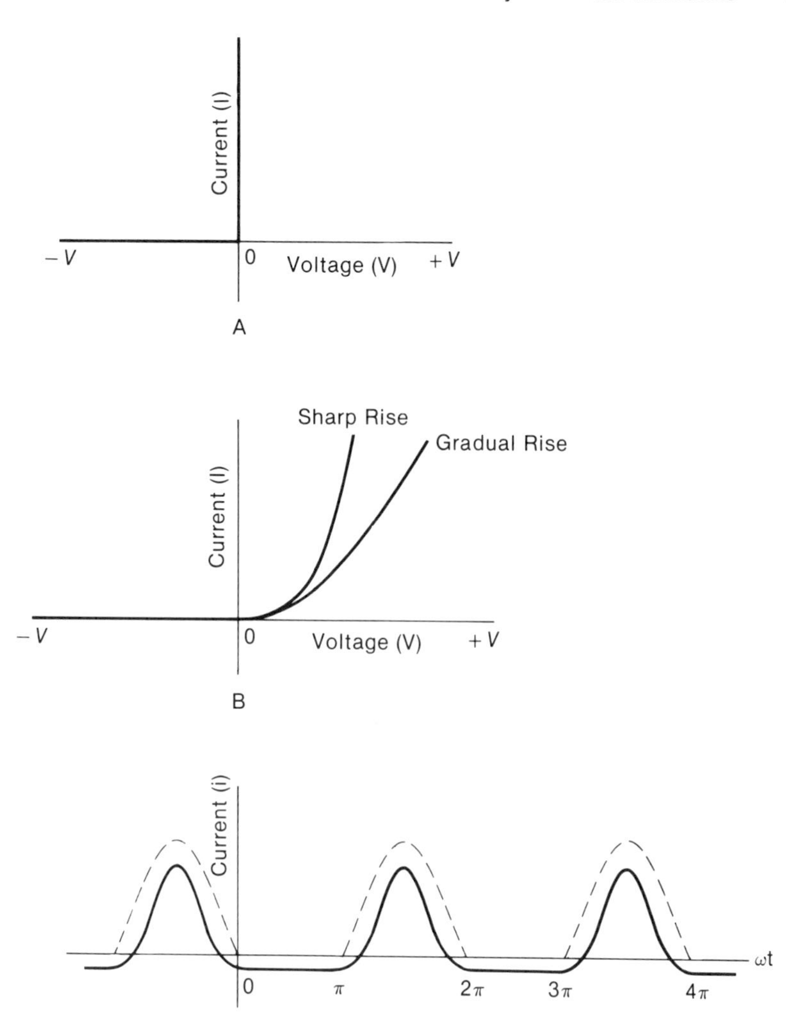

Fig. 10.3 Comparison of ideal and practical diodes. A. Cut-in voltage for an ideal diode. B. Cut-in voltage for a practical diode. C. Diode current when sinusoidal voltage is applied. Ideal diode current is indicated by the dashed curve. Practical diode current is indicated by the solid curve.

For example, to generate a narrow square pulse train as shown in Fig. 10.4 A, an avalanche diode, Zener diode, back diode, or PIN diode—all of which show a sharp breakdown characteristic under a reverse bias voltage of a desirable value—is useful. From Eq. 10.1.10, the amplitude of the fundamental frequency current in the diode is

$$a_1 = \frac{2i_o}{\pi} \sin \frac{\omega\tau}{2} \tag{10.1.15}$$

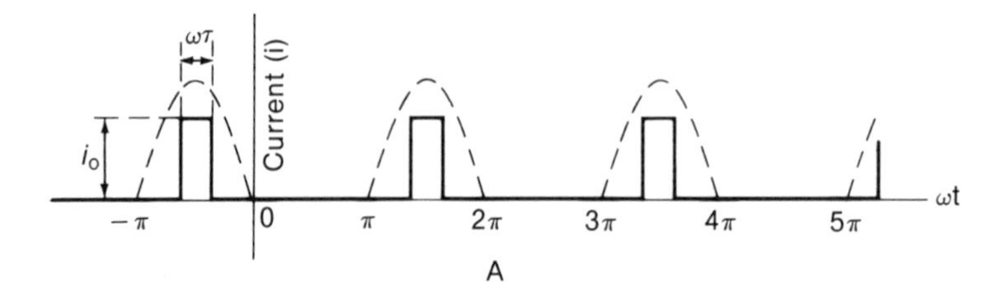

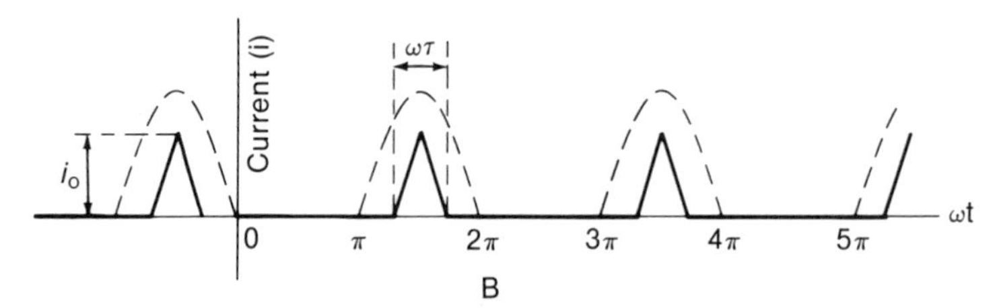

Fig. 10.4 Sinusoidal wave application. A. Diode with deep cut-in and sharp rise produces rectangular pulses. B. Diode with deep cut-in and gradual rise produces triangular pulses. Ideal diode current is indicated by the dashed curve. Idealized practical diode current is indicated by the solid curve.

If the frequency is to be multiplied four times, then

$$a_4 = \frac{2i_o}{4\pi}\sin\frac{4\omega\tau}{2} = \frac{i_o}{2\pi}\sin 2\omega\tau \tag{10.1.16}$$

To maximize the harmonic amplitude a_4,

$$\sin 2\omega\tau = 1 \quad \text{or} \quad 2\omega\tau = \frac{\pi}{2} + 2N\pi \tag{10.1.17}$$

where N is zero or positive integer.

$$\omega\tau = \frac{\pi}{4} + N\pi \tag{10.1.18}$$

However, as seen from Fig. 10.4 A, $\omega\tau < \pi$, therefore $N = 0$. As a result, $\omega\tau = \pi/4$ and $a_4 = i_o/2\pi$. If $\omega\tau$ decreases, then a_4 decreases following $\sin(2\omega\tau)$. The fourth harmonic of a rectangular pulse train is smaller than the ideal half-wave rectified sinusoidal wave. However, at the twentieth harmonic

$$a_{20} = \frac{2i_o}{20\pi}\sin\frac{20\omega\tau}{2} = \frac{i_o}{10\pi}\sin 10\omega\tau \tag{10.1.19}$$

To maximize a_{20}, $10\omega\tau = \pi/2$ or $\omega\tau = \pi/20$ and thus $a_{20} = i_o/10\pi$. If this is an ideal diode, from Eq. 10.1.7 $a_{20} = 2i_o/(20 - 1)(20 + 1)\pi \approx i_o/200\pi$. The rectangular pulse train therefore produces stronger harmonics. If $f = 9.3$ GHz, $f_{20} = 186$ GHz. For 9.3 GHz, $\omega T = 2\pi$ or $T = 2\pi/2\pi f = 1/f = 1/9.3 \times 10^9\ \mathrm{s}^{-1} = 0.11$ ns. $\omega\tau = \pi/20$ means $\tau = \pi/(20 \times 2\pi f) = 1/40f = 1/(40 \times 9.3 \times 10^9\ \mathrm{s}^{-1}) = 2.69$ ps.

When $a_{20} = i_o/10\pi$ the conversion loss is $a_{20}/i_o = 0.318 = 20\log_{10}0.0318$ dB $= -29.95$ dB. If an ideal diode is used and $\omega\tau = \pi$, then $a_{20}/i_o = 1/200\pi = 0.000159 = 20\log_{10}0.000159$ dB $= -75.97$ dB. This means that, in terms of the voltage, if the input fundamental frequency signal voltage $v_o = 1$ V, then for the pulse train diode, $v_{20} = 0.0318 \times 1$ V $= 31.8$ mV and for the ideal diode, $v_{20} = 0.000159 \times 1$ V $= 159$ μV. These are viable values in practical microwave engineering.

10.2 Step Recovery Diodes

A step recovery diode is a junction diode specifically designed to store carrier charge at the junction during the forward cycle and discharge it during the reverse cycle with a discharge current in the form of a sharp step function. A step recovery diode is made of a p–n junction with a reasonable amount of junction capacitance [2], a p^+–p–n diode with the p region used for charge storage [3], or PIN structure with the i region used for the storage of charge [4].

When the square wave voltage is applied to the diode as shown in Fig. 10.5 the diode current does not shut off immediately. As seen from Fig. 10.5 the reverse current will flow for a short period of time due to the stored charge accumulated in the junction during the forward cycle. The reverse current lasts only for a short period of time until the stored charge is exhausted. When the charge is exhausted, the diode opens circuits or suddenly goes into the off-state, as illustrated in Fig. 10.5. The off-state is the state that the diode is supposed to be in under the reverse bias condition. This is the reason that this type of diode is called the "step recovery diode" or the "snap off diode". This step recovery transient is very fast. Recovery times in the picosecond range have been reported [3]. This fast transient is useful for short pulse generation and microwave harmonic generation [5]. The discharge time of typical microwave diodes are reported to be 0.5–10 ns [2].

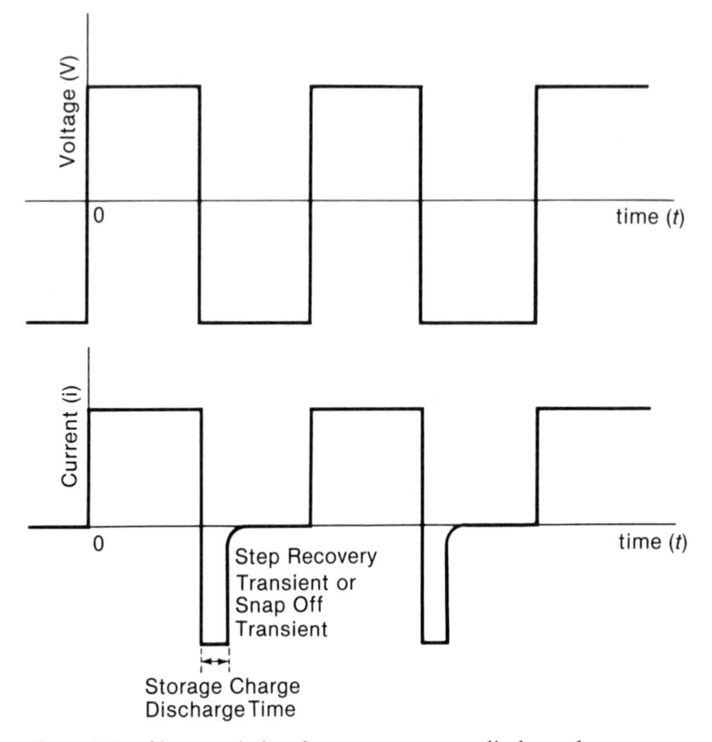

Fig. 10.5 Characteristic of a step recovery diode under square wave voltage application. The storage charge discharge time is between 0.5 and 10 ns. The step recovery (snap off) transient is on the order of picoseconds.

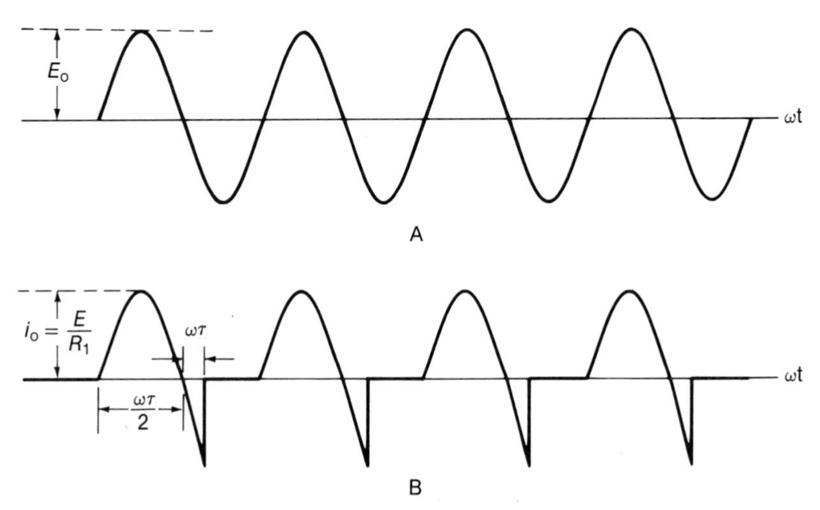

Fig. 10.6 Current waveform in a step recovery diode. A. Sinusoidal emf applied to the diode. B. Current in the diode.

10.3 Harmonic Generation by Step Recovery Diodes

If a sinusoidal emf

$$E = E_o \sin \omega t \qquad \text{(V)} \tag{10.3.1}$$

is applied to a step recovery diode as shown in Fig. 10.6 A, the current in the diode will have a waveform as shown in Fig. 10.6 B. The current waveform for the positive half-cycle is similar to Fig. 10.1 C. As a result, the frequency spectrum will be, from Eqs. 10.1.6, 10.1.7, and 10.1.8,

$$\left.\begin{aligned} \frac{a_o}{2} &= \frac{i_o}{\pi} \\ b_1 &= \frac{i_o}{2} \\ a_n &= \frac{2i_o}{(n-1)(n+1)\pi} \\ &\qquad n = 2, 4, 6, \ldots \\ b_n &= 0 \end{aligned}\right\} \tag{10.3.2}$$

where i_o is the peak forward current. In the reverse half-cycle of the applied emf, the reverse current's waveform resembles Fig. 10.4 B. According to Eqs. 10.1.12 10.1.13, and 10.1.14,

$$\left.\begin{aligned} \frac{a_o}{2} &= \frac{i_{or}\omega\tau}{4\pi} \\ a_n &= \frac{4i_{or}\left(1 - \cos\dfrac{n\omega\tau}{2}\right)}{n^3\omega\tau} \\ &\qquad n = 1, 2, 3, 4, \ldots \\ b_n &= 0 \end{aligned}\right\} \tag{10.3.3}$$

where i_{or} is the reverse peak current and

$$i_{or} \simeq i_o \sin \omega\tau \tag{10.3.4}$$

The current for the step recovery diode is therefore, as shown in

Fig. 10.6 B, and the spectrum is a proper combination of Eqs. 10.3.2 and 10.3.3

$$\left.\begin{aligned}
\frac{a_o}{2} &= \frac{i_o}{\pi}\left(1 - \frac{\omega\tau}{4}\right) \\
b_1 &= \frac{i_o}{2} \\
a_n &= 2\frac{1}{(n-1)(n+1)\pi} - \frac{4\left(1 - \cos\left(\frac{n\omega\tau}{2}\right)\right)}{n^3\omega\tau} i_o \\
& \qquad n = 2, 4, 6, 8, \ldots \\
a_m &= -\frac{4\left(1 - \cos\left(\frac{m\omega\tau}{2}\right)\right)}{m^3\omega\tau} i_o \\
& \qquad m = 3, 5, 7, 9, \ldots \\
b_n &= 0
\end{aligned}\right\} \quad (10.3.5)$$

Comparing Eqs. 10.3.2. and 10.3.3, the step recovery diode produces richer harmonics than an ordinary nondegenerate p–n junction diode.

10.4 Harmonic Generation by Backward Diodes

By inspection of the current–voltage characteristics of a backward diode as shown in Fig. 7.6, the current waveform can be sketched as shown in Fig. 10.7. In these drawings, it is assumed that the fundamental frequency signal voltage is sufficient to go over the threshold forward voltage. For the positive half-cycle, the diode conducts for one short period of time when the input signals exceed the forward threshold voltage. At the reverse half-cycle, the diode is a good conductor. Therefore, the current waveform of a single backward diode takes a form as sketched in Fig. 10.7 B. This is a combination of Eqs. 10.1.9, 10.1.10, 10.1.11, and

Eqs. 10.1.6, 10.1.7, 10.1.8. As a result,

$$
\left.
\begin{aligned}
\frac{a_o}{2} &= \frac{i_o}{\pi}\left(\frac{\omega\tau}{2} - 1\right) \\
b_1 &= \frac{i_o}{2} \\
a_n &= -\frac{2i_o}{\pi}\left(\frac{1}{n}\sin\left(\frac{n\omega\tau}{2}\right) - \frac{1}{(n-1)(n+1)}\right) \\
&\qquad n = 2, 4, 6, 8, \ldots \\
a_m &= -\frac{2i_o}{m\pi}\sin\left(\frac{m\omega\tau}{2}\right) \\
&\qquad m = 1, 3, 5, 7, \ldots \\
b_n &= 0 \\
&\qquad n = 2, 3, 4, 5, \ldots
\end{aligned}
\right\} \quad (10.4.1)
$$

When two backward diodes are connected in series back to back with each other, the current waveform as illustrated in Fig. 10.7 C will appear. The Fourier coefficients of this type of waveform are known to be [1]

$$
\left.
\begin{aligned}
\frac{a_o}{2} &= 0 \\
a_n &= 0 \\
b_n &= 2i_o \frac{\sin\left(\frac{n\omega\tau}{2}\right)\sin\left(\frac{n\pi}{2}\right)}{\frac{n\pi}{2}} \\
&\qquad n = 1, 2, 3, \ldots
\end{aligned}
\right\} \quad (10.4.2)
$$

10.5 Harmonic Generation by Nondegenerate p – n Junction Diodes

As was explained in Fig. 10.3, when a sinusoidal emf is applied across a nondegenerate p–n junction diode, the diode current

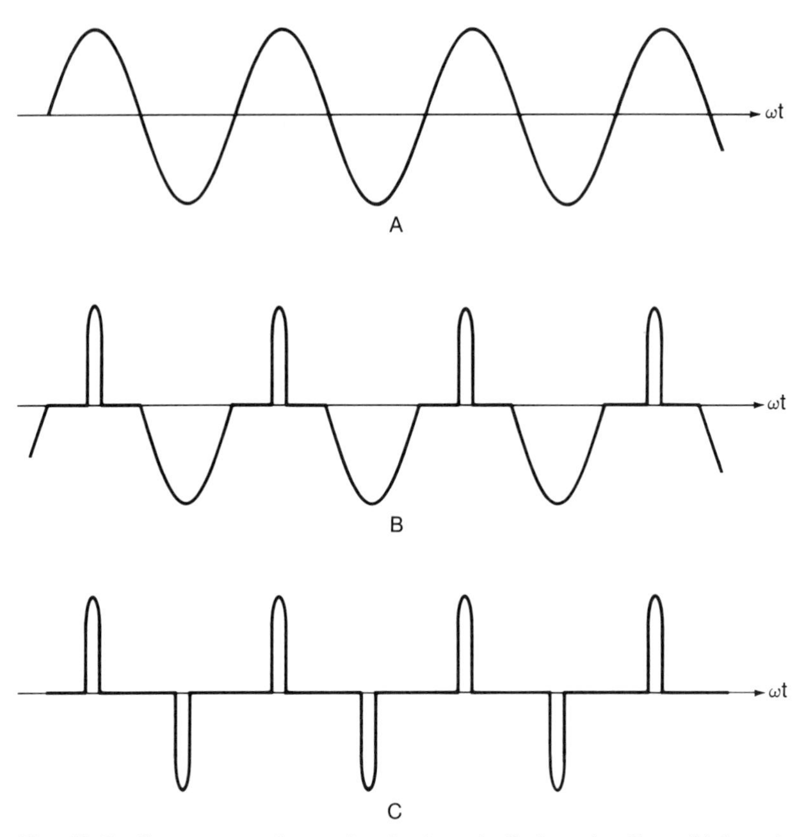

Fig. 10.7 Current waveforms for backward diodes. A. Sinusoidal emf applied to a backward diode. B. Current waveform in a backward diode. C. Current waveform when two backward diodes are connected in series back to back.

takes a waveform illustrated by Fig. 10.3 C. This may be approximated by a train of triangular pulse currents as shown by Fig. 10.4 B. The Fourier coefficients are then given by Eq. 10.1.12, Eq. 10.1.13, and Eq. 10.1.14. A variable capacitance diode, or varactor, which is presented in Chapter 9, is one of the most common nondegenerate p–n junction diodes used for microwave harmonic generation. If a combination of a sinusoidal voltage $v = v_i \sin \omega t$ and a dc bias voltage V_{ro} is applied to the diode, then the total reverse bias voltage across the diode V_r is

$$V_r = V_o + V_{ro} + v = V_o + V_{ro} + v_i \sin \omega t \qquad (10.5.1)$$

where V_o is the junction contact potential. The capacitance of a varactor C is then, as in Eq. 9.2.4,

$$C = S\sqrt{\frac{qN_d\varepsilon}{2(V_o + V_{ro})}}\left(1 + \frac{v_i}{V_o + V_{ro}}\sin\omega t\right)^{-1/2} \quad (10.5.2)$$

As shown in Eq. 9.2.9, the diode's time varying current is

$$i = \frac{d}{dt}(CV_r) = C\frac{dV_r}{dt} + V_r\frac{dC}{dt} \quad (10.5.3)$$

Substituting Eqs. 10.5.1 and 10.5.2 in Eq. 10.5.3,

$$i = \frac{C_o v_i \omega \cos\omega t}{\sqrt{1 + \frac{v_i}{V_o + V_{ro}}\sin\omega t}} - \frac{C_o}{2}\frac{v_i^2\omega \sin\omega t \cos\omega t}{\left(1 + \frac{v_i}{V_o + V_{ro}}\sin\omega t\right)^{3/2}(V_o + V_{ro})} \quad (10.5.4)$$

where C_o is the dynamic capacitance of the p–n junction without microwave signals and given by Eq. 9.2.7. Applying a binomial expansion to the denominators of Eq. 10.5.4,

$$\begin{aligned} i &= C_o v_i \omega \cos\omega t - \frac{C_o v_i^2 \omega}{2(V_o + V_{ro})}\cos\omega t \sin\omega t \\ &\quad + \frac{C_o v_i^2 \omega}{1(V_o + V_{ro})}\cos\omega t \sin\omega t \\ &\quad - \frac{3C_o v_i^3 \omega}{4(V_o + V_{ro})^2}\sin\omega t \cos\omega t \sin\omega t \\ &= C_o v_i \omega \cos\omega t - \frac{3C_o v_i^3}{8(V_o + V_{ro})^2}\sin 2\omega t \sin\omega t \\ &= C_o v_i \omega \cos\omega t - \frac{3C_o v_i^3 \omega}{16(V_o + V_{ro})^2}(\cos\omega t - \cos 3\omega t) \\ &= C_o v_i \omega\left(1 - \frac{3v_i^2}{16(V_o + V_{ro})^2}\right)\cos\omega t \\ &\quad + \frac{3C_o v_i^3 \omega}{16(V_o + V_{ro})^2}\cos 3\omega t \end{aligned} \quad (10.5.5)$$

As seen from this result, the varactor is useful for third harmonic generation. Inspecting Eqs. 9.2.7 and 10.5.5, respectively, it is wise to operate with the zero dc bias voltage for a larger harmonic output if it is possible. Usually, the self-bias is developed across the diode when the fundamental frequency power is fed. As a result, it is impossible to create a condition $V_{ro} = 0$, unless a small capacitor is connected in series with the diode to block the flow of dc current. In an actual varactor diode, for the forward half-cycle, the diode behaves like an ordinary nondegenerate p–n junction diode, and for the reverse half-cycle, it is a varactor. The actual frequency spectrum is therefore a combination of Eqs. 10.1.12, 10.1.13, 10.1.14, and 10.5.5.

10.6 Microwave Harmonic Generator Circuit

Microwave harmonic generation is needed when

1. A high microwave frequency signal is needed and there is no available economical fundamental frequency oscillator at the required frequency.
2. An accurate crystal controlled high microwave frequency signal is needed but no high frequency crystal at the required frequency is available, or it is expensive.

A schematic diagram of a microwave harmonic generator circuit is shown in Fig. 10.8. In this figure, microwave power of a fundamental frequency f is fed through a coaxial line to a harmonic generating diode. The current in the diode contains microwave harmonic current and launches microwave harmonic fields into a waveguide cavity resonator which consists of a shorting plunger and a waveguide iris. The resonator is tuned at a

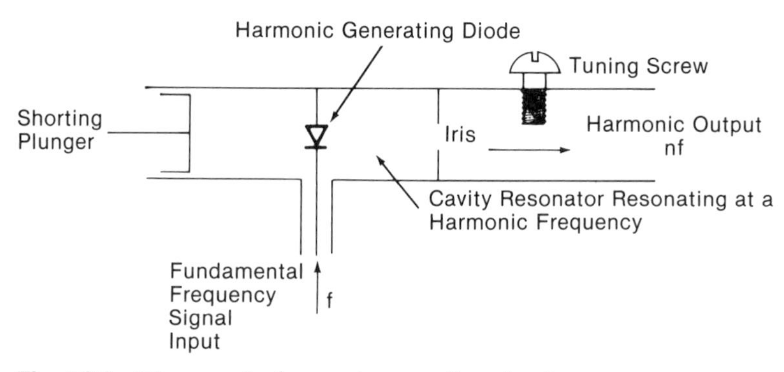

Fig. 10.8 Diagram of a harmonic generation circuit.

desired microwave harmonic frequency. Through the iris, the harmonic signals are launched into the output waveguide. A screw tuner is used for matching the impedance of the harmonic. In this arrangement, the fundamental frequency signals do not enter the waveguide because the waveguide is made too small and is cut off to the fundamental signal wavelength [6,8–10].

Since the harmonic current is generated in the semiconductor, if the junction capacitance is significant and the lead inductance is not negligible, then the generated harmonic current is internally bypassed or short circuited and may not even be radiated into the waveguide cavity resonator. In the diode current equations in previous sections, the order of the harmonic n can mathematically be any desirable large number. This simply means the harmonic current may be generated in the junction but does not mean that it actually radiates the particular harmonic. Except for custom-made laboratory diodes, discharge times of commercial homojunction step recovery diodes are 50–450 ps and the minimum cutoff frequencies $f_c = 1/2\pi R_s C_J$ are 100–350 GHz, yet the actual operating frequency range is 1–20 GHz, with an operating output power range of 0.3–10 W, an order of harmonic $n = 5$–10, a junction capacitance $C_J = 0.1$–4.7 pF, and a reverse breakdown voltage of 20–100 V dc [3,7].

For example, at $f = 300$ GHz, if $C_J = 0.25$ pf, the impedance of the junction capacitance is $1/2\pi f C_J = 1/(2\pi\ 300 \times 10^9\ \mathrm{s}^{-1} \times 0.25 \times 10^{-12}\ \mathrm{F} = 2.1\ \Omega$. In order to couple 100 GHz energy out of the diode, the coupling of the diode to the cavity must be adjusted so that the diode couples at the low impedance location in the cavity. This is usually an extreme corner of a cavity resonator [6]. At any rate, the effective and efficient coupling of the diode to the cavity resonator with low junction capacitive impedance is not easy, if not impossible.

10.7 Microwave Harmonic Generating Diodes

Degenerate p–n junction diodes, nondegenerate p–n junction diodes, Schottky barrier diodes, heterojunction diodes, step recovery diodes, tunnel diodes, varactor diodes, and any other nonlinear diodes produce harmonics. The principle of harmonic generation is explained as follows. When pure sinusoidal voltage of a fundamental frequency is applied across a nonlinear diode, the current is a nonsinusoidal periodic function of the fundamental frequency. Fourier expansion of a nonsinusoidal function contains inherent harmonics of the fundamental frequency. Harmonics are therefore generated. In addition to harmonics, as stated in Chapter 8, the nonlinear impedance can mix different frequencies. The fundamental and the harmonics can mix, and the harmonics themselves can mix with each other. At any rate, the beat frequencies coincide with the fundamental and its harmonics. The original harmonics and the beat frequencies usually have different phase angles with each other [1]. Therefore, if mixing occurs there is interference in phase angle in each spectrum of harmonics. At any rate, if a particular harmonic signal of good quality is desired, a good reliable filter is needed to reject undesired signals.

Problems

10.1 Define harmonic frequencies.

10.2 Define a nonlinear p–n junction.

10.3 Explain rectification by an ideal p–n junction diode.

10.4 Using the Fourier coefficient formulas, find the spectrum of $\sin \omega t$.

10.5 Point out the differences between actual diodes and ideal diodes.

10.6 Assuming a diode's characteristic follows the "square law", find a frequency spectrum of the diode using Fourier coefficients.

10.7 Sketch a current–voltage characteristic of a p–n junction diode which produces rectangular pulses in the rectified current.

10.8 Sketch a current–voltage characteristic of a p–n junction diode which produces triangular pulses.

10.9 Define a step recovery diode.

10.10 Explain the function of a step recovery diode.

10.11 Try to sketch a current–voltage curve of a step recovery diode. Point out any difference in the characteristics from an ordinary nondegenerate p–n junction diode. Explain the reason for the difference.

10.12 Relate the snap off or step recovery time to the strength of the harmonic analytically.

10.13 Compare the third harmonics of an ordinary nondegenerate p–n junction diode and a step recovery diode.

10.14 Try to generate a strong fifth harmonic. Select a type of harmonic generating diode and justify the selection.

References

1 J. J. Tuma, "Engineering Mathematics Handbook." McGraw-Hill, New York, 1970.
2 E. S. Yang, "Fundamentals of Semiconductor Devices." McGraw-Hill, New York, 1978.
3 B. G. Streetman, "Solid-State Electron Devices." Prentice-Hall, Englewood Cliffs, New Jersey, 1972.
4 A. van der Ziel, "Solid-State Physical Electronics." 2nd ed. Prentice-Hall, Englewood Cliffs, New Jersey, 1976.
5 J. L. Moll and S. A. Hamilton, Physical modeling of the step recovery diode for pulse and harmonic generation circuit. *Proc. IEEE* **57**(7), 1250–1259 (1969).
6 T. K. Ishii, "Microwave Engineering." Ronald Press, New York, 1966.
7 Hewlett Packard, "Diode and Transistor Designer's Catalog." Hewlett Packard, San Jose, California, 1980.
8 D. Roddy, "Microwave Technology." Prentice-Hall, Englewood Cliffs, New Jersey, 1986.
9 H. E. Thomas, "Handbook of Microwave Techniques and Equipment." Prentice-Hall, Englewood Cliffs, New Jersey, 1972.
10 J. T. Coleman, "Microwave Devices." Reston Publ., Reston, Virginia, 1982.
11 U. Lott, Simultaneous magnitude and phase measurement of harmonics in non-linear microwave two ports. *IEEE MTT-S Int. Microwave Symp. Dig.*, **1**, 225–228 (May 1988).

11

Microwave Switching Semiconductor Devices

11.1 Microwave Binary States of Diodes and Transistors

By controlling the bias voltage properly, any nonlinear semiconductor diode can be placed in a binary state. For example, in a conventional nondegenerate PN junction rectifier diode, if the diode is forward biased, the diode is considered to be in an "on" state. When the bias is reversed, the same diode is considered to be in an "off"state. There is a particular diode which is specifically designed for switching on and off with microwave signals. One such diode is called the PIN diode. When this diode is forward biased the diode conducts microwave current. When forward bias is removed or reverse bias is given the diode is turned to the open, or off, state for microwaves [5–8].

Variable capacitance diodes or varactors which are described in Chapter 9 present relatively large capacitances at lower reverse voltages. At high reverse voltages, the capacitance becomes small and the diode is considered to be open. In this chapter, PIN diodes and varactor diodes are presented as microwave power controlling devices by putting them in the binary states of "off" and "on". Applications of these devices for microwave switches, diplexers, step attenuators, and phase shifters are presented [10–13]. Microwave transistors can also be used for microwave switching [6–9]. The switching mechanisms and the applications are presented in the following sections.

There are two categories of microwave switches: the on–off switch and the routing switch. No matter what category, the switching function depends on the binary state of the nonlinear component of the switch. The binary states are the low impedance state and high impedance state. These binary states are created by a bias in one way or the other. When the device is in the low impedance state, the microwave current flows through the device. When the device is in the high impedance state, very little microwave current can flow through the device.

Some important characterizing parameters of on–off microwave switches are, besides the working frequency range and other factors, the insertion loss and the isolation.

In the microwave transmission line, if the output of the transmission line is P_o (W), keeping the input power constant, as well as when an on–off switch is inserted with the "on" state in the transmission line, and the power received drops down to P_{on} (W), then the insertion loss of this switch is defined as

$$L_{ins} = 10 \log_{10}\left(\frac{P_o}{P_{on}}\right) \text{dB} \tag{11.1.1}$$

Under the same conditions, if the switch is turned off and the received transmission power drops down to P_{off} (W), then the isolation of this switch is

$$L_{iso} = 10 \log_{10}\left(\frac{P_o}{P_{off}}\right) \text{dB} \tag{11.1.2}$$

Typical insertion loss of a commercial switching diode is 0.9 dB per diode, and the isolation is 25 dB per diode, depending on the bias [3].

Another important switching parameter is the switching speed. The shortest transition time between the binary states is called the switching speed. A typical switching speed of commercial switching diodes is less than 2 ns [3]. This is not so fast if signals on the order of GHz are to be immediately and completely stopped. For example, during the switching time of 2 ns, a 10 GHz microwave signal will pass through 20 cycles. In other words, the period T of the frequency $f = 10$ GHz microwave is $T = 1/(10 \times 10^9\ \text{s}^{-1}) = 0.1$ ns. If the switching time is $\tau = 2$ ns, then the number of cycles of microwave signals passing through before the switch opens is $N = \tau/T = 2\ \text{ns}/0.1\ \text{ns} = 20$ cycles.

If the incident microwave power P_o is 1 mW or 0 dBm, and when the switch is on P_{on} is 0.9 mW or $10\log_{10}(0.9\text{ mW}/1\text{ mW})$ dBm $= -0.46$ dBm, then the insertion loss of this switch is, from Eq. 11.1.1,

$$L_{ins} = 10\log_{10}\left(\frac{1\text{ mW}}{0.9\text{ mW}}\right) = 0.46\text{ dB}$$

or

$$L_{ins} = 0\text{ dBm} - (-0.46\text{ dBm}) = 0.46\text{ dB}$$

If the switch is open and $P_{off} = -30$ dBm $= 10^{-3}$ mW, then from Eq. 11.1.2, the isolation is $L_{iso} = 10\log_{10}(1\text{ mW}/10^{-3}\text{ mW}$ $= 30$ dB. $P_{off} = -30$ dBm $= 10^{-3}$ mW is still significant power in many cases. The sensitivity of many microwave receivers is -50 to -120 dBm or 10^{-5} to 10^{-12} mW. In order to obtain desirable isolation, multiple switching diodes or transistors are combined in a well-thought-out transmission circuit impedance design. If the switch is a routing switch rather than an on–off switch then, in Eqs. 11.1.1 and 11.1.2, P_{on} is the switch output of the on-route output and P_{off} is the switch output of the off-route output.

11.2 Microwave Binary States of PIN Diodes

A doping profile of a PIN diode is shown in Fig. 11.1 A. A simplified microwave equivalent circuit under the forward bias condition is shown in Fig. 11.1 B and under the reverse bias conditions shown in Fig. 11.1 C. R_F is the microwave resistance

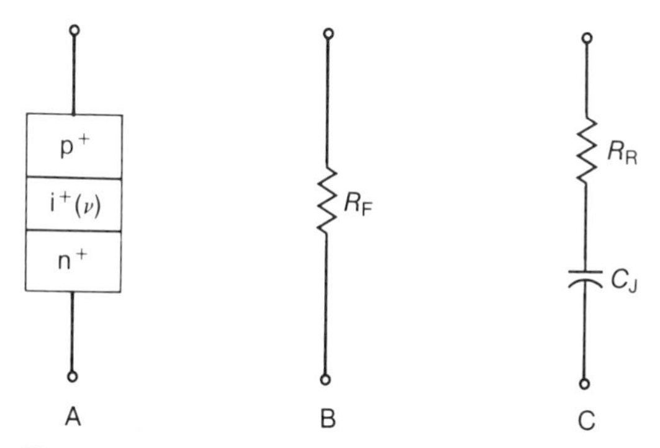

Fig. 11.1 A. PIN diode. B. Forward bias (microwave binary state). C. Reverse bias (microwave binary state).

and is considered to be reasonably small at microwave frequencies. There is no ideal i-type semiconductor. Most practical PIN diodes are made of silicon which is not 100% pure. Most practical PIN diode i-layers contain a small amount of n-type impurity, and are identified as n-type i-layers. It therefore conducts a small amount of dc current under forward bias and an appreciable amount of microwave frequency current. Under reverse bias conditions, the depletion layer punches through the i-layer and the junction capacitance is minimized. The conduction, dependent upon the minority carrier and the microwave reverse resistance R_R, is considered to be mostly made of bulk p^+- and n^+– layer resistance and ohmic contact resistance. The value of R_R and the small value of the reverse junction capacitance C_J make the microwave impedance under reverse bias high.

If the thickness of the i-layer is L (m), the cross section is S (m^2), and the conductivity is σ (S-m) under forward bias, then the microwave resistance of the i-layer R_I is given by

$$R_I = \frac{L}{S\sigma} \tag{11.2.1}$$

Under forward bias superimposed by a microwave voltage, the carriers will be injected into the i-layer from both the n^+-layer and the p^+-layer. Electrons and holes injected into the i-layer will recombine there. If the mobility of electrons in the i-layer is μ_n (m^2/V-s), the electron density is n (m^{-3}), the mobility of holes is μ_p (m^2/V-s), and the hole density is p (m^{-3}), then the conductivity is

$$\sigma = q(\mu_n n + \mu_p p) \tag{11.2.2}$$

In the i-layer, by the definition of holes, statistically

$$\mu_n \approx \mu_p \equiv \mu \tag{11.2.3}$$

where μ is the average mobility of the carrier.

The charge neutrality law requires that

$$n = p$$

then

$$\mu_n n \approx \mu_p p \tag{11.2.4}$$

Substituting Eq. 11.2.4 in Eq. 11.2.2,

$$\sigma = 2q\mu n \tag{11.2.5}$$

Substituting Eq. 11.2.5 in Eq. 11.2.1,

$$R_I = \frac{L}{2Sq\mu n} \quad (\Omega) \tag{11.2.6}$$

The electron charge in the i-layer is then

$$Q = LSqn \quad (C) \tag{11.2.7}$$

If the lifetime of the electron in the i-layer is τ seconds and the diode current is I_o (A), then

$$Q = I_o\tau \quad (C) \tag{11.2.8}$$

Combining Eqs. 11.2.7 and 11.2.8,

$$LSqn = I_o\tau \tag{11.2.9}$$

Substituting Eq. 11.2.9 in Eq. 11.2.6, the microwave resistance of the i-layer under forward bias is then

$$R_I = \frac{L^2}{2\mu n I_o \tau} \tag{11.2.10}$$

In practical microwave PIN diodes, the value of R_I is on the order of 10^{-1} Ω [1]. This is on the same order as R_b, the magnitude of the resistance in the bulk of p^+- and n^+-regions. The total microwave resistance under the forward bias is then

$$R_F = R_b + R_{I_F} + R_o \tag{11.2.11}$$

The carrier mobility is on the order of 10^{-1} m^2/V-s. The diode current on the order of 10^{-3} A, and R_o is the combined contact resistance and lead resistance. The thickness of the i-layer is on the order of 10^{-4} m. The contribution of the skin effect is considered to be small for forward resistance [1].

When the dc bias is reversed, the depletion layer spreads throughout the i-layer and there will be reverse resistance in series with a small capacitance caused by the i-layer as shown in Fig. 11.1. The microwave resistance under the reverse bias, R_R, consists of the bulk resistance of the p^+- and n^+-layers R_b, the ohmic contact resistance R_o, and a small amount of R_{I_R}, if any. The equivalent series resistance is due to dissipation loss in the carrier depleted i-layer

$$R_R = R_b + R_o + R_{I_R} \tag{11.2.12}$$

This resistance is essentially linear. Therefore, R_F in Eq. 11.2.11

and R_R in Eq. 11.2.12 are quite different. Inspection of specification sheets for commercially available microwave PIN diodes reveals that, in most diodes, $R_b + R_o$ are equal for forward and reverse and the values of these are on the order of 10^{-3}–10^0 Ω. Some high loss diodes have R_R much higher than R_F [1–3].

The junction capacitance under reverse bias can be calculated based on the geometry of the i-layer.

$$C_J = \varepsilon \frac{S}{L} = \varepsilon_o \varepsilon_r \frac{S}{L} \tag{11.2.13}$$

Again, inspection of commercially available microwave PIN diodes reveals that the value of junction capacitance ranges from 10^{-1} to 10^0 pF.

The microwave impedance of a reverse biased PIN diode is then

$$Z_R = R_R - j\frac{1}{\omega C_J} \tag{11.2.14}$$

The order of magnitude computation of the impedance Z_R at 10 GHz is $10^3 - j10^3$ Ω. The binary states of a microwave PIN diode are then 10^{-1} Ω when forward biased and $10^3 - j10^3$ Ω when reverse biased. These switching characteristics are useful for microwave circuit switching, attenuation, and phase control [5,10,13].

For example, the conductivity of intrinsic silicon is known to be $\sigma = (1/2.5 \times 10^5)$ S/μm. If the thickness of the i-layer is $L = 100$ μm and the cross-sectional area is $S = 100$ μm × 100 μm = 10^4 μm², then from Eq. 11.2.1, the resistance of the intrinsic layer R_I is $R_I = (100\ \mu\text{m})(2.5 \times 10^5\ \Omega\text{-}\mu\text{m})/10^4\ \mu\text{m}^2 = 2.5 \times 10^3\ \Omega$ at zero bias or reverse bias.

Under the same condition, the electron mobility in intrinsic silicon is known to be $\mu_n = 1350\ \text{cm}^2/\text{V-s} = 0.135\ \text{m}^2/\text{V-s}$ and the electronic charge $q = 1.6 \times 10^{-19}$ C. The resistance of the i-layer under reverse bias is, from Eq. 11.2.6,

$$R_I = \frac{100\ \mu\text{m}}{2 \times 10^4\ \mu\text{m}^2 \times 1.6 \times 10^{-19}\ \text{C} \times 0.135\ \text{m}^2/\text{V-s}}$$

$$= 2.3 \times 10^3\ \Omega$$

When the PIN diode is forward biased, the electrons in the i-layer are discharged to the anode and the diode current I_o flows.

The lifetime of electrons in the i-layer is known to be 100 ns in commercial PIN diodes [3]. The forward current is 50 mA [3]. Then, from Eq. 11.2.10,

$$R_{I_F} = \frac{(100 \times 10^{-6}\ \text{m})^2}{[(2 \times 0.135\ \text{m}^2/\text{V-s})(50 \times 10^{-3}\ \text{A})(100 \times 10^{-9}\ \text{S})]}$$

$$= 0.074\ \Omega$$

As a result, according to Eqs. 11.2.11 and 11.2.12, the forward resistance R_F can be on the order of $10^0\ \Omega$ but the reverse resistance can be on the order of $10^3\ \Omega$ [3].

11.3 Microwave Switching and Multiplexing

The principle of microwave switching by the use of a PIN diode is illustrated by Fig. 11.2 A. This waveguide switch is basically an E-plane T [2,12]. The PIN diode is mounted on the E-arm, one half wavelength from the junction. According to transmission line theory, the impedance of a lossless transmission line repeats the same value every half-wavelength [2,12]. The PIN diode mounted physically one half-wavelength from the junction is therefore electrically the same as mounting the PIN diode right at the joint. The shorting plunger behind the PIN diode is a tuner that produces an inductance to create a parallel resonance with the diode's junction capacitance when the diode is reverse biased. The parallel resonance produces a high impedance at the junction and the waveguide is therefore open circuited at the junction. When the diode is forward biased, the waveguide is practically short circuited making microwave transmission possible.

The input impedance of a short circuited transmission line of length l is given by

$$Z_i = jZ_o \tan\left(\frac{2\pi l}{\lambda_g}\right) \tag{11.3.1}$$

where Z_o is the characteristic impedance of the waveguide and λ_g is the waveguide wavelength [2,12]. If the length l is chosen to produce a positive input impedance, then Z_i is inductive,

$$Z_i = j\omega L_i \tag{11.3.2}$$

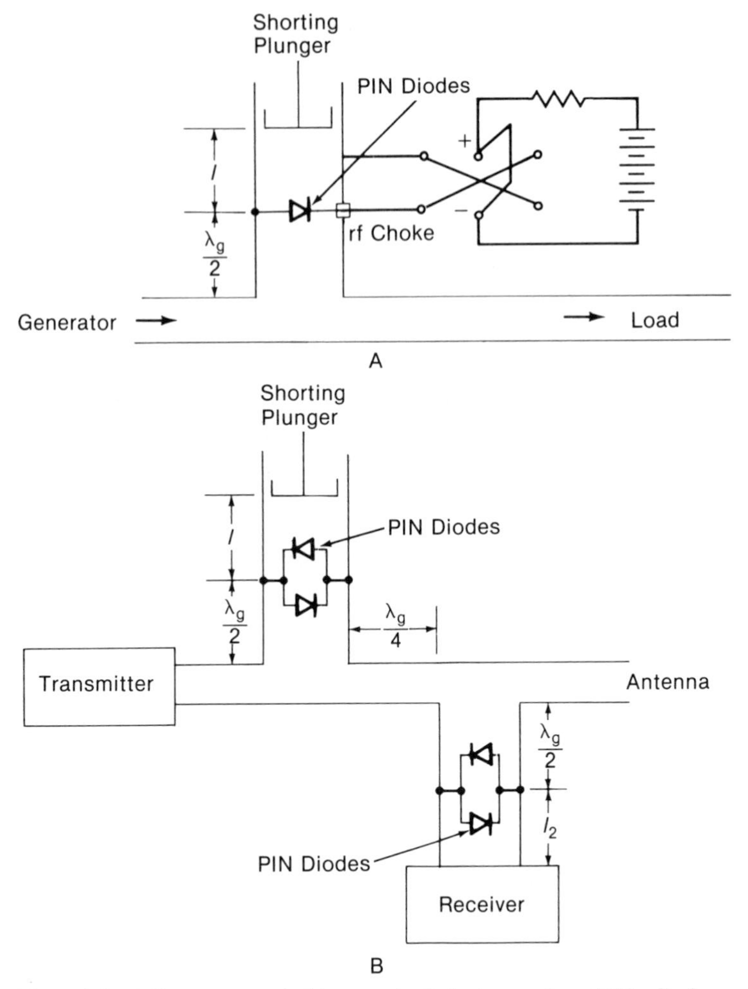

Fig. 11.2 Microwave switching and diplexing using PIN diodes. A. E-plane T-waveguide switch. B. E-plane T-waveguide diplexer.

The input equivalent inductance is represented by L_i. Combining Eqs. 11.3.1 and 11.3.2,

$$L_i = \frac{Z_o}{\omega} \tan\left(\frac{2\pi l}{\lambda_g}\right) \tag{11.3.3}$$

The impedance of a reverse biased PIN diode is given by Eq. 11.2.14. The parallel impedance of the reverse biased PIN diode and the short circuited waveguide is, as seen in Fig. 11.2 A,

$$\dot{Z}_L = \frac{\dot{Z}_i \dot{Z}_R}{\dot{Z}_i + \dot{Z}_R} \tag{11.3.4}$$

Substituting 11.2.14 and 11.3.1 in Eq. 11.3.4,

$$\dot{Z}_L = \frac{jZ_o \tan\left(\dfrac{2\pi l}{\lambda_g}\right)\left(R_R - j\dfrac{1}{\omega C_J}\right)}{R_R + j\left(Z_o \tan\dfrac{2\pi l}{\lambda_g} - \dfrac{1}{\omega C_J}\right)}$$

$$= \frac{jZ_o \tan\left(\dfrac{2\pi l}{\lambda_g}\right)\left[R_R - j\left(Z_o \tan\left(\dfrac{2\pi l}{\lambda_g}\right) - \dfrac{1}{\omega C_J}\right)\right]\left(R_R - j\dfrac{1}{\omega C_J}\right)}{R_R^2 + \left(Z_o \tan\left(\dfrac{2\pi l}{\lambda_g}\right) - \dfrac{1}{\omega C_J}\right)^2} \quad (11.3.5)$$

then,

$$|\dot{Z}_L| = \frac{Z_o \tan\left(\dfrac{2\pi l}{\lambda_g}\right)\sqrt{\left[R_R^2 - \dfrac{1}{\omega C_J}\left(Z_o \tan\left(\dfrac{2\pi l}{\lambda_g}\right) - \dfrac{1}{\omega C_J}\right)\right]^2 + R_R^2\left[\dfrac{1}{\omega C_J} + \left(Z_o \tan\left(\dfrac{2\pi l}{\lambda_g}\right) - \dfrac{1}{\omega C_J}\right)\right]^2}}{R_R^2 + \left(Z_o \tan\left(\dfrac{2\pi l}{\lambda_g}\right) - \dfrac{1}{\omega C_J}\right)^2} \quad (11.3.6)$$

The maximum value of $|\dot{Z}_L|$ or open circuit condition is obtained when

$$Z_o \tan\left(\frac{2\pi l}{\lambda_g}\right) - \frac{1}{\omega C_J} = 0 \quad (11.3.7)$$

or

$$l = \frac{\lambda_g}{2\pi}\tan^{-1}\left(\frac{1}{\omega Z_o C_J}\right) \quad (11.3.8)$$

and the open circuit impedance is

$$|\dot{Z}_L|_{max} = \frac{1}{\omega C_J}\sqrt{1 + \frac{1}{\omega C_J}} \tag{11.3.9}$$

If $\omega \approx 10^9\ s^{-1}$, $C_J \approx 10^{-13}$ F, and $R_R \approx 10^3\ \Omega$, then $|\dot{Z}_L|$ is on the order of $10^5\ \Omega$.

When the diode is forward biased, the forward resistance is $10^0\ \Omega$. This is usually negligible with respect to the matched load which is on the order of $10^2\ \Omega$. Thus the waveguide switch is closed when the PIN diode is forward biased.

When the PIN diode is switched electronically using a switching circuit for a waveguide junction with many waveguide branches, multiplexing is possible by sequentially switching from one waveguide to another.

When two PIN diodes are connected back to back as shown in Fig. 11.2 B, the combination provides a limiter, or limiting switch. When a microwave signal is weak, as small as 10^{-18} W, the diode is practically open circuited as demonstrated by Eq. 11.3.9. The carrier depletion is throughout the i-layer of the PIN diode. Therefore, for either the positive half-cycle or the negative half-cycle of small input signals, the diodes are practically open circuited. When the microwave signal is strong, as large as 10^4 W, the diodes are self-biased forward and the forward resistance R_F is now in the range of 10^{-1} to $10^0\ \Omega$. The diodes are now switched to the closed state.

In Fig. 11.2, when weak signals come in from the antenna, the PIN diodes are open. Then the signals reach the receiver. When the transmitter is on, high power from the transmitter closes the diode switch, but the power goes to the antenna, not to the receiver. The receiver is protected by the PIN diodes, which act as limiters. This system is a microwave diplexer.

For example, a commercial 5082–3202 microwave switching PIN diode has a switching time of less than 5 ns, a forward resistance $R_F = 0.8\ \Omega$, a reverse resistance $R_R = 8\ \Omega$, and an isolation $L_{iso} > 25$ dB. The total capacitance, including the package capacitance and the junction capacitance, is $C_T = 0.32$ pF. This diode is designed to operate up to 8 GHz [3]. If the operating frequency of 7 GHz is chosen, then the free space wavelength is $\lambda_o = 3 \times 10^8\ m/s/f = 3 \times 10^8\ m/s/7 \times 10^9\ s^{-1} = 0.042$ m = 4.2 cm. If a microstripline of characteristic impedance $Z_o = 50\ \Omega$ on an alumina substrate of $\varepsilon_r = 10$ is used, then the transmission line wavelength $\lambda_l = \lambda_o/\sqrt{\varepsilon_r} = 1.33$ cm. The tuning transmission line length l is now obtained from Eq. 11.3.8.

$$l = \frac{\lambda_l}{2\pi}\tan^{-1}\left(\frac{1}{\omega Z_o C_J}\right)$$

$$= \frac{1.33\text{ cm}}{2\pi}\tan^{-1}\left(\frac{1}{2\pi(7\times 10^9\text{ s}^{-1})(50\ \Omega)(0.32\times 10^{-12}\text{ F})}\right)$$

$$= 0.2\text{ cm},$$

or $\lambda_l/2$ more can be added for ease of fabrication with the same impedance. Transmission line theory states that the impedance value repeats itself every $\lambda_l/2$ [2,12]. In this case, $\lambda_l/2 = 1.33$ cm/2 = 0.665 cm. Thus, $l = 0.2$ cm or $l = 0.2$ cm + 0.665 cm = 0.865 cm.

In reality, the short circuited transmission line is inconvenient for the microstripline configuration. According to transmission line theory, if a $\lambda_l/4$ long open circuited line is added, this configuration works the same as the original short circuited line [2,12]. In this case, $\lambda_l/4 = 1.33$ cm/4 = 0.33 cm. As a result, if the open circuited microstripline is used, then the tuning line length is either $l = 0.2$ cm + 0.33 cm = 0.53 cm or $l = 0.865$ cm + 0.33 cm = 1.195 cm. At any rate, the maximum impedance at the diode is, from Eq. 11.3.9,

$$Z_{L_{max}} = \sqrt{1 + \frac{1}{\omega C_J}}\cdot\frac{1}{\omega C_J}$$

$$= \sqrt{1 + 1/\left[2\pi(7\times 10^9\text{ s}^{-1})(0.32\times 10^{-12}\text{ F})\right]}$$

$$\times 1/\left[2\pi(7\times 10^9\text{ s}^{-1})(0.32\times 10^{-12}\text{ F})\right] = 602\ \Omega.$$

This is considered to be high in comparison with $Z_o = 50\ \Omega$. If this is not high enough for some reason, then more than one PIN diode may be connected in series. This is easily done using monolithic microwave integrated circuit (MMIC) technology [5]. Now this single PIN diode presents 602 Ω and opens the transmission line of $Z_o = 50\ \Omega$. This means that, from the input side, the input 50 Ω line is terminated by an impedance of 602 Ω + 50 Ω = 652 Ω. According to transmission line theory [2,12], the voltage reflection coefficient at this open state at the input is then

$$\rho = \frac{(Z_{L_{max}} + Z_o) - Z_o}{(Z_{L_{max}} + Z_o) + Z_o} = \frac{652 - 50}{652 + 50} = 0.85$$

The voltage transmission coefficient at this point is

$$\tau = 1 - \rho = 1 - 0.85 = 0.15$$

This voltage is further divided into 652 Ω and 50 Ω. At the output transmission line, the voltage transfer coefficient across the 50 Ω output line is therefore

$$\tau_v = \tau \frac{Z_o}{Z_{L_{max}} + Z_o} = 0.15 \times \frac{50\ \Omega}{602\ \Omega + 50\ \Omega} = 0.012$$

As a result, the power transfer coefficient is

$$\tau_p = \tau_v^2 = (0.012)^2 = 0.000144$$

Indeed, not much power can go through.

From the definition of Eq. 11.1.2, the isolation of the PIN diode switch is

$$L_{iso} = 10 \log_{10}\left(\frac{1}{\tau_p}\right) = 10 \log_{10}\left(\frac{1}{0.000144}\right) = 38.4\ \text{dB}$$

This is quite adequate for a single diode switch [5].

Now if the diode is in the "on" state, the diode presents $R_F = 0.8\ \Omega$ to the $Z_o = 50\ \Omega$ output transmission line. The voltage reflection coefficient at the input side is then

$$\rho = \frac{(R_F + Z_o) - Z_o}{(R_F + Z_o) + Z_o} = \frac{0.8\ \Omega + 50\ \Omega - 50\ \Omega}{0.8\ \Omega + 50\ \Omega + 50\ \Omega} = 0.0079$$

The voltage transmission coefficient is then

$$\tau = 1 - \rho \approx 1 - 0.00179 = 0.9921$$

The voltage transfer coefficient from the input to the output of the switch is then

$$\tau_v = \tau \frac{Z_o}{R_F + Z_o} = 0.9921 \frac{50\ \Omega}{0.8\ \Omega + 50\ \Omega} = 0.98$$

The power transfer coefficient is

$$\tau_p = \tau_v^2 = 0.957$$

This means that 95.7% power goes through and only $1 - \tau_v^2 = 1 - 0.957 = 0.043$ or 4.3% power is lost in the switch. According to the definition of Eq. 11.1.1, the insertion loss of this PIN diode switch is therefore

$$L_{ins} = 10\log_{10}\left(\frac{1}{\tau_p}\right) = 10\log_{10}\left(\frac{1}{0.957}\right) = 0.191 \text{ dB}$$

This is considered an acceptable insertion loss for a single PIN diode.

11.4 Microwave Step Attenuators and Phase Shifters

The switching structure shown in Fig. 11.2 A can be applied to switch attenuators and phase shifters in a microwave system. Examples are shown in Fig. 11.3. In Fig. 11.3 A, when the PIN diode Switches 1 and 2 are on while Switches 3 and 4 are off, Attenuator 1 is in the circuit. When the PIN diode Switches 1 and 2 are off while Switches 3 and 4 are on, Attenuator 2 is in the circuit. Attenuators 1 and 2 can be replaced by phase shifters, and the device becomes a step phase shifter.

A short circuited waveguide stub acts as a reactance. When it is inserted in a waveguide, it therefore works as a step phase shifter. A PIN diode controlled step phase shifter is schematically illustrated in Fig. 11.3 B. The phase shifter, or the reactance, is active when the diode is reverse biased. When the diode is forward biased, the phase shifter is out of the circuit.

In these attenuators and phase shifters, PIN diodes are employed merely for switching. The PIN diodes are also used as attenuation elements or phase shifting elements. The amount of attenuation or phase shift across the PIN diode depends on the bias voltage. By controlling the bias voltage, it is therefore possible to make a digital step attenuator or phase shifter or an analog attenuator or phase shifter. When a PIN diode is forward biased, the diode is an attenuator [13]. When the PIN diode is reverse biased, the diode is a capacitive phase shifter [10]. If a quarter-wavelength transmission line is attached to this reverse biased PIN diode, the PIN diode quarter-wavelength loaded transmission line becomes an inductive phase shifter.

For example, with a commercial HPND–4165 PIN diode, the forward resistance changes from $R_F = 1000\ \Omega$ to $R_F = 3\ \Omega$, when the forward bias current changes from 0.01 mA to 3 mA [3]. If this diode is mounted in series in a microstripline of characteristic

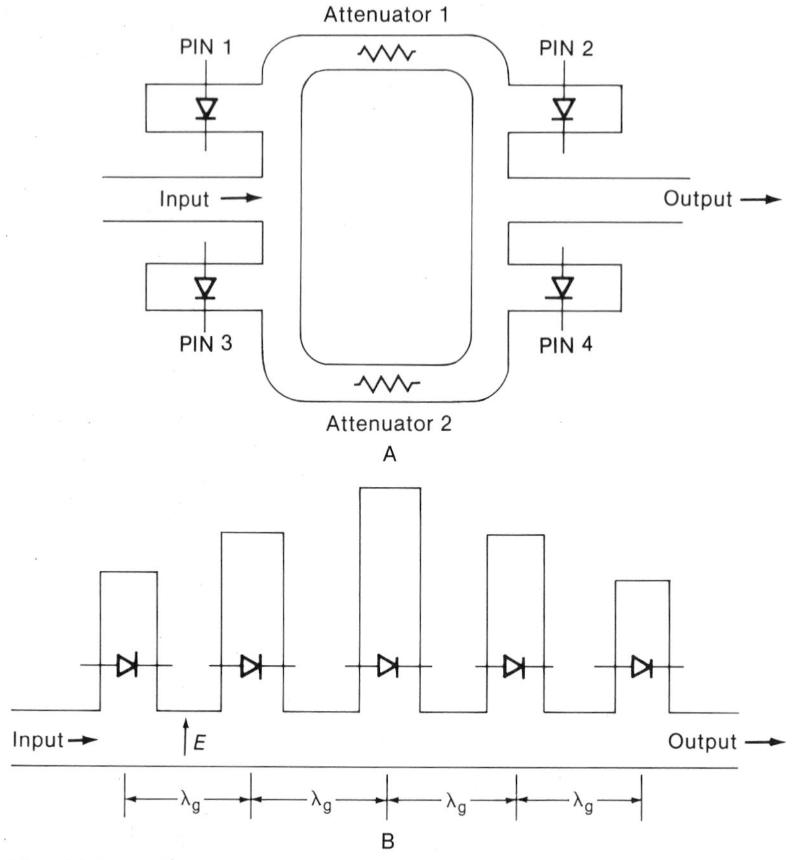

Fig. 11.3 A. Step attenuator. B. Step phase shifter.

impedance $Z_o = 50\ \Omega$, then the insertion loss of the PIN diode or the attenuation due to the diode is calculated in the same way as shown in Section 11.3.

The voltage reflection coefficient at the switch is

$$\rho = \frac{R_F + Z_o - Z_o}{R_F + Z_o + Z_o} = \frac{R_F}{R_F + 2Z_o} \tag{11.4.1}$$

The voltage transmission coefficient at the switch is

$$\tau = 1 - \rho = \frac{2Z_o}{R_F + 2Z_o} \tag{11.4.2}$$

The voltage transfer coefficient at the output side of the diode is

$$\tau_v = \tau \frac{Z_o}{R_F + Z_o} = \frac{2Z_o^2}{(R_F + 2Z_o)(R_F + Z_o)} \tag{11.4.3}$$

The power transfer coefficient from the input side of the PIN diode to the output side is

$$\tau_p = \tau_v^2 = \left[\frac{2Z_o^2}{(R_F + 2Z_o)(R_F + Z_o)}\right]^2 \tag{11.4.4}$$

The attenuation or the insertion loss due to the PIN diode is

$$\text{Att.} = 10\log\left(\frac{1}{\tau_p}\right) = 20\log\frac{(R_F + 2Z_o)(R_F + Z_o)}{2Z_o^2}\ \text{dB} \tag{11.4.5}$$

In this particular case using the HPND–4165 with a forward bias current of 0.01 mA,

$$\text{Att.} = 20\log\frac{(1000\ \Omega + 2 \times 50\ \Omega)(1000\ \Omega + 50\ \Omega)}{2 \cdot (50\ \Omega)^2}\ \text{dB}$$

$$= 47.3\ \text{dB}$$

If the bias current is increased to 3 mA, then

$$\text{Att.} = 20\log\frac{(3\ \Omega + 2 \times 50\ \Omega)(3\ \Omega + 50\ \Omega)}{2 \cdot (50\ \Omega)^2}\ \text{dB}$$

$$= 0.75\ \text{dB}$$

By changing the dc forward bias current between 0.01 mA and 3 mA, continuously or in a stepwise fashion, the diode itself can be used as a microwave attenuator [13].

According to the specification of the same PIN diode, the diode capacitance changes from $C = 0.5$ pF to $C = 0.22$ pF when the reverse bias changes from 0 V to 10 V [3]. The phase shift due to the diode inserted in series with $Z_o = 50\ \Omega$ microstripline is

$$\phi = \tan^{-1}\left(\frac{1}{\omega C Z_o}\right) \tag{11.4.6}$$

If the operating frequency is $f = 7$ GHz, for a dc reverse voltage of 0 V,

$$\phi = \tan^{-1}\left(\frac{1}{2\pi \times 7 \times 10^9\ \text{Hz} \times 0.5 \times 10^{-12}\ \text{F} \times 50\ \Omega}\right)$$

$$= 0.738\ \text{rad} = 42.3°$$

when the dc reverse biased voltage is 10 V,

$$\phi = \tan^{-1}\left(\frac{1}{2\pi \times 7 \times 10^9\ \text{Hz} \times 0.22 \times 10^{-12}\ \text{F} \times 50\ \Omega}\right)$$

$$= 1.12\ \text{rad} = 64.2°$$

Examples of the way to actually realize these attenuators and phase shifters in microstrip configuration are shown in Fig. 11.4.

A surface mountable 5082–0001 PIN diode chip has a size of 0.38 mm × 0.38 mm and is 0.11 mm thick [3]. The diode is mounted in series with the microstripline of $Z_o = 50\ \Omega$, as shown in Fig. 11.4 A. The diode is biased through thin microstripline $\lambda_l/4$ rf chokes. The $\lambda_l/4$ rf choke consists of a high impedance thin microstrip $\lambda_l/4$ in length and a capacitor pad. The capacitor pad forms an rf ground while the $\lambda_l/4$ thin microstrip of high impedance (as high as 200 Ω) presents an extremely high impedance at the T-junction to the 50 Ω microstripline. This prevents the escape of microwaves into the dc bias circuit. The distance between the T-junction and the diode chip is chosen to be $\lambda_l/4$ to minimize reflections [2,12]. When the diode is forward biased, it is an attenuator. When the diode is reverse biased, it is a capacitive phase shifter. By changing the bias voltage stepwise,

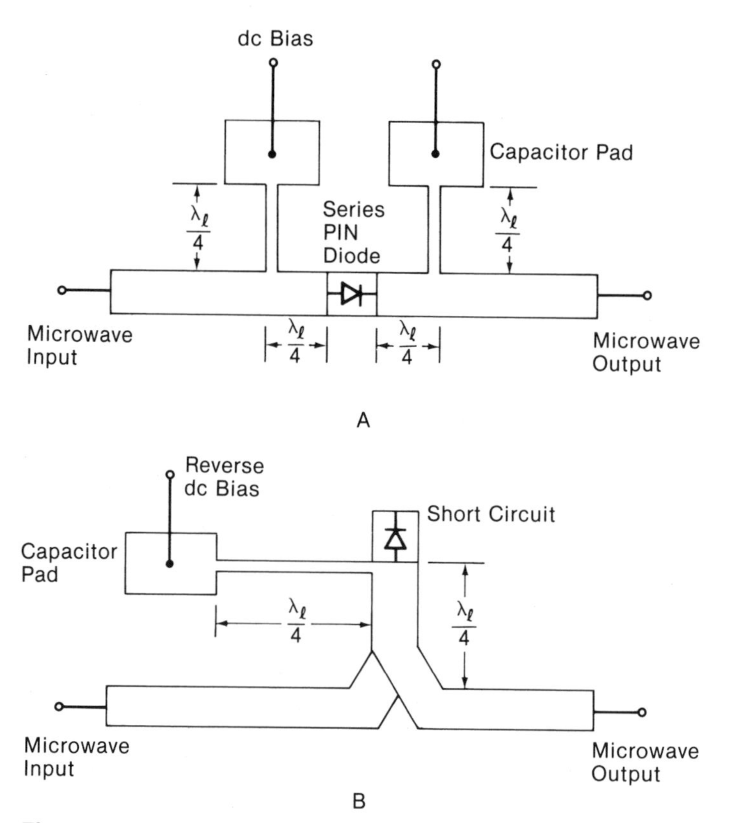

Fig. 11.4 A. PIN diode attenuator or capacitive phase shifter. B. PIN diode inductive phase shifter.

either a step attenuator or step phase shifter is made. At 7 GHz, if the substrate is alumina with $\varepsilon_r = 10$, $\lambda_l/4 = 3.3$ mm, as explained in Section 11.3.

To make an inductive phase shifter, a $\lambda_l/4$ microstrip section must be attached to the diode and must be connected in series with the 50 Ω microstripline as shown in Fig. 11.4 B. The $\lambda_l/4$ microstripline transforms the capacitance of the diode to inductance at the microstripline input [2,12]. The diode is reverse biased through an rf choke composed of a capacitor pad and a thin high impedance $\lambda_l/4$ line as seen in Fig. 11.4 B. The other side of the diode must be short circuited to the low side of the microstrip. The ground plate near the low side microstrip of the $\lambda_l/4$ transformer line must be etched away. The lower microstrip must be connected to the upper microstrip of the input microwave line through the substrate as shown in Fig. 11.4 B.

In these designs, if the impedance of the high impedance thin $\lambda_l/4$ microstripline is 200 Ω, a capacitive reactance of 5 Ω would be sufficient for rf grounding. This means that $1/\omega C = 20\ \Omega$ or $C = 1/(20\ \Omega)(2\pi \times 7 \times 10^9\ \text{Hz}) = 1.13$ pF. If the substrate is alumina with $\varepsilon_r = 10$, the thickness is $H = 0.5$ mm, and the area of the capacitor pad is S m^2, then $C \approx \varepsilon_o \varepsilon_r S/H$. So the area must be $S = CH/\varepsilon_o \varepsilon_r = 1.13 \times 10^{-12}\ \text{F} \times 0.5 \times 10^{-3}\ \text{m}/8.854 \times 10^{-12}\ \text{F/m} \times 10 = 0.0635 \times 10^{-4}\ \text{m}^2$. This means that $S = 0.25 \times 10^{-2}\ \text{m} \times 0.25 \times 10^{-2}\ \text{m} = 2.5\ \text{mm} \times 2.5\ \text{mm}$ or larger will fulfill the function of rf grounding.

The PIN diode can be mounted in parallel to the microstripline to make an attenuator or the phase shifter as illustrated in Fig. 11.5. If the diode is forward biased then it is an attenuator and if reverse biased, it is a phase shifter. As seen in Fig. 11.5, microwaves are fed through the bypass chip capacitors and the diode [13]. The diode is biased through a $\lambda_l/4$ microwave choke and a capacitor pad. If the microstrip's characteristic admittance is Y_o and the forward conductance of the diode is G_F, then the current reflection coefficient at the input of the diode is

$$\rho = \frac{Y_o + G_F - Y_o}{Y_o + G_F + Y_o} = \frac{G_F}{G_F + 2Y_o} \tag{11.4.7}$$

The current transmission coefficient is $1 - \rho$ and the current transfer coefficient τ is

$$\tau = (1 - \rho)\frac{Y_o}{G_F + Y_o} = \frac{2Y_o}{G_F + 2Y_o}\frac{Y_o}{G_F + Y_o} \tag{11.4.8}$$

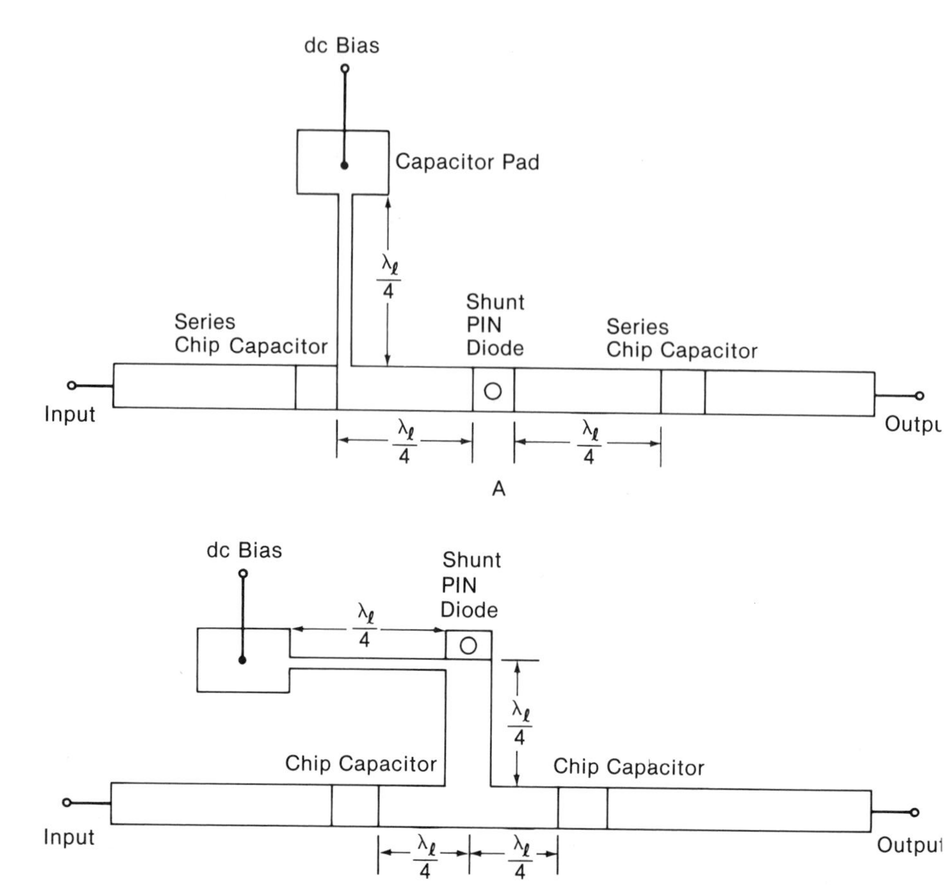

Fig. 11.5 Single shunt PIN diode attenuator or capacitive phase shifter. B. Inductive phase shifter.

The power transfer coefficient is then

$$\tau_p = \tau^2 = (1-\rho)^2\left(\frac{Y_o}{G_F + Y_o}\right)^2 \tag{11.4.9}$$

The attenuation is

$$\text{Att.} = 10\log\frac{1}{\tau_\rho} = 20\log\frac{1}{\tau}$$

$$= 20\log\frac{(G_F + 2Y_o)(G_F + Y_o)}{2Y_o^2}\ \text{dB} \tag{11.4.10}$$

For example, if $Y_o = 1/50$ S $= 0.02$ S and $G_F = 1/10$ S $= 0.1$ S, then from Eq. 11.4.10,

$$\text{Att.} = 20\log\frac{(0.1\ \text{S} + 2\times 0.01\ \text{S})(0.1\ \text{S} + 0.02\ \text{S})}{2\times(0.02\ \text{S})^2}\ \text{dB}$$

$$= 26.4\ \text{dB}$$

If the bias is reversed, the diode becomes a variable capacitor. The admittance toward the load at the diode is $j\omega C + Y_o$. The transmission line current at the input will therefore have a phase shift of

$$\phi = \tan^{-1}\frac{\omega C}{Y_o} \tag{11.4.11}$$

For example, if $f = 7$ GHz, $Y_o = 1/50$ S, and $C = 0.4$ pF, then

$$\phi = \tan^{-1}\left(\frac{2\pi\times 7\times 10^9\ \text{s}^{-1}\times 0.4\times 10^{-12}\ \text{F}}{1/50\ \text{S}}\right)$$

$$= 0.72\ \text{rad} = 41.3°$$

For simplicity, the effect of reverse diode conductance is neglected in this analysis, since it is small in comparison with Y_o.

When a $\lambda_l/4$ microstripline is attached to the diode as shown in Fig. 11.5 B, the diode presents an inductive admittance $1/\omega L$ at the T-junction, where L is an equivalent shunt inductance created at the T-junction by the loaded $\lambda_l/4$ microstripline stub. According to transmission line theory with $\lambda_l/4$ apart [2,12],

$$1/\omega C = 1/\omega L \tag{11.4.12}$$

This produces an admittance of $Y_o - j1/\omega L$. The phase shift is therefore

$$\phi = \tan^{-1}\left(-\frac{1}{\omega L Y_o}\right) = -\tan^{-1}\left(\frac{1}{\omega L Y_o}\right)$$

or using Eq. 11.4.12

$$\phi = -\tan^{-1}\left(\frac{1}{\omega C Y_o}\right) \tag{11.4.13}$$

Like before, if $f = 7$ GHz, $C = 0.4$ pF, and $Y_o = 1/50$ S, then

$$\phi = -\tan^{-1}\left(\frac{1}{2\pi\times 7\times 10^9\ \text{s}^{-1}\times 0.4\times 10^{-12}\ \text{F}\times 1/50\ \text{S}}\right)$$

$$= -1.57\ \text{rad} = -89.997°.$$

11.5 Microwave Switching by Transistors

Almost all types of microwave transistors are used for microwave switching. Both bipolar junction transistors (BJT) and field effect transistors (FET) are used for microwave switching. HEMTs are useful for high speed switching. For BJTs, the collector current is controlled by the base voltage. If the base–emitter is forward biased, then the transistor is on, but if the base–emitter is reverse biased, the carrier depletion region created at the base–emitter junction shuts off the collector current.

For an n-channel FET, when the gate is positively biased or zero biased with respect to the source, the channel is wide open and the drain current flows. The transistor is therefore on. When the gate is deeply negative, the channel gets narrow and pinched off by the negative potential on the gate. This makes the flow of drain current impossible. The drain circuit of the transistor is therefore open. By controlling the gate voltage, the FET can be turned off and on. The "on" state of microwave transistors can be connected in microwave circuits either in series or parallel and controls microwave current. The action of microwave transistor switches in microwave circuits is basically identical to the action of microwave diode switches, but for transistors the base voltage or gate voltage controls the action of the switch. Most commercial transistor switches use more than one transistor and are developed in the form of MIC or MMIC [6,9]. These MMIC type microwave switches are used as on–off switches, routing switches, and microwave multiplexers and are commercially available. The isolation and the insertion loss of microwave switching transistors can be treated similarly to switching diodes. The treatment is therefore not repeated here.

As a practical example, a KSW–2–46 MMIC GaAs transistor SPDT microwave switch has an insertion loss of 1 dB and an isolation of 50 dB between dc and 4.6 GHz [6]. The entire switch is packaged in 0.185 in. × 0.185 in. × 0.06 in. or 4.7 mm × 4.7 mm × 1.55 mm. The switching speed is specified as 2–3 ns, with microwave power handling capacity up to 27 dBm. The microwaves are turned on or off by a dc control voltage of 0 V and −8.0 V [6].

As another practical example, a single GaAs MESFET used as a microwave switch will be presented. Using a transistor curve tracer, the output characteristic curves of the MESFET (V_{ds} versus I_{ds}) are obtained. From those curves, the differential resistance $r_{ds} = (dV_{ds}/dI_{ds})$ is calculated for various values of V_{ds} and V_{gs}.

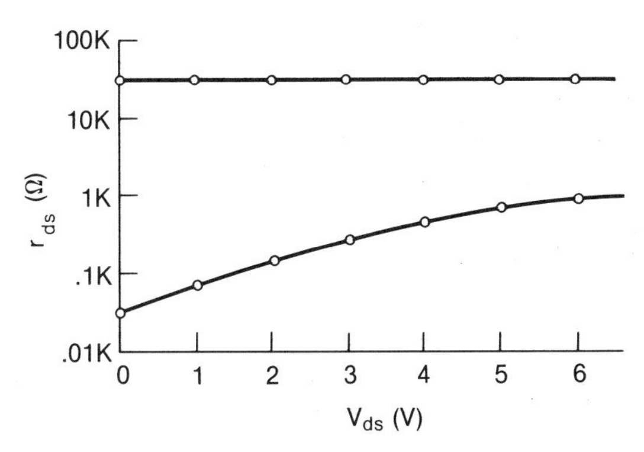

Fig. 11.6 Differential drain–source resistance, $r_{ds} = dV_{ds}/dI_{ds}$, of a microwave GaAs MESFET for various values of the drain–source dc bias V_{ds}. $V_{gs} = -2.5$ V (*top*) and $V_{gs} = 0$ V (*bottom*).

An example is shown in Fig. 11.6. According to the curves of this n-channel GaAs MESFET, to turn it on $V_{gs} = 0$ V is sufficient and to turn it off $V_{gs} = -2.5$ V is sufficient. As seen from this curve, though this particular FET is common for microwave amplification, the resistance is high. This type of transistor is therefore useful only for high impedance circuit switching. For low impedance circuit switching, transistors of low differential resistance must be used.

11.6 Microwave Switching

Microwave diodes and transistors are useful for microwave switching [7,8]. Switching can be simple on–off, routing, or multiplexing. The switching devices can be mounted in a microwave circuit in series or parallel depending on the desired switching characteristics. The switching characteristics are controlled by a dc bias on the switching device. The switching speed of most commercial solid-state microwave switches is on the order of ns. A number of cycles of microwave signals can therefore pass through before the switch is fully off when operating in high GHz microwave frequencies. The switching devices can be applied to attenuators and phase shifters [10,11]. The attenuation or phase shifting can be done within the semiconductor device, or the semiconductors are simply used for switching while the actual attenuation or phase shifting is done by a separate set of attenuators or phase shifters.

A microwave semiconductor switching device is characterized by the insertion loss, isolation, VSWR, operating frequency, frequency bandwidth, bias voltage, bias current, switching speed, power handling capacity, and packaging configuration.

Problems

11.1 Define the microwave binary states of a nonlinear semiconductor junction.

11.2 Create "*n*-nary" states of a nonlinear semiconductor junction.

11.3 Describe what closes and opens the PIN diode switch at microwave frequencies.

11.4 Point out the reason for which a PIN diode conducts under forward bias.

11.5 Under reverse and forward bias conditions, in spite of the existence of the intrinsic i-layer, the PIN diode presents relatively low forward or reverse microwave resistance. Explain the reason for this.

11.6 The dc resistance of a biased PIN diode differs greatly under forward and reverse bias. Yet if the same diode is properly biased, the microwave impedance for the positive half-cycle is not much different from the impedance for the negative half-cycle. Explain the reason for this. Also, explain what is meant by "properly biased."

11.7 State the reason most PIN diodes are made of silicon.

11.8 Explain the depletion layer "punch through" under reverse bias conditions.

11.9 Sketch the current–voltage curve of a PIN diode.

11.10 Sketch the microwave equivalent circuit of a PIN diode.

11.11 Explain what makes the microwave impedance of a PIN diode high under proper reverse bias.

11.12 Derive a relationship between the resistivity and the mobility of the carrier.

11.13 Identify the elements which form the microwave forward and reverse resistance, respectively.

11.14 Assuming 50 kW transmitter pulse modulated microwaves of pulse width 8 μs, 800 pulses per second (pps) hits a PIN diode limiter; estimate the limiter voltage. Assume that the waveguide characteristic impedance is 400 Ω.

11.15 Sketch a schematic diagram of a microwave diplexer using H-plane junctions with PIN diodes.

11.16 A reverse biased PIN diode has a microwave impedance of approximately $j10^3\ \Omega$. Show that by adding an inductance, the impedance will increase.

11.17 Design a microstrip version of a PIN diode switch.

11.18 Show analytically that a short circuited waveguide stub acts as a reactance.

11.19 Show analytically that if a reactance is inserted in a transmission line in series, it acts as a step phase shifter.

11.20 Describe a procedure to produce a differential drain–source resistance r_{ds} versus the drain–source voltage V_{ds} characteristic curve from the output characteristic curve of a microwave FET.

References

1 J. F. White, "Semiconductor Control." Artech House, Dedham, Massachusetts, 1977.

2 T. K. Ishii, "Microwave Engineering." Ronald Press, New York, 1966.

3 Hewlett Packard, "Diode and Transistor Designer's Catalog." Hewlett Packard, San Jose, California, 1980.

4 B. G. Streetman, "Solid-State Electronic Devices," 2nd ed. Prentice-Hall, Englewood Cliffs, New Jersey, 1980.

5 M/ACOM, Monolithic PIN diode SPDT switches. *Microwave J.*, **31**(1), 99 (1989).

6 Mini-Circuit, Tiny SPDT switches. *Microwave J.*, **31**(1), (1989).

7 S. Y. Liao, "Microwave Solid-State Devices." Prentice-Hall, Englweood Cliffs, New Jersey, 1985.

8 H. E. Thomas, "Handbook of Microwave Techniques and Equipment." Prentice-Hall, Englewood Cliffs, New Jersey, 1972.

9 D. T. Bryant, GaAs FET single-pole double-throw (SPDT) monolithic microwave integrated circuit. *IEEE MTT-S Int. Microwave Symp. Dig.* **1**, 371–374 (1988).

10 D. M. Krafcsik, MMIC dual-varactor analog reflection phase shifter for 6 to 18 GHz operation. *IEEE Trans. Microwave Theory Tech.* **MTT-36**(12), 1938–1941 (1988).

11 H. Kondoh, DC-50 GHzMMIC variable attenuator with 30 dB dynamic range. *IEEE MTT-S Int. Microwave Symp. Dig.*, **1**, 499–502 (1988).

12 S. Y. Liao, "Microwave Devices and Circuits," 2nd ed. Prentice-Hall, Englewood Cliffs, New Jersey, 1985.

13 R. Baeten, T. K. Ishii, and J. S. Hyde, PIN diode attenuator with small phase shift. *IEEE Trans. Microwave Theory Tech.* **MTT-36**(4), 789–791 (1988).

12

Quantum Electron Devices

12.1 Microwave Amplification by Stimulated Emission of Radiation

In a paramagnetic crystal such as a ruby or in an ionically bonded gas such as ammonia, it is possible to artificially create a situation in which the molecular density of high energy state molecules is greater than the molecular density of low energy state molecules [1,8,9]. This state is called the density inversion. Creating the state of density inversion is called pumping. If the difference in energy of the higher energy state and the lower energy state of the molecules is ΔE_p, the energy difference can be supplied by the pump oscillator by feeding or pumping photons into the system. If the photon frequency is f_p then, according to quantum theory,

$$\Delta E_p = hf_p \tag{12.1.1}$$

where h is Planck's constant and

$$h = 6.63 \times 10^{-34} \text{ J-s} \tag{12.1.2}$$

For photon energy at microwave frequencies of approximately 10^9 Hz, approximately 10^{-25} J energy is needed to pump the energy of a particle from the lower state to the upper state of the system. The particle can be an electron, an atom, or a molecule depending on the type of maser.

In practice, the pump oscillator power is on the order of $P = 10^{-2}$ W. If the energy transition from the lower level to the upper level takes $\tau = 10^{-12}$ s, and the number of particles making the transition in τ is N, then

$$P = Nhf_{\mathrm{p}}/\tau \qquad (12.1.3)$$

$$N = P\tau/hf_{\mathrm{p}} \qquad (12.1.4)$$

In this example

$$N \approx (10^{-2}\ \mathrm{W})(10^{-12}\ \mathrm{s})/10^{-25}\ \mathrm{J} = 10^{11}$$

particles that take upward transition in 10^{-12} s. This means that 10^{23} particles take upward transition in 1 s. The particles of higher energy states always have a tendency to make the downward transition, provided that such an energy state is quantum mechanically allowable. Suppose that such a state is available at ΔE_{s} lower than the higher state. If the system is in a microwave cavity resonator of frequency f_{s} with a small input power which satisfies the following quantum mechanical relationship:

$$\Delta E_{\mathrm{s}} = hf_{\mathrm{s}} \qquad (12.1.5)$$

Then the downward energy transition is triggered by the stimulation from the input signal. This stimulation input signal and the microwave cavity resonator trigger a massive downward transition [9]. The energy of transition is small, but the number of particles are sufficiently large and the transition takes place within a short period of time. When the particles in the higher energy state make the downward transition by stimulation given by Eq. 12.1.3, photon energy will be emitted. The *energy* of emission may be as small as 10^{-25} J but the number of particles which make the downward transition is large and the transition occurs within a short period of time. The emitted *power* is therefore sufficiently large to be used as a microwave amplifier. The emitted power is on the order of 10^{-10} W. Small stimulation power of approximately 10^{-13} W can produce the amplified output. This is the *m*icrowave *a*mplification by *s*timulated *e*mission of *r*adiation, or the *maser*.

If the emission of radiation from the maser is $P_{\mathrm{e}} = 10^{-10}$ W, and the number of particles making the downward transition is N_{d} in $\tau = 10^{-12}$ s then,

$$P_{\mathrm{e}} = N_{\mathrm{d}}hf_{\mathrm{s}}/\tau \qquad (12.1.6)$$

$$N_{\mathrm{d}} = P_{\mathrm{e}}\tau/hf_{\mathrm{s}} \qquad (12.1.7)$$

In this example,

$$N_d \approx (10^{-10}\ \text{W})(10^{-12}\ \text{s})/10^{-25}\ \text{J} = 10^3$$

particles make the downward transition in 10^{-12} s or 10^{15} particles make the downward transition in 1 s to produce emission of 10^{-10} W.

For stimulation, if $P_s = 10^{-13}$ W is absorbed in the maser, then

$$P_s = N_s h f_s/\tau \tag{12.1.8}$$

$$N_s \approx P_s \tau / h f_s \tag{12.1.9}$$

In this example,

$$N_s = (10^{-13}\ \text{W})(10^{-12}\ \text{s})/10^{-25}\ \text{J} = 1$$

particle makes an upward transition in 10^{-12} s, or 10^{12} particles make upward transitions for stimulation to produce the emission of radiation each second.

In this example, 10^{23} particles are pumped up from the lowest level to the highest level in one second. During the same time, 10^{12} particles are stimulated from the second lowest level to the highest level. Also, during the same time, 10^{15} particles make downward transitions from the highest level to the second lowest level and emit amplified radiation. Thus, in this case, 10^{12} photons are fed into the system and 10^{15} photons are emitted at signal frequency in one second. The power gain is therefore $10^{15}/10^{12} = 1000$ or $10 \log 1000$ dB $= 30$ dB.

Note that these examples show only an order of magnitude calculation rather than a precise, accurate calculation.

12.2 Pumping

If the energy level of the high energy state of the particles of the maser material is E_h and the energy level of the low energy state is E_{l_p}, then

$$E_h > E_{l_p} \tag{12.2.1}$$

If the particle density in E_h is N_h^* and the particle density in E_{l_p} is $N_{l_p}^*$ under the condition that neither pumping or stimulation is given, then

$$N_h^* < N_{l_p}^* \tag{12.2.2}$$

This relation is described by the Boltzmann relation as [1]

$$\frac{N_{\mathrm{h}}^{*}}{N_{l_{\mathrm{p}}}^{*}} = \frac{w_{\mathrm{h}} e^{-(E_{\mathrm{h}}/\mathrm{k}T_{\mathrm{o}})}}{w_{l_{\mathrm{p}}} e^{-(E_{l_{\mathrm{p}}}/\mathrm{k}T_{\mathrm{o}})}} = \frac{w_{\mathrm{h}}}{w_{l_{\mathrm{p}}}} e^{-(E_{\mathrm{h}} - E_{l_{\mathrm{p}}}/\mathrm{k}T_{\mathrm{o}})} = \frac{w_{\mathrm{h}}}{w_{l_{\mathrm{p}}}} e^{-(\Delta E_{\mathrm{p}}/\mathrm{k}T_{\mathrm{o}})} \tag{12.2.3}$$

where w_{h} is the statistical weight of level E_{h}, $w_{l_{\mathrm{p}}}$ is the statistical weight of level $E_{l_{\mathrm{p}}}$, k is the Boltzmann constant 1.38×10^{-23} J, T_{o} is the temperature of the maser material, and

$$\Delta E_{\mathrm{p}} \equiv E_{\mathrm{h}} - E_{l_{\mathrm{p}}}$$

Equation 12.2.3 is simply stating that if $w_{\mathrm{h}} \approx w_{l_{\mathrm{p}}}$, then the number of particles in the higher energy state is less than the number of particles in the lower energy state.

Under the thermal equilibrium, some particles in the lower energy state make the upward transition spontaneously and some molecules in the higher energy state make the downward transition spontaneously [10].

Statistically, if the probability of occurrence of the upward transition is $\omega_{l_{\mathrm{p}}\mathrm{h}}^{*}$ per unit of time and the probability of occurrence of the downward transition is $\omega_{\mathrm{h}l_{\mathrm{p}}}^{*}$ per unit of time, then under thermal equilibrium, the number of particles which make the upward transition $N_{l_{\mathrm{p}}}^{*}\omega_{l_{\mathrm{p}}\mathrm{h}}$ must be statistically equal to the number of particles making the downward transition, which is $N_{\mathrm{h}}^{*}\omega_{\mathrm{h}l_{\mathrm{p}}}$.

$$N_{l_{\mathrm{p}}}^{*}\omega_{l_{\mathrm{p}}\mathrm{h}}^{*} = N_{\mathrm{h}}^{*}\omega_{\mathrm{h}l_{\mathrm{p}}}^{*} \tag{12.2.4}$$

When the maser material is pumped at an exact frequency given by Eq. 12.1.1, then

$$\omega_{l_{\mathrm{p}}\mathrm{h}} > \omega_{l_{\mathrm{p}}\mathrm{h}}^{*} \tag{12.2.5}$$

The upward transition probability under pumping can be represented by assuming a linear relationship as

$$\omega_{l_{\mathrm{p}}\mathrm{h}} = \xi P_{\mathrm{p}} + \omega_{l_{\mathrm{p}}\mathrm{h}}^{*} \tag{12.2.6}$$

where ξ is a proportionality constant and P_{p} is the pump power [2].

Under steady state, with pumping and without stimulation,

$$N_{\mathrm{l_p}}\omega_{l_{\mathrm{p}}\mathrm{h}} = N_{\mathrm{h}}\omega_{\mathrm{h}l_{\mathrm{p}}}^{\mathrm{s}} \tag{12.2.7}$$

where $\omega_{\mathrm{h}l_{\mathrm{p}}}^{\mathrm{s}}$ is the spontaneous downward transition probability per unit time under pumping. Since

$$\omega_{l_p h} > \omega^{s}_{h l_p} \tag{12.2.8}$$

under pumping, to satisfy Eq. 12.2.7

$$N_{l_p} < N_h \tag{12.2.9}$$

This is the density inversion which means that, while pumping, the number of particles in the higher energy state is greater than the number of particles in the lower energy state. This relationship is opposite to Eq. 12.2.2. The density inversion is created by pumping. Substituting Eq. 12.2.6 in Eq. 12.2.7

$$N_h = \frac{\omega_{l_p h}}{\omega^{s}_{h l_p}} N_{l_p} = \frac{\xi P_p + \omega^{*}_{l_p h}}{\omega^{s}_{h l_p}} N_{l_p} \tag{12.2.10}$$

With greater pumping, the number of particles in the higher energy state that are available for maser action is greater.

For example, if $\Delta E_p = 9.5 \times 10^{-25}$ J, then the pump oscillator frequency f_p must be adjusted to

$$f_p = \frac{\Delta E_p}{h} = \frac{9.5 \times 10^{-25}\ \text{J}}{6.63 \times 10^{-34}\ \text{J-s}} = 1.4 \times 10^{9}\ \text{Hz}$$

This is 1.4 GHz. Since the energy of one electron-volt (1 eV) is 1.6×10^{-19} J,

$$\Delta E_p = \frac{9.5 \times 10^{-25}\ \text{J}}{1.6 \times 10^{-19}\ \text{J/eV}} = 5.9 \times 10^{-6}\ \text{eV} = 5.9\ \mu\text{eV}$$

In practice, the value of w_h and w_{l_p} are in the neighborhood of unity. If $w_h = 1.01$ and $w_{l_p} = 0.99$, then $(w_h/w_{l_p}) = 1.02$. If the maser material is cooled to liquid helium temperature which is $T_o = 2$ K, then the thermal energy level is $kT_o = (1.38 \times 10^{-23}$ J/K)(2 K) $= 2.76 \times 10^{-23}$ J $= 1.725 \times 10^{-4}$ eV $= 172.5\ \mu$eV. As a result, the ratio $\Delta E_p/kT_o = 5.9\ \mu\text{eV}/172.5\ \mu\text{eV} = 0.034$. The pump energy gap ΔE_p is very small in comparison with the thermal energy kT_o. It is wise to choose E_{l_p} at a value far above the thermal energy kT_o. If E_{l_p} is chosen to be $E_{l_p} = 10\ kT_o = 10 \times 172.5\ \mu\text{eV} = 1725\ \mu\text{eV}$, then $E_h = E_{l_p} + \Delta E_p = 1725\ \mu\text{eV} + 5.9\ \mu\text{eV} = 1730.9\ \mu\text{eV}$.

According to Eq. 12.2.3,

$$\frac{N^{*}_h}{N^{*}_{l_p}} = \frac{\omega_h}{\omega_{l_p}} e^{-(\Delta E_p/kT_o)} = 1.02\ e^{-0.034} = 0.986$$

So N_h^* and $N_{l_p}^*$ are nearly equal, though $N_h^* < N_{l_p}^*$ in this case. If $N_{l_p}^* = 10^{30}$ particles. Then $N_h^* = 0.986 N_{l_p}^* = 0.986 \times 10^{30}$ particles. The difference is therefore $N_{l_p}^* - N_h^* = (1 - 0.986) \times 10^{30} = 1.4 \times 10^{28}$ particles. The $(N_h^*/N_{l_p}^*)$ ratio is not appreciable but the difference $(N_{l_p}^* - N_h^*)$ is still a large quantity.

In terms of transition probabilities, from Eq. 12.2.4,

$$\frac{\omega_{l_p h}^*}{\omega_{h l_p}^*} = \frac{N_h^*}{N_{l_p}^*} = 0.986$$

If $N_{l_p}^* = 10^{30}$ particles and $N_h = 10^{23}$ particles are pumped up in 1 s by the pump power of $P_p = 10^{-2}$ W as shown in Section 12.1, then the transition probability

$$\omega_{l_p h} = \frac{N_h}{N_{l_p}^*} = \frac{10^{23}}{10^{30}} = 10^{-7}\ \text{s}^{-1}$$

If a poor case is considered, $\omega_{l_p h} = 10\omega_{l_p h}^*$, then $\omega_{l_p h}^* = \omega_{l_p h}/10 = 10^{-8}\ \text{s}^{-1}$. Now from Eq. 12.2.6

$$\omega_{l_p h} = \xi P_p + \omega_{l_p h}^*$$

$$\xi = \frac{\omega_{l_p h} - \omega_{l_p h}^*}{P_p} = \frac{10^{-7}\ \text{s}^{-1} - 10^{-8}\ \text{s}^{-1}}{10^{-2}\ \text{W}} = 9 \times 10^{-6}\ (\text{W-s})^{-1}$$

In the example of Section 12.1, the number of particles which make upward transitions in 1 sec is $10^{23}\ \text{s}^{-1} = \omega_{l_p h} N_{l_p}$. Now $\omega_{l_p h} = 10^{-7}\ \text{s}^{-1}$, so $N_{l_p} = 10^{23}\ \text{s}^{-1}/\omega_{l_p h} = 10^{23}\ \text{s}^{-1}/10^{-7}\ \text{s}^{-1} = 10^{30}$. The spontaneous transition is at most $\omega_{h l_p}^s = \omega_{l_p h}/10 = 10^{-8}\ \text{s}^{-1}$. From Eq. 12.2.7,

$$N_h = \frac{N_{l_p}\omega_{l_p h}}{\omega_{h l_p}^s} = \frac{10^{23}\ \text{s}^{-1}}{10^{-8}\ \text{s}^{-1}} = 10^{31}$$

In summary of this example, before pumping, $N_h^* = 0.986 \times 10^{30}$, $N_{l_p}^* = 1 \times 10^{30}$ particles, and after pumping started, $N_h = 1 \times 10^{31}$, $N_{l_p} = 1 \times 10^{30}$. These sample calculations are to gain a quantitative understanding rather than to show accurate and exact cases.

12.3 Stimulation

Suppose that there is an allowable quantum mechanical energy state at E_{l_s} for the maser material and E_{l_s} is lower than E_h. The energy difference between E_h and E_{l_s} is represented by

$$\Delta E_s = E_h - E_{l_s} \tag{12.3.1}$$

If the microwave energy of Eq. 12.1.3 is fed to this maser material, some of the particles in E_{l_s} will make the upward transition to E_h. The upward transition probability per unit time $\omega_{l_s h}$ is under a principle similar to Eq. 12.2.6,

$$\omega_{l_s h} = \eta P_s + \omega^*_{l_s h} \tag{12.3.2}$$

where η is a proportionality constant, P_s is the microwave input signal power, and $\omega^*_{l_s h}$ is the spontaneous upward transition probability from energy state E_{l_s} to E_h under the thermal equilibrium condition.

The number of particles which make the downward transition from E_h to E_{l_s} per unit time is then, under a steady state condition with pump power on the input signal power feeding,

$$N_h \omega_{h l_s} = N_{l_s} \omega_{l_s h} \tag{12.3.3}$$

Substituting Eq. 12.3.2 in Eq. 12.3.3,

$$N_h \omega_{h l_s} = N_{l_s}\left(\eta P_s + \omega^*_{l_s h}\right) \tag{12.3.4}$$

As seen from this equation, the presence of the signal power increases the number of particles which make downward transitions per unit time. This process is called stimulation.

For example, if the frequency of a signal to be amplified by the maser is $f_s = 1$ GHz $= 1 \times 10^9$ Hz $= 1 \times 10^9$ s^{-1}, then $\Delta E_s = E_h - E_{l_s} = hf_s = (6.63 \times 10^{-34}$ J-s$)(1 \times 10^9$ s$^{-1}) = 6.63 \times 10^{-25}$ J $= (6.63 \times 10^{-25}/1.6 \times 10^{-19})$ eV $= 4.14$ μeV. In the example of Section 12.2, $E_h = 1730.9$ μeV. As a result, $E_{l_s} = E_h - \Delta E_s = 1730.9$ μeV $- 4.14$ μeV $= 1726.25$ μeV. This is a three level maser: $E_h = 1730.9$ μeV, $E_{l_s} = 1726.25$ μeV, and $E_{l_p} = 1725$ μeV with a thermal noise level of $kT_o = 172.5$ μeV.

Using the Boltzmann relation Eq. 12.2.3,

$$\frac{N_{l_s}}{N_{l_p}} = \frac{w_s}{w_p} e^{-(E_{l_s} - E_{l_p}/kT_o)}$$

If $w_s \approx w_p \approx 1$, then

$$\frac{N_{l_s}}{N_{l_p}} = e^{-(1726.25\ \mu\text{eV} - 1725\ \mu\text{eV}/172.5\ \mu\text{eV})} = 0.991$$

From the example presented in Section 12.2, if $N_{l_p} = 1 \times 10^{30}$, then $N_{l_s} = 0.991 \times N_{l_p} = 0.991 \times 10^{30}$. As shown in Section 12.2, when the maser is pumped, $N_h = 1 \times 10^{31}$. If this maser emits 10^{-10} W as is the case shown in Section 12.1, then $N_d = 10^{15}\ s^{-1}$. The downward transition probability in one second is therefore $\omega_{hl_s} = N_d/N_h = 10^{15}\ s^{-1}/1 \times 10^{31} = 10^{-16}\ s^{-1}$.

From Eq. 12.3.3,

$$\omega_{l_sh} = \frac{N_h\omega_{hl_s}}{N_{l_s}} = \frac{10^{30} \times 10^{-16}\ s^{-1}}{0.991 \times 10^{30}} = 1.009 \times 10^{-16}\ s^{-1}$$

If $\omega^*_{l_sh} = \omega_{l_sh}/10 = 1.009 \times 10^{-17}\ s^{-1}$, then from Eq. 12.3.2

$$\eta = \frac{\omega_{l_sh} - \omega^*_{l_sh}}{P_s} = \frac{1.009 \times 10^{-16}\ s^{-1} - 1.009 \times 10^{-17}\ s^{-1}}{10^{-10}\ W}$$

$$= 9.08 \times 10^{-7}\ \text{W-s}^{-1}$$

12.4 Emission

When one particle in the upper energy level E_h makes the downward transition to E_{l_s}, it loses energy ΔE_s in Eq. 12.1.3 and emits a photon hf_s. If $N_h\omega_{hl_s}$ particles as in Eq. 12.3.4 make the downward transition from E_h to E_{l_s}, the total number of photons emitted in a unit time is $N_h\omega_{hl_s}$ and the power emitted is

$$P_o = N_h\omega_{hl_s}\Delta E_s = N_h\omega_{hl_s}hf_s \quad (W) \tag{12.4.1}$$

substituting Eq. 12.3.4 in Eq. 12.4.1,

$$P_o = N_{l_s}(\eta P_s + \omega^*_{l_sh})hf_s \quad (W) \tag{12.4.2}$$

This shows a linear relationship between the input signal and the output signal.

Substituting Eq. 12.2.10 in Eq. 12.4.1

$$P_o = N_{l_p}\frac{\xi P_p + \omega^*_{l_ph}}{\omega^s_{hl_p}}\omega_{hl_s}hf_s \tag{12.4.3}$$

This shows a linear relationship between the output power and the pump power. In this equation, ω_{hl_s} is the coherent downward transition probability per unit time for E_h to E_{l_s}.

By observation of Eq. 12.3.4, ω_{hl_s} is linearly proportional to the input signal power. If the proportionality constant is η then, in comparison with Eq. 12.4.2,

$$\omega_{hl_s} = \eta P_s + \omega^{s}_{hl_s} \tag{12.4.4}$$

Substituting Eq. 12.4.4 in Eq. 12.4.3

$$P_o = N_{l_p} \frac{\xi P_p + \omega^{*}_{l_p h}}{\omega^{s}_{hl_p}} \left(\eta P_s + \omega^{s}_{hl_s} \right) hf_s \tag{12.4.5}$$

This equation shows the contribution of the pump power and the input signal power to the output signal power. The output power can be increased by increasing the pump power and the input signal power.

12.5 Gain

The gain of a maser is the ratio of the signal output power to the signal input power

$$G = \frac{P_o}{P_s} \tag{12.5.1}$$

Substituting Eq. 12.4.6 in Eq. 12.5.1,

$$G = N_{l_p} \frac{\xi P_p + \omega^{*}_{l_p h}}{P_s \omega^{s}_{hl_p}} \left(\eta P_s + \omega^{s}_{hl_s} \right) hf_s \tag{12.5.2}$$

This equation shows that the gain G is a function of input signal level P_s. This means that the maser is inherently a nonlinear amplifier, especially at an extremely small signal level such that

$$\eta P_s \ll \omega^{s}_{hl_s} \tag{12.5.3}$$

If the input signal level is sufficiently high so that

$$\eta P_s \gg \omega^{s}_{hl_s} \tag{12.5.4}$$

Then, omitting $\omega^{s}_{hl_s}$ in Eq. 12.5.2,

$$G = N_{l_p} \frac{\xi P_p + \omega^{*}_{l_p h}}{\omega^{s}_{hl_p}} \eta hf_s \tag{12.5.5}$$

This is not a function of P_s. The maser is therefore linear. The gain depends on P_p.

If the P_s is too high, η in Eq. 12.5.5 becomes a function of P_s. In a large P_s, the available vacancies in E_h are quickly filled up and hence $\omega_{l_s h}$ decreases. This causes a decrease in η as seen from Eq. 12.4.5. This results in the saturation of the maser.

In the example shown in Section 12.3, it is shown that $\eta = 9.08 \times 10^{-7}$ (W-s)$^{-1}$ and $\omega^{s}_{hl_s} = \omega^{*}_{hl_s} = 1.009 \times 10^{-17}$ s^{-1}. So to satisfy condition 12.5.3,

$$\eta P_s = 0.01\omega^{s}_{hl_s}$$

or

$$P_s = \frac{0.01\omega^{s}_{hl_s}}{\eta} = \frac{0.01 \times 1.009 \times 10^{-17}\ \mathrm{s}^{-1}}{9.08 \times 10^{-7}\,(\mathrm{W\text{-}s})^{-1}} = 0.1 \times 10^{-12}$$

$$= 0.1\ \mathrm{pW} \approx -130\ \mathrm{dBW} = -100\ \mathrm{dBm}.$$

For this example maser, when the input power becomes less than -100 dBm, the gain is approximately, from Eqs. 12.5.2 and 12.2.10,

$$G = \frac{\xi P_p + \omega^{*}_{l_p h}}{P_s \omega^{s}_{hl_p}} N_{l_p} \omega^{s}_{hl_s} h f_s = \frac{N_h \omega^{s}_{hl_s} h f_s}{P_s} \qquad (12.5.6)$$

In this example, for $P_p = 10^{-2}$ W $= 10$ mW $= 10$ dBm, $\omega_{l_p h} = \xi P_p + \omega^{*}_{l_p h} = 10^{-7}$ s^{-1}, $N_{l_p} = 10^{30}$, $\Delta E_s = h f_s = 6.63 \times 10^{-25}$ J for $f_s = 1 \times 10^{9}$ s^{-1} $= 1$ GHz, $\eta = 9.08 \times 10^{-7}$ (Ws)$^{-1}$, $N_h = 10^{31}$, $\omega_{hl_p} = 10^{-8}$ s^{-1}, $\omega_{hl_s} = 1.009 \times 10^{-17}$ s^{-1}, and $P_s = -100$ dBm $= 0.1 \times 10^{-12}$ W, then

$$G = \frac{10^{-7}\ \mathrm{s}^{-1}}{(0.1 \times 10^{-12}\ \mathrm{W}) \times (10^{-8}\ \mathrm{s}^{-1})} \times 10^{30}$$

$$\times 1.009 \times 10^{-17}\ \mathrm{s}^{-1} \times 6.63 \times 10^{-25}\ \mathrm{J}$$

$$= 6.63 \times 10^{2} = 28.2\ \mathrm{dB}$$

or

$$G = \frac{10^{31} \times 6.63 \times 10^{-25}\ \mathrm{J}}{0.1 \times 10^{-12}\ \mathrm{W}} \times 1.009 \times 10^{-17}\ \mathrm{s}^{-1}$$

$$= 6.63 \times 10^{2} = 28.2\ \mathrm{dB}$$

However, this will not happen in reality. Condition 12.5.3 means the downward transition loses coherency and spontaneous transition takes over. There is no maser action when this happens.

If condition 12.5.4 is attained and if

$$\eta P_s = 100\omega^s_{hl_s} \quad \text{or} \quad P_s = \frac{100\omega^s_{hl_s}}{\eta} = \frac{100 \times 1.009 \times 10^{-17}\ \text{s}^{-1}}{9.08 \times 10^{-7}\ (\text{W-s})^{-1}}$$

$$= 0.1 \times 10^{-8}\ \text{W} = 1\ \text{nW} = -90\ \text{dBW} = -87\ \text{dBm}$$

then, by Eqs. 12.5.5 and 12.2.10, $G = N_h \eta h f_s = 10^{31} \times 9.08 \times 10^{-7}\ (\text{W-s})^{-1} \times 6.63 \times 10^{-25}\ \text{J} = 6.02 = 10 \log_{10} 6.02\ \text{dB} = 7.8\ \text{dB}$

12.6 Noise

The reason for using a maser as an amplifier is due to low noise. Noise power emitted from a maser due to maser action can be obtained from Eq. 12.4.6. By letting the input signal power $P_s = 0$, any radiation coming out of the maser is considered to be noise power P_N.

From Eq. 12.4.6 with $P_s = 0$,

$$P_N = N_{l_p} \frac{\xi P_p + \omega^*_{l_p h}}{\omega^s_{hl_p}} \omega^s_{hl_s} h f_s \tag{12.6.1}$$

As seen from this equation, spontaneous transitions represented by $\omega^*_{l_p h}$, $\omega^s_{hl_s}$ and $\omega^s_{hl_p}$ are the cause of noise power emission [10]. Besides this quantum mechanical noise, the maser emits thermal noise given by

$$P_t = kTB \tag{12.6.2}$$

where k is the Boltzmann constant, 1.38×10^{-23} J/K.

T is the absolute temperature of the maser in K, and B is the frequency bandwidth in Hz. The frequency bandwidth of a maser is extremely narrow. For example, it is common that B is on the order of 10^5 Hz, at 10^{10} Hz operating signal frequency. Some masers are cryogenically refrigerated to 2 K, the temperature of liquid helium. This makes the thermal noise of the maser as low as 10^{-18} W. If it is not refrigerated and instead operated at room temperature, 300 K, then the thermal noise is on the order of 10^{-16} W.

The maser is a quantum mechanical device. The operating frequency must exactly fit into the quantum mechanical energy gap of the maser material as shown in Eq. 12.3.1. If E_h and E_{l_s} are discretely defined, the operating frequency must therefore be exactly equal to

$$f = \frac{E_{\mathrm{h}} - E_{l_{\mathrm{s}}}}{h} \tag{12.6.3}$$

The energy levels of an isolated single molecule in low pressure gas masers are discretely defined. The bandwidth is therefore practically zero.

From Eq. 12.6.3,

$$B = \Delta f = (1/h)\Delta(E_{\mathrm{h}} - E_{l_{\mathrm{s}}})$$

$$= \frac{(E_{\mathrm{h}} + \frac{1}{2}\Delta E_{\mathrm{h}}) - (E_{l_{\mathrm{s}}} - \frac{1}{2}\Delta E_{l_{\mathrm{s}}})}{h}$$

$$- \frac{(E_{\mathrm{h}} - \frac{1}{2}\Delta E_{\mathrm{h}}) - (E_{l_{\mathrm{s}}} + \frac{1}{2}\Delta E_{l_{\mathrm{s}}})}{h} = \frac{\Delta E_{\mathrm{h}} + \Delta E_{l_{\mathrm{s}}}}{h} \tag{12.6.4}$$

The spread of the energy level $\Delta E_{l_{\mathrm{s}}}$ is called the line width. For an isolated single molecule, the linewidth is zero. The frequency bandwidth is consequently zero.

If the maser molecules are packed together in a crystal, these molecules interact with each other. The energy levels are no longer strictly discrete. The linewidth ΔE_{h}, and $\Delta E_{l_{\mathrm{s}}}$ appear by the mutual interaction of maser molecules.

In most practical masers, the thermal noise power is 10^{-18}–10^{-16} W or $P_{\mathrm{t}} = -150$–-130 dBm depending on the temperature.

In the maser investigated in Section 12.5, Eq. 12.6.1 is rewritten as

$$P_{\mathrm{N}} = N_{\mathrm{h}} \omega^{\mathrm{s}}_{\mathrm{h}l_{\mathrm{s}}} h f_{\mathrm{s}} \tag{12.6.5}$$

For the maser example presented in this chapter, $N_{\mathrm{h}} = 10^{31}$, $\omega^{\mathrm{s}}_{\mathrm{h}l_{\mathrm{s}}} = 1.009 \times 10^{-17}\ \mathrm{s}^{-1}$ and $\Delta E = hf_{\mathrm{s}} = 6.63 \times 10^{-25}$ J. Then $P_{\mathrm{N}} = 10^{31} \times 1.009 \times 10^{-17}\ \mathrm{s}^{-1} \times 6.63 \times 10^{-25}\ \mathrm{J} = 6.69 \times 10^{-11}$ W $= -102$ dBW $= -72$ dBm.

This amount is not considered low noise for a maser. The quantum mechanical noise is much higher than the thermal noise. This is an example of hard pumping which makes N_{h} large, accounting for high quantum mechanical noise. If the pump power is reduced 20 dB, then $P_{\mathrm{N}} = -92$ dBm.

This is still unacceptably high noise for a maser, which is supposed to amplify low level signals. Cutting the pump power down will cut the quantum mechanical noise down, but at the same time, it will cut down the gain. As a result, the gain and the noise have to be simultaneously considered to obtain optimum performance.

Now if $B = \Delta f = 10$ MHz, for example, from Eq. 12.6.4 $\Delta E_{\text{h}} + \Delta E_{l_s} = h\Delta f = 6.63 \times 10^{-34}$ J-s $\times 10 \times 10^6$ s$^{-1} = 6.63 \times 10^{-28}$ J. This should be compared with the transition energy gap $\Delta E_{\text{s}} = 6.63 \times 10^{-25}$ J. The total linewidth of combined linewidths on E_{h} and E_{l_s} is only 1/1000 of the transitional energy gap ΔE_{s}.

If the power gain of the maser is G, the total noise output is the result of the amplified thermal noise kTBG, and the quantum mechanical noise P_{N} is P_{NT}, then

$$P_{\text{NT}} = \text{k}TBG + P_{\text{N}} \tag{12.6.6}$$

The noise figure of this maser is defined as

$$F = \frac{P_{\text{NT}}}{\text{k}TBG} = 1 + \frac{P_{\text{N}}}{\text{k}TBG} \tag{12.6.7}$$

For example, if a maser has a gain $G = 30$ dB $= 10^3$, $T = 2$ K, $B = 10^7$ Hz and $P_{\text{N}} = 10^{-12}$ W, then

$$\text{F} = 1 + \frac{10^{-12}\ \text{W}}{1.38 \times 10^{-23}\ \text{J/K} \times 2\ \text{K} \times 10^7\ \text{s}^{-1} \times 10^3}$$
$$= 4.6 = 10 \log 4.6\ \text{dB} = 6.62\ \text{dB}.$$

If P_{N} is reduced to $P_{\text{N}} = 10^{-13}$ W, then $F = 1.36 = 1.33$ dB. If P_{N} is reduced to $P_{\text{N}} = 10^{-14}$ W, then $F = 1.036 = 0.15$ dB. If the relationship

$$P_{\text{NT}} = \text{k}T_{\text{N}}BG \tag{12.6.8}$$

is forced, the equivalent temperature T_{N} is called the noise temperature of the maser. Then

$$T_{\text{N}} = \frac{P_{\text{NT}}}{\text{k}GB} = T + \frac{P_{\text{N}}}{\text{k}GB} \tag{12.6.9}$$

P_{N} is also called the additional noise. The quantity $(P_{\text{N}}/\text{k}GB) = T_{\text{E}}$ is called the excess noise temperature of the maser.

For example, in a maser, if $P_{\text{N}} = 10^{-13}$ W, $B = 10^7$ Hz and $G = 30$ dB $= 10^3$, then $T_{\text{E}} = 10^{-13}$ W/(1.38 $\times 10^{-23}$ J/K)($10^3 \times 10^7$ sec^{-1} = 0.72 K. If $T = 2$ K, then $T_{\text{N}} = 2$ K + 0.72 K = 2.72 K.

12.7 Frequency Standard

As seen from Eq. 12.6.3, the energy gap between E_{h} and E_{l_s} determines the maser signal frequency f. There are characteristic values for $E_{\text{h}} - E_{l_s}$ for each maser material. Thus, each maser

Table 12.1 Characteristic Frequencies

Maser material	Characteristic frequency
Ammonia	23.870 GHz
Cesium	9.192631830 GHz
Sodium	1.772 GHz
Hydrogen	1.420 GHz

material has its characteristic frequency. The characteristic frequency is distinctly defined by Eq. 12.6.3. Some examples of maser materials and the characteristic frequencies of maser materials are listed in Table 12.1 [2]. The characteristic frequencies are so accurate that they are used for frequency standards and time standards, as exemplified by an atomic clock.

Most maser output is very small and is not used unless it is amplified. Straightforward amplification requires an extremely stable and accurate signal source to begin with. Since such an accurate signal source is not usually available, the straightforward amplifier is not generally utilized. Instead, a maser controlled oscillator, as shown in the block diagram in Fig. 12.1, is used as a standard frequency signal source. In this diagram, a small portion of the voltage controlled microwave oscillator signal is sampled through a coupler and is fed into the maser. The voltage controlled oscillator can be any voltage controlled solid-state active diode or transistor oscillator. When the oscillator frequency coincides with the characteristic frequency of the maser, the signals are

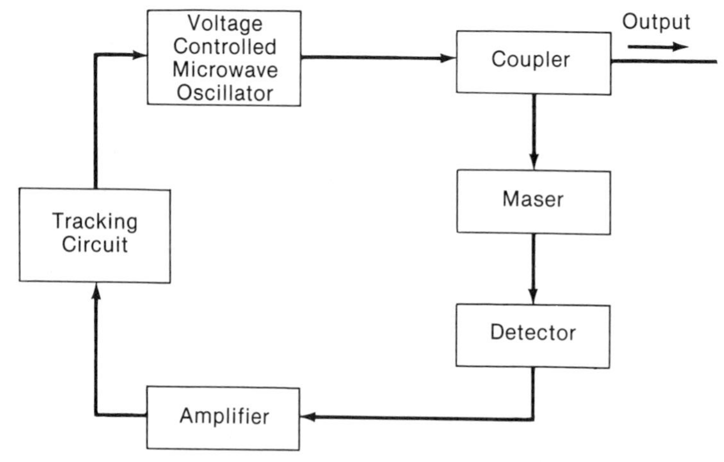

Fig. 12.1 Maser controlled standard frequency generator block diagram.

amplified by the maser and the output is detected, amplified, and fed into a tracking circuit to maintain the power supply voltage of the voltage controlled microwave oscillator as well as to maintain coincidence of the microwave oscillator frequency and maser characteristic frequency [3].

12.8 Gas Masers

As seen from Table 12.1, the majority of frequency standard masers are gas masers. Molecules of low pressure gas are reasonably isolated from each other without much interaction or coupling among themselves, producing a well defined single line frequency spectrum. As an example of a gas maser, an ammonia maser has been selected [4]. A schematic diagram of the ammonia gas maser is shown in Fig. 12.2. Low pressure ammonia gas at few mm Hg is shot out of a nozzle to an evacuated environment. The gas molecular beam is led to a molecular beam focuser. The ammonia molecule is a dipolar molecule which consists of one "positive" nitrogen atom and three "negative" hydrogen atoms. The electrical center of the nitrogen atom and the electrical center of the three hydrogen atoms do not coincide; thus, they form a molecular dipole. These electrical centers vibrate with each other at 23.870 GHz, creating an oscillating dipole. Some molecules will oscillate strongly and some molecules not quite as strongly, depending on the quantum state of the molecular energy. In other words, some molecules are in a state of E_{h} and other molecules are in a state of E_{l_s}.

When the ammonia molecules are led to the focuser, the focuser separates the E_{h} molecules and the E_{l_s} molecules by exposing the ammonia molecules to a nonuniform high electric field. The high electric field is created by a quadrupole structure.

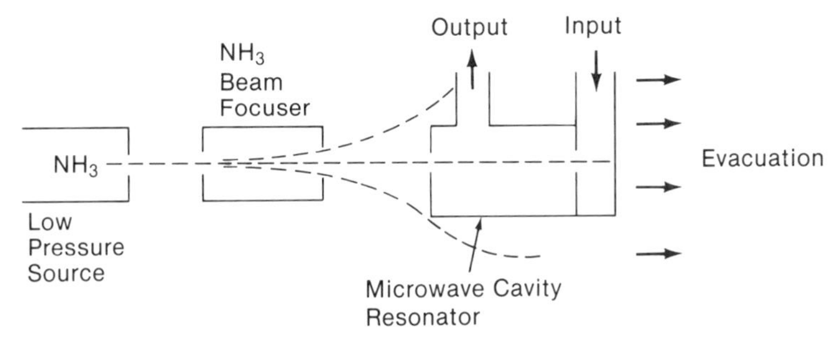

Fig. 12.2 Diagram of an ammonia maser. The microwave cavity resonates at 23.870 GHz.

Molecules in the E_h state are pushed inward by the nonuniform field of the quadrupole structure and the E_{l_s} state molecules are pushed outward [2]. This causes the E_h molecules to focus and the E_{l_s} molecules to diverge. The focused molecules are received by a cavity resonator shown in Fig. 12.2 and the diverged molecules are evacuated by a vacuum pump. The microwave cavity resonator is adjusted to resonate with the characteristic frequency of the ammonia molecules, 23.870 GHz. The majority of the molecules inside the cavity resonator is in the E_h state and fewer molecules are found to be in the E_{l_s} state. This is the density inversion state which was explained in Eq. 12.2.10 [7,8]. In this case, pumping is achieved by natural spontaneous thermal pumping while the gas is still in the tank. In fact, when the ammonia molecules are coming out of the nozzle, the density is not inverted. It follows the Boltzmann relation as shown by Eq. 12.2.3. The beam focuser selects the molecules and allows molecules in the E_h state into the cavity resonator [7,8]. The cavity resonator is excited by the input signal. The input signals stimulate the molecules in the cavity resonator as explained in Section 12.3. The emitted radiation power expressed by Eq. 12.4.2 is enhanced by the presence of the cavity resonator [5].

Without the cavity resonator, e_m, the maximum microwave electric field strength associated with the emitted microwave radiation P_o, is, by Ohm's law

$$e_m = \sqrt{2\eta_o P_o} \text{ (V/m)} \tag{12.8.1}$$

where η_o is the wave impedance of the maser gas medium. If the quality factor of the cavity resonator is Q then, by definition, the maximum electric field strength is

$$e_{fm} = \sqrt{Q}\sqrt{2\eta_o P_o} \text{ (V/m)} \tag{12.8.2}$$

In many cases, masers do not work without a properly tuned cavity resonator. Most often the emitted field strength given by Eq. 12.8.1 is very small in comparison with the noise given by Eq. 12.6.1 for quantum mechanical noise and by Eq. 12.6.2 for thermal noise. The emitted fields are masked by noise and no coherent maser action is possible. On the other hand, if the cavity resonator is used, the emitted field strength at the quantum frequency given by Eq. 12.6.3 is strengthened, as seen from Eq. 12.8.2. The cavity resonator is considered to be a very narrow band pass filter. It therefore suppresses noise frequencies other than the resonance frequency of the cavity resonator.

For example, in a microwave cavity resonator, if the quality factor $Q = 10{,}000$ is not uncommon, then $\sqrt{Q} = 100$. If a maser emitted to the vacuum a wave impedance of $\eta_o = 377\ \Omega$, with emission of $P_e = 10^{-10}$ W $= -100$ dBW $= -70$ dBm without a cavity resonator, the electric field associated with this emission is, from Eq. 12.8.1, $e_m = \sqrt{2 \times 377\ \Omega \times 10^{-10}\ \text{W}} = 8.68 \times 10^{-3}$ V/m = 8.68 mV/m. If the cavity resonator is present, this field strength is stepped up by a factor of $\sqrt{Q} = 100$ or $e_m = 8.68$ mV/m $\times$ 100 = 868 mV/m. For stimulation, if the input power of this maser is $P_s = 10^{-13}$ W $= -130$ dBW $= -100$ dBm then, without the cavity resonator, the maximum field strength is

$$e_m = \sqrt{2 \times 377\ \Omega \times 10^{-13}\ \text{W}} = 8.68 \times 10^{-6}\ \text{V/m} = 8.68\ \mu\text{V/m}$$

If the cavity resonator is present, this is boosted up to

$$e_m = \sqrt{10{,}000}\, 8.68\ \mu\text{V/m} = 868\ \mu\text{V/m}$$

In summary, the cavity resonator acts like a step-up transformer. Like a transformer, it can step-up the field strength but it cannot increase the power. The power is conserved but the energy associated with the field strength is increased. A resonator is capable of storing electromagnetic energy. This illustrates how effectively a cavity resonator helps stimulate maser action and strengthens the radiating electromagnetic fields of emission.

According to circuit theory, the quality factor of a resonator can be calculated from $Q = f/\Delta f$, where f is the resonance frequency and Δf is the frequency bandwidth of the resonator. If $f = 1$ GHz $= 10^9$ Hz and $Q = 10{,}000$, then $\Delta f = f/Q = 10^9$ Hz/10^4 = 10^5 Hz = 0.1 MHz = 100 kHz. This means that the cavity allows electrical noise within 100 kHz of the resonance frequency of 1 GHz but it rejects the rest of the noise spectrum. For example, if the emission frequency bandwidth without cavity resonator is $\Delta f_o = 10$ MHz $= 10 \times 10^6$ Hz and the cavity frequency bandwidth is $\Delta f = 10^5$ Hz = 0.1 MHz, then the noise power within the cavity frequency bandwidth will be suppressed by the ratio $\Delta f/\Delta f_o = 0.1$ MHz/10 MHz = 0.01 = -20 dB. If the noise power emission without cavity resonator is $P_N = -72$ dBm in the 10 MHz bandwidth, then with the cavity resonator the noise power is reduced to $P_N = -72$ dBm -20 dB $= -92$ dBm in 100 kHz bandwidth. The cavity resonator therefore helps to strengthen stimulation and emission of radiation and to suppress noise.

These calculations are again merely order of magnitude calculations to illustrate some quantitative aspects and calculation procedures. It is therefore not necessarily exact, accurate, nor specific to any particular type of maser.

The emitted microwave power from the ammonia molecules in the E_h state into the cavity resonator re-stimulates the gas molecules by joining the newly fed input signals. This process repeats as long as the signals and the pump feed the activated ammonia molecules in the cavity resonator. The amplified microwave signals may be taken out from a waveguide which is coupled to the cavity resonator as shown in Fig. 12.2, or the output can be taken from the same waveguide as is used for the input. After microwave power emission, the ammonia molecules transfer from the energy state E_h to the state E_{l_s}. Molecules in the lower energy state in the cavity will be evacuated from the cavity resonator by a vacuum pump. New high energy molecules are continuously fed into the cavity resonator through the nozzle and the molecular beam focuser. In this type of maser, the pumping is done thermally at room temperature and the molecular density inversion is accomplished by selectively focusing the beam of high energy molecules into a microwave cavity resonator.

12.9 Solid Masers

Paramagnetic crystals such as ruby or rutile are used for solid maser material [1]. An example of a solid maser, a ruby crystal maser, is presented in this section. A schematic diagram of a ruby crystal maser is shown in Fig. 12.3. A ruby crystal is basically a crystal of Al_2O_3 with a small amount of Cr in it. A crystal of pure Al_2O_3 is an alumina crystal which is white and transparent. When a small amount of Cr is introduced, the color ranges from pink to dark red, depending upon the amount of Cr impurity in the host crystal of Al_2O_3. The more Cr atoms, the darker the color is. The color of ruby crystals used for the maser is usually pink. These Cr atoms, as an impurity in the pink ruby, are responsible for maser action. The Cr atoms in the host Al_2O_3 crystal structure are paramagnetic. The total magnetic dipole moment of the Cr atom is specified by its quantum numbers [2]. The natural quantum energy state separation ΔE_s of Cr atoms in the ruby crystal to satisfy Eq. 12.1.5 is in the optical wavelength of pink to red. For microwave frequency applications, it is necessary to create an artificial energy level separation ΔE_s to fit the operating microwave frequency. This is accomplished by applying a magnetic field to the ruby crystal. As seen from Fig. 12.3, a magnet is

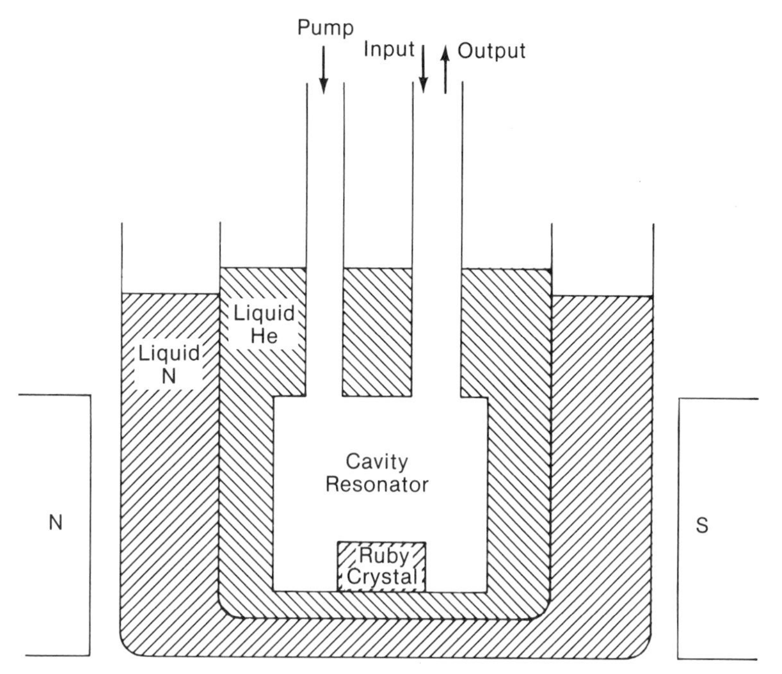

Fig. 12.3 Diagram of a ruby maser.

needed for the maser. When the Cr atoms in the Al_2O_3 host crystal are magnetized, new energy levels are generated by the Zeeman effect as follows [1].

$$\Delta E = M_J g_J \beta B \tag{12.9.1}$$

M_J is the angular momentum quantum number of Cr atoms in the Al_2O_3 host crystal, g_J is the spectroscopic splitting factor which is related to the energy loss in the crystal, β is the Bohr magneton which is a magnetic dipole moment due to the spin of an electron [6], and B is the applied magnetic flux density. Generation of energy levels by the magnetic field is called the Zeeman effect. These generated energy levels are called Zeeman energy levels. As seen from Eq. 12.9.1, the Zeeman level expressed by this equation is the relative energy increase or decrease with respect to the energy level with zero magnetic field. These energy splits are caused by the magnetic dipole moment of the Cr atoms in the Al_2O_3 host crystal with respect to the applied magnetic fields. If the magnetic dipole moment of a Cr atom is parallel to the applied magnetic field, the atom is in the lowest possible energy state. If the magnetic dipole moment is antiparallel to the applied magnetic field, the atom is in the highest possible energy state. Between these two extreme states there are a number of energy states available for the application of maser action. As seen from Eq. 12.9.1, the energy separation ΔE_s to satisfy Eq. 12.1.5 at microwave frequencies is available by simply adjusting the magnetic flux density B to produce ΔE_s to satisfy Eq. 12.1.5 for the signal frequency.

Since there are a number of other energy levels available, it is not difficult to find a third energy level which satisfies Eq. 12.1.1 in conjunction with Eq. 12.2.3 and Eq. 12.3.1. The frequency of the pump oscillator must be tuned to satisfy Eq. 12.1.1 with pump power sufficient to produce sufficient density inversion between the level E_h and the level E_{l_s}. The pump power is fed through the waveguide shown in Fig. 12.3 into the cavity resonator. The signal power is fed through another waveguide as shown in the same figure. The cavity resonator is therefore a dual mode resonator, resonating at the pump frequency and also at the signal frequency in different modes [2,5]. The cavity is designed so that the two modes are orthogonal to each other. The two modes interact with each other only in the ruby crystal when the maser action is activated. The input signals stimulate the density inverted pumped Cr atoms in the ruby crystal to make a massive downward transition. This downward transition emits the radiation of microwaves in the cavity in strength. The emission and incoming signals join to re-stimulate the Cr atoms in the crystal, which causes more emission of radiation. This process repeats and an equilibrium state is reached. The amplified signal is taken out from the same waveguide into which the input signals are being fed, as seen from Fig. 12.3.

In the maser, the useful energy gap ΔE_s is smaller than the thermal noise energy kT at room temperature. At room temperature, ΔE_s is masked by the noise and maser action is impossible. As a result, in a practical maser, as seen from Fig. 12.3, the maser is placed in a double dewer which contains liquid helium and liquid nitrogen. The liquid helium cools the maser as low as 2 K. To achieve this low temperature and to prevent evaporation and boiling loss of liquid helium, the maser is also cooled by liquid nitrogen at 77 K, as seen in Fig. 12.3.

For example, the Bohr magneton β is known to be $\beta = 9.27 \times 10^{-24}$ A-m^2 [6]. If $f_s = 1$ GHz, the frequency of microwave signals to be amplified, $\Delta E_s = hf_s = 6.63 \times 10^{-34}$ J-s $\times 1 \times 10^9$ s^{-1} $= 6.63 \times 10^{-25}$ J. In Eq. 12.9.1, the quantum number of a Cr ion in Al_2O_3 is known to be $M_J = 1/2$ [6]. The necessary magnetic flux density B to produce $\Delta E_s = 6.63 \times 10^{-25}$ J is dependent on the direction of magnetic flux with respect to the crystal axis. In practice, $B = 0.2$ T is found to be adequate for 1 GHz amplification [6]. Then, from Eq. 12.9.1,

$$g_J = \frac{\Delta E}{M_J \beta B} = \frac{6.63 \times 10^{-25}\ \text{J}}{\frac{1}{2} \times 9.27 \times 10^{-24}\ \text{A-m}^2 \times 0.2\ \text{T}}$$

$$= 0.715\ \text{J/A-T-m}^2$$

12.10 Masers

A feature of microwave quantum electron devices is that the interaction of electrons and microwaves takes place on bound electrons rather than free electrons. In transistors and semiconductor diodes or, for that matter, in various thermionic devices, microwaves interact with moving carriers which are free electrons and holes. In quantum electron devices, electrons are bound. This is true even in gas masers. In gas masers, the gaseous atoms or molecules are moving, yet the interacting electrons are attached to the atom or molecule and never separate from their home particles. The orientation of the electron, atom, or molecule with respect to a reference axis determines the relative energy level of the electron, atom, or molecule. The transition of energy, upward or downward, simply means the orientation of the electron, atom, or molecule is changing to high energy state or low energy state. A maser can be an electronic maser, an atomic maser, or a molecular maser, depending on what kind of particles are emitting coherent radiation. The coherency of microwave radiation is created by a cavity resonator which resonates at the frequency of the emitted radiation due to quantum transition of the active particles. Inside the cavity resonator, there is a standing wave of microwaves. In standing waves, all electromagnetic fields are in phase or change in unison with respect to time, establishing coherency. The cavity resonator helps establish coherency and helps increase the efficiency of stimulation and emission of radiation. The process of preparing particles for emission of radiation by feeding energy to them is called pumping. Pumping can be done coherently or incoherently. Coherent pumping is done by irradiating the particles by coherent radiation of a proper wavelength. Incoherent pumping can be done thermally by simply heating the particles involved.

Maser amplification depends on the particle density, inversion and dependency of the downward transition probability on stimulation.

Problems

12.1 Define a paramagnetic crystal.

12.2 State the procedure and the principle of the density inversion.

12.3 Find the photon energy of 10 GHz microwaves.

12.4 A maser requires 300 mW of pump power at 7.5 GHz. Compute the number of photons per second to pump this maser.

12.5 Explain the mechanism of quantum mechanical stimulation.

12.6 Compute the value of ΔE_s necessary to amplify 4 GHz microwave signals.

12.7 Compute the required number of photons per unit time to produce 10^{-10} W at 10 GHz. At the same time, the input signal power absorbed by a maser is 10^{-7} W at 10 GHz. Find the number of photons fed per unit time.

12.8 Define a maser.

12.9 Under thermal equilibrium conditions at room temperature, a maser has a molecular density of 1.2×10^{10} molecules/cm^3 for $N_{l_p}^*$. The pump frequency needed is 7.5 GHz. Assuming that the ratio of the statistical weights (w_h/w_{l_p}) is unity, find N_h^*. If (w_h/w_{l_p}) is 1.5, then what is N_h^*?

12.10 In most cases, under thermal equilibrium, $N_h^* < N_{l_p}^*$ by the Boltzmann relation. But sometimes the Boltzmann relation under thermal equilibrium states that $N_h^* > N_{l_p}^*$. Find a condition which makes $N_h^* > N_{l_p}^*$ under thermal equilibrium.

12.11 A molecule does not usually stay in the same energy state permanently. A particular energy state has a statistical lifetime. Try to relate the statistical lifetime of a particular energy state to the molecular density of the energy state.

12.12 If a maser of $N_h^* = 10^{10}$ molecules/cm^3 is emitting 10^{-13} mW at 8.5 GHz, find the downward transition probability per unit time. Is room temperature of $T_o = 300$ K enough to sustain the emission of radiation by simple thermal pumping?

12.13 Show that $N_h > N_{l_p}$ under pumping.

12.14 Show that feeding input signal power to a maser induces larger signal power emission. Also show that the amount of emission is controlled by the amount of the input signal power.

12.15 Propose methods to increase the power gain of a maser.

12.16 Explain the reason for the saturation of the maser gain for high input signals.

12.17 Discuss the advantages and disadvantages of the maser over other types of microwave amplifiers.
12.18 Show that maser noise increases with pump power.
12.19 Estimate the quantum mechanical noise of a maser with respect to the thermal noise.
12.20 Derive the noise bandwidth of a maser.
12.21 Point out the reason most frequency or time standards are obtained by a gas maser rather than a solid maser.
12.22 Draw a schematic diagram of a maser controlled electronic oscillator.
12.23 Sketch a schematic diagram of an ammonia gas maser. Explain how it works.
12.24 Explain the principle of pumping in a ruby maser.
12.25 For a maser, explain the objectives of the microwave cavity resonator.
12.26 If a maser emits 10^{-15} mW into a space where $\eta_o = 377\ \Omega$ find the electric field strength. Calculate the noise field strength assuming the medium temperature is 300 K and the maser has a 3 kHz frequency bandwidth at 30 GHz signal frequency. Calculate the emission electric field if a cavity resonator of $Q = 10{,}000$ is properly attached.
12.27 Assume that a ruby maser operates with the pump frequency of 10.525 GHz and the signal frequency is 2.45 GHz. Sketch an energy level diagram of this maser and identify E_h, E_{l_s} and E_{l_p}.
12.28 In a ruby maser, the microwave emission is coming from the Cr atoms in an Al_2O_3 host crystal. Is it possible to make a maser using Cr crystals alone?
12.29 There is an optimum concentration of Cr atoms in the Al_2O_3 host crystal for best maser performance. State the reason for this.
12.30 Show numerically that, at room temperature, ΔE_s of a maser is masked by the thermal noise.
12.31 Show numerically that refrigeration to liquid helium temperature is sufficient to discern the signal due to the ΔE_s transition at microwave frequencies.

References

1 A. E. Siegman, "Microwave Solid-State Masers." McGraw-Hill, New York, 1964.
2 T. K. Ishii, "Maser and Laser Engineering." Krieger, New York, 1980.
3 F. J. Tischer, "Microwellen Mesztechnik." Springer-Verlag, Berlin, 1958.

4 J. P. Gordon, H. J. Ziegler, and C. H. Townes, Molecular microwave oscillator and new hyperfine structure in the microwave spectrum of NH_3. *Phys. Review* **95**, 282–284 (July 1, 1954).
5 T. K. Ishii, "Microwave Engineering." Ronald Press, New York, 1967.
6 A. E. Siegman, "An Introduction to Lasers and Masers." McGraw-Hill, New York, 1971.
7 M. Brune, J. M. Raimond, P. Goy, L. Davidovich, and S. Haruche, The two-photon Rydberg atom micromaser. *IEEE J. Quantum Electron.* **24**(7), 1323–1330 (1988).
8 N. Nayak, R. K. Bullough, B. V. Thompson, and G. S. Agarwal, Quantum collapse and revival of Rydberg atoms in cavities of arbitrary Q at finite temperature. *IEEE J. Quantum Electron.* **24**(7), 1331–1337 (1988).
9 A. Bandilla and H. H. Ritze, Physical interpretation of operator dynamics in Jaynes–Cummings model. *IEEE J. Quantum Electron.* **24**(7), 1338–1345 (1988).
10 F. W. Cummings, On spontaneous emission. *IEEE J. Quantum Electron.* **24**(7), 1346–1350 (1988).

13

Ferrimagnetic Electron Devices

13.1 Ferrimagnetic Materials

Ferrimagnetic materials provide important parts in various microwave electronic systems [1]. Ferrimagnetic materials are used in tuners, filters, attenuators, isolators, circulators, modulators, switches, and shutters. In this chapter, practical aspects and basic theory of the interaction of ferrimagnetic materials with microwaves are studied, and principles of practical ferrimagnetic microwave circuit components are investigated.

Common ferrimagnetic materials used for microwave applications are ferrites and garnets. When ferrites and garnets are dc magnetized and the microwave magnetic field **h** is applied perpendicularly to the magnetizing dc magnetic field, the induced microwave flux density **b** is not parallel to the magnetic field **h**. On the other hand, for ferromagnetic material, the microwave magnetic field **h** and microwave magnetic flux density **b** are parallel to each other whether the material is magnetized or not. In the magnetized ferrimagnetic material, the microwave magnetic field **h** and microwave magnetic flux density **b** are not parallel to each other. In the soft ferrite, if the magnetizing dc magnetic field is removed and no residual magnetism exists, it then turns into pseudo-ferromagnetic material, or the microwave magnetic flux density **b** becomes parallel to microwave magnetic fields **h** [2]. It is interesting to investigate why **b** is not parallel to **h** in dc magnetized ferrimagnetic material.

13.2 Gyromagnetic Equations

The ferrimagnetic phenomenon is due to the precessing magnetic dipole moment of bound unpaired electrons in the ferrite material. For example, if the ferrite is Fe_3O_4, it contains unpaired electrons in Fe^{3+} ions. These unpaired spinning electrons are responsive to applied magnetic fields, both dc and microwave. Since electrons are negatively charged, if they are spinning they can be considered tiny magnets. Each spinning unpaired electron has a magnetic dipole moment. For simplicity, if homogeneous ferrite is considered, there is a number of spinning magnetic dipole moments in the ferrite. When a dc magnetic field is applied to the ferrite, the spinning magnetic dipole moments precess around the direction of applied dc magnetic fields the way a tipping and spinning top precesses around gravitational fields. If the total magnetic dipole moment of precessing electrons per unit volume of the ferrite material is **M** A/m under the applied dc magnetic field **H** A/m then, for a lossless system, the torque per unit of the magnetization **M**, which is the total magnetic dipole moment per unit volume, is by the classical magnetodynamic principle

$$\tau = \mu_o \mathbf{M} \times \mathbf{H} \qquad (\text{m-N/m}^3) \tag{13.2.1}$$

When the magnetization **M** is introduced, the permeability of the medium is considered to be μ_o as $\mathbf{B} = \mu_o(\mathbf{H} + \mathbf{M})$. According to classical mechanics, the torque is equal to the time rate of change of the total mechanical angular momentum **J**.

$$\tau = \frac{d\mathbf{J}}{dt} \qquad (\text{m-N/m}^3) \tag{13.2.2}$$

where **J** (kg/s^3) is the total mechanical angular momentum of precessing electrons per unit volume of ferrite material. The magnetization vector **M** and the angular momentum vector **J** are considered to be related by definition

$$\mathbf{M} = \gamma \mathbf{J} \tag{13.2.3}$$

The proportionality constant γ is termed the gyromagnetic ratio.

Combining Eqs. 13.2.3 and 13.2.2,

$$\tau = \frac{1}{\gamma}\frac{d\mathbf{M}}{dt} \tag{13.2.4}$$

Equating Eqs. 13.2.1 and 13.2.4,

$$\frac{d\mathbf{M}}{dt} = \mu_o\gamma(\mathbf{M} \times \mathbf{H}) \tag{13.2.5}$$

This is the gyromagnetic equation for the magnetization of an ideal lossless ferrite. For a realistic ferrite with losses, modifying the Landau–Lifschitz equation, Lax and Button [1] introduced a gyromagnetic equation for ferrites with slight losses; modifying it further in the MKS unit system,

$$\frac{d\mathbf{M}}{dt} = \mu_o\gamma(\mathbf{M} \times \mathbf{H}) - \frac{\alpha}{M}\mathbf{M} \times \frac{d\mathbf{M}}{dt} \tag{13.2.6}$$

where α is a damping constant for loss associated with the precession and M is the magnitude of the magnetization vector $\mathbf{M}$. Equation 13.2.6 states that the loss term is proportional to the torque, which is a reasonable assumption for the first approximation.

Since the electron charge is considered to be negative, the magnetization vector $\mathbf{M}$ and the angular momentum density $\mathbf{J}$ are antiparallel to each other. The gyromagnetic ratio γ, as defined in Eq. 13.2.3, is therefore a negative scalar quantity. The unit of the gyromagnetic ratio γ is $[(\mathrm{A/m})/(\mathrm{kg/s^3})] = [\mathrm{A\text{-}kg/m\text{-}s^3}]$.

As far as the precession of the electron spin is concerned, if the angular velocity of the precession is ω_o (rad/s) then, according to classical mechanics of precession,

$$\tau = \omega_o \times \mathbf{J} \tag{13.2.7}$$

With Eqs. 13.2.3 and 13.2.6,

$$\omega_o \times \mathbf{J} = \mu_o\mathbf{M} \times \mathbf{H} - \frac{\alpha}{\gamma}\frac{\mathbf{M}}{M} \times \frac{d\mathbf{M}}{dt}$$

$$\omega_o \times \frac{\mathbf{M}}{\gamma} = \mu_o\mathbf{M} \times \mathbf{H} - \frac{\alpha}{\gamma}\frac{\mathbf{M}}{M} \times \frac{d\mathbf{M}}{dt} \tag{13.2.8}$$

This means that for the lossless ferrite, $\alpha = 0$ and ω_o is parallel to $\mathbf{H}$ since γ is negative. For a lossy ferrite, $\alpha \neq 0$ and ω_o is still parallel to $\mathbf{H}$. The reason is that $-d\mathbf{M}/dt$ is parallel to $d\mathbf{J}/dt$

and $\mathrm{d}\mathbf{J}/\mathrm{d}t$ is parallel to $\boldsymbol{\omega}_o$. For a lossless ferrite,

$$\boldsymbol{\omega}_o \times \frac{\mathbf{M}}{\gamma} = \mu_o \mathbf{M} \times \mathbf{H}$$

$$\mathbf{M} \times \boldsymbol{\omega}_o = -\gamma\mu_o \mathbf{M} \times \mathbf{H} = \mathbf{M} \times (-\gamma\mu_o \overline{\mathbf{H}}) \quad (13.2.9)$$

Therefore

$$\boldsymbol{\omega}_o = -\gamma\mu_o \mathbf{H} \quad (13.2.10)$$

Taking magnitude only

$$f_o = \frac{1}{2\pi}|\gamma|\mu_o H \quad (13.2.11)$$

where f_o is the precession frequency (Hz).

Alternatively,

$$|\gamma| = \frac{2\pi f_o}{\mu_o H} \qquad (\mathrm{Hz/T}) \quad (13.2.12)$$

This is an alternative unit of the gyromagnetic ratio; it is simpler than (A-kg/m-s^3), previously presented. The unit (Hz/T) is more widely used in practice in MKS unit system.

For a lossy ferrite, Eq. 13.2.8 is modified as

$$\boldsymbol{\omega}_o \times \frac{\mathbf{M}}{\gamma} = -\mu_o \mathbf{H} \times \mathbf{M} + \frac{\alpha}{M}\frac{\mathrm{d}\mathbf{M}}{\mathrm{d}t} \times \frac{\mathbf{M}}{\gamma}$$

$$= \left(-\gamma\mu_o \mathbf{H} + \frac{\alpha}{M}\frac{\alpha\mathbf{M}}{\mathrm{d}t}\right) \times \frac{\mathbf{M}}{\gamma} \quad (13.2.13)$$

Then

$$\boldsymbol{\omega}_o = -\gamma\mu_o \mathbf{H} + \frac{\alpha}{M}\frac{\mathrm{d}\mathbf{M}}{\mathrm{d}t} \quad (13.2.14)$$

Since $\mathbf{H}$ is antiparallel to $\mathrm{d}\mathbf{M}/\mathrm{d}t$, depending on the sign of α, ω_o can be greater or less than $\gamma\mu_o H$. If $\alpha < 0$, then $\omega_o > |\gamma|\mu_o H$, noting that $\gamma < 0$. If $\alpha > 0$, then $\omega_o < |\gamma|\mu_o H$.

For microwave operation, ω_o must be on the order of 10^{10} Hz to have some interaction with microwave frequency signals. For practical purposes, $\mu_o H$ should be on the order of 10^{-1} T. This means that in most practical microwave ferrite devices

$$|\gamma| = \frac{\omega_o}{\mu_o H} = \frac{10^{10}\ \mathrm{Hz}}{10^{-1}\ \mathrm{T}} = 10^{11}\ \mathrm{Hz/T} \approx 100\ \mathrm{GHz/T}$$

This is an approximate order of magnitude calculation. In a specific device, the actual value changes greatly depending on the specific type of ferrite material used, the geometry employed, and the magnitude of dc magnetic bias and its direction to the crystal axis if it is a single crystal ferrite. For example, if a ferrite device is operated at 8 GHz under a dc magnetic bias of 0.6 T, then the gyromagnetic ratio of this particular ferrite under this specific operation is

$$\gamma = \frac{\omega_o}{\mu_o H} = \frac{2\pi \times 8 \times 10^9 \text{ Hz}}{0.6 \text{ T}} = 81 \text{ GHz/T}$$

13.3 Tensor Permeability

In Eq. 13.2.6, if both the dc magnetic field $\mathbf{H}_o$ (A/m) and the microwave field $\mathbf{h}$ (A/m) are applied to the ferrite material at the same time, it can be expressed that

$$\mathbf{M} = \mathbf{M}_o + \mathbf{m}e^{j\omega t} \quad \text{(A/m)} \qquad (13.3.1)$$

$$\mathbf{H} = \mathbf{H}_o + \mathbf{h}e^{j\omega t} \quad \text{(A/m)} \qquad (13.3.2)$$

where $\mathbf{M}_o$ is the dc magnetization. The corresponding microwave frequency components are represented by $\mathbf{m}e^{j\omega t}$ and $\mathbf{h}e^{j\omega t}$ respectively. Substituting Eqs. 13.3.1 and 13.3.2 in Eq. 13.2.6 and equating the microwave frequency components only,

$$j\omega\mathbf{m} = \mu_o\gamma(\mathbf{M}_o \times \mathbf{h}) - \mu_o\gamma(\mathbf{H}_o \times \mathbf{m}) - j\omega\alpha(\mathbf{u}_z \times \mathbf{m}) \quad (13.3.3)$$

where $\mathbf{u}_z$ is the unit vector in the direction of $\mathbf{M}_o$ and $\mathbf{H}_o$ which is usually in the z-direction [1].

The magnetization vector $\mathbf{m}$ and the magnetizing magnetic field vector $\mathbf{h}$ are related by

$$\mathbf{m} = \overleftrightarrow{\mathbf{X}}\mathbf{h} \qquad (13.3.4)$$

and $\overleftrightarrow{\mathbf{X}}$ is called the susceptibility tensor with a dimensionless unit. The Cartesian coordinate representation of Eq. 13.3.4 is

$$\begin{pmatrix} m_x \\ m_y \\ m_z \end{pmatrix} = \begin{pmatrix} X_{xx} & X_{xy} & X_{xz} \\ X_{yx} & X_{yy} & X_{yz} \\ X_{zx} & X_{zy} & X_{zz} \end{pmatrix} \begin{pmatrix} h_x \\ h_y \\ h_z \end{pmatrix} \quad \text{(A/m)} \quad (13.3.5)$$

By comparing Eqs. 13.3.3 and 13.3.5 component by component [1]

$$X_{xx} = X_{yy} = \frac{(\omega_o + j\omega\alpha)\,\omega_M}{(\omega_o + j\omega\alpha)^2 - \omega^2} \tag{13.3.6}$$

$$X_{yx} = -X_{xy} = \frac{j\omega\omega_M}{(\omega_o + j\omega\alpha)^2 - \omega^2} \tag{13.3.7}$$

where

$$\omega_o \equiv \omega_o \gamma H_o \text{ (rad/s)} \tag{13.3.8}$$

and

$$\omega_M \equiv \mu_o \gamma M_o \text{ (rad/s)} \tag{13.3.9}$$

For lossless ferrites, $\alpha = 0$. When the operating frequency ω approaches ω_o, the susceptibility approaches infinity. This is the condition for gyromagnetic resonance and ω_o is called the gyromagnetic resonance (angular) frequency. The loss term prevents the susceptibility from approaching infinity.

Comparison of Eqs. 13.3.3 and 13.3.5 with assumptions based on practical applications yields

$$(M_o, H_o) \gg (m, h) \tag{13.3.10}$$

$$X_{xz} = X_{yz} = X_{zz} = X_{zx} = X_{zy} = 0 \tag{13.3.11}$$

In other words, the ferrite is saturated in the z-direction. As seen from Eqs. 13.3.6 and 13.3.7, the susceptibility is a complex quantity. It is therefore presentable in a general form

$$X = X' - jX'' \tag{13.3.12}$$

The real term represents the dispersive susceptibility and the imaginary part represents the dissipative susceptibility.

According to the definition of magnetic susceptibility, the susceptibility $\overleftrightarrow{\mathbf{X}}$ is related to the magnetic permeability of vacuum μ_o by

$$\mathbf{b} = \mu_o(\overleftrightarrow{\mathbf{I}} + \overleftrightarrow{\mathbf{X}})\mathbf{h} \tag{13.3.13}$$

With consideration of Eqs. 13.3.6, 13.3.7, 13.3.11, and 13.3.13,

$$\begin{pmatrix} b_x \\ b_y \\ b_z \end{pmatrix} = \mu_o \begin{pmatrix} 1 + X_{xx} & -X_{yx} & 0 \\ X_{yx} & 1 + X_{xx} & 0 \\ 0 & 0 & 1 \end{pmatrix} \begin{pmatrix} h_x \\ h_y \\ h_z \end{pmatrix} \tag{13.3.14}$$

The magnetic permeability is then

$$\overleftrightarrow{\mu} = \begin{pmatrix} \mu_o(1 + X_{xx}) & -\mu_o X_{yx} & 0 \\ \mu_o X_{yx} & \mu_o(1 + X_{xx}) & 0 \\ 0 & 0 & \mu_o \end{pmatrix} \quad (13.3.15)$$

It is interesting to note that the microwave magnetic field h_z will not be affected by the presence of the ferrite material at all. This is mainly due to the assumption of Eq. 13.3.10. For a small dc magnetic bias or zero dc magnetic bias, there is appreciable coupling of h_z to **b** unless it is intentionally made so that $h_z = 0$ [2].

In order to obtain some practical and quantitative concepts, the following examples are presented. If a microwave power of P = 100 mW is concentrated in the crosssectional area $S = 1$ cm $\times$ 2 cm $= 2 \times 10^{-4}$ m^2, then the magnitude of the Poynting vector **P** is $P/S = 0.1$ W$/2 \times 10^{-4}$ m$^2 = 0.5 \times 10^3$ W/m^2. If the wave impedance $\eta = 200\ \Omega$, then

$$\mathrm{P} = \tfrac{1}{2}\eta h^2,$$

or

$$h = \sqrt{\frac{2\mathrm{P}}{\eta}} = \sqrt{\frac{2 \times 0.5 \times 10^3\ \mathrm{W/m^2}}{200\ \Omega}} = 2.24\ \mathrm{A/m} \quad (13.3.16)$$

If the ferrite shows gyromagnetic resonance at $f_o = 9$ GHz with a dc saturation flux density of $\mu_o H_o = 0.3$ T, then the gyromagnetic ratio of this ferrite is, from Eq. 13.3.8,

$$|\gamma| = \frac{\omega_o}{\mu_o H_o} = \frac{2\pi f_o}{\mu_o H_o} = \frac{2\pi \times 9 \times 10^9\ \mathrm{Hz}}{0.3\ \mathrm{T}}$$

$$= 188.5 \times 10^9\ \mathrm{Hz/T} \quad (13.3.17)$$

If M_o is the saturation magnetization and $\mu_o M_o = 0.4$ T, then

$$M_o = \frac{0.4\ \mathrm{T}}{\mu_o} = \frac{0.4\ \mathrm{T}}{4\pi \times 10^{-7}\ \mathrm{H/m}} = 3.18 \times 10^5\ \mathrm{A/m} \quad (13.3.18)$$

Then from Eq. 13.3.8

$$\omega_o = |\gamma|\mu_o H_o = 188.5 \times 10^9\ \mathrm{Hz/T} \times 0.3\ \mathrm{T}$$

$$= 5.66 \times 10^{10}\ \mathrm{rad/s} \quad (13.3.19)$$

From Eq. 13.3.9

$$\omega_M = |\gamma|\mu_o M_o = 188.5 \times 10^9 \text{ Hz/T} \times 0.4 \text{ A/m}$$
$$= 7.54 \times 10^{10} \text{ rad/s} \quad (13.3.20)$$

If this is a low loss ferrite and $\alpha \approx 0$, then from Eq. 13.3.6,

$$X_{xx} = X_{yy} = \frac{\omega_o \omega_M}{\omega_o^2 - \omega^2} \quad (13.3.21)$$

From Eq. 13.3.7,

$$X_{yx} = -X_{xy} = \frac{j\omega\omega_M}{\omega_o^2 - \omega^2} \quad (13.3.22)$$

If the operating microwave frequency is $f = 8.5$ GHz-8.5×10^9 Hz, then

$$\omega = 2\pi f = 2\pi \times 8.5 \times 10^9 \text{ Hz} = 53.4 \times 10^9 \text{ rad/s} \quad (13.3.23)$$

Therefore,

$$X_{xx} = X_{yy} = \frac{5.66 \times 10^{10} \text{ rad/s} \times 7.54 \times 10^{10} \text{ rad/s}}{(5.66 \times 10^{10} \text{ rad/s})^2 - (5.34 \times 10^{10} \text{ rad/s})^2}$$
$$= 12.1 \quad (13.3.24)$$

Then

$$X_{yx} = -X_{xy} = j\frac{5.34 \times 10^{10} \text{ rad/s} \times 7.54 \times 10^{10} \text{ rad/s}}{(5.66 \times 10^{10} \text{ rad/s})^2 - (5.34 \times 10^{10} \text{ rad/s})^2}$$
$$= j11.4 \quad (13.3.25)$$

The susceptibility tensor is

$$\overleftrightarrow{\mathbf{X}} = \begin{pmatrix} 12.1 & -j11.4 & 0 \\ j11.4 & 12.1 & 0 \\ 0 & 0 & 0 \end{pmatrix} \quad (13.3.26)$$

and the permeability tensor is

$$\overleftrightarrow{\mu} = 4\pi \times 10^{-7} \begin{pmatrix} 13.1 & -j11.4 & 0 \\ j11.4 & 13.1 & 0 \\ 0 & 0 & 1 \end{pmatrix} \quad (13.3.27)$$

If $\mu_o H_o = 0.3$ T, then

$$H_o = \frac{0.3\ \text{T}}{4\pi \times 10^{-7}\ \text{H/m}} = 2.39 \times 10^5\ \text{A/m} \quad (13.3.28)$$

This is much greater than $h = 2.24$ A/m.

In this case, $\mu_o M_o = 0.4$ T, then

$$M_o = \frac{0.4\ \text{T}}{4\pi \times 10^{-7}\ \text{H/m}} = 3.18 \times 10^5\ \text{A/m} \quad (13.3.29)$$

This is compared to

$$m_x = X_{xx} h_{x_{max}} = 12.1 \times 2.24\ \text{A/m}$$
$$= 27.1\ \text{A/m} \quad (13.3.30)$$

This is much smaller than $M_o = 3.18 \times 10^5$ A/m. Thus the conditions of Eq. 13.3.10 are met.

Since the dc magnetic flux density B_o is

$$B_o = \mu_o(H_o + M_o) = \mu_o H + \mu_o M_o$$
$$= 0.3\ \text{T} + 0.4\ \text{T} = 0.7\ \text{T} \quad (13.3.31)$$

If the dc permeability in the direction of $\mathbf{H}_o$ is μ_{dc}, then

$$B_o = \mu_{dc} H_o \quad (13.3.32)$$

$$\mu_{dc} = \frac{B_o}{H_o} = \frac{0.7\ \text{T}}{2.39 \times 10^5\ \text{A/m}} = 2.93 \times 10^{-6}\ \text{H/m} \quad (13.3.33)$$

The relative permeability of this ferrite in the direction of $\mathbf{H}_o$ is

$$\mu_r = \frac{\mu_{dc}}{\mu_o} = \frac{2.93 \times 10^{-6}\ \text{H/m}}{4\pi \times 10^{-7}\ \text{H/m}} = 2.33 \quad (13.3.34)$$

13.4 Faraday Rotation

When linearly polarized electromagnetic waves are propagating in a dc magnetically biased ferrite in the direction of the biasing dc magnetic field, the direction of the polarization rotates as the waves propagate. The direction of rotation is the same regardless of the direction of propagation of the waves with respect to the direction of applied dc magnetic bias field. The angle of rotation of the polarization is determined by the ferrite material, the amount of dc magnetic bias field, and the operating frequency. On the other hand, the rotation angle is independent of the propagation direction. This phenomenon was originally discovered by Michael Faraday on linearly polarized light propagating along an axially magnetized glass rod. This phenomenon of rotation of polarization due to an axially magnetized medium is called Faraday rotation. Faraday rotation can be explained using the permeability tensor which is represented by Eq. 13.3.15. For simplicity of explanation, if the ferrite considered is lossless, then the damping parameter in Eq. 13.3.6 and 13.3.7 is zero,

$$\alpha = 0 \tag{13.4.1}$$

$$X_{xx} = X_{yy} = \frac{\omega_o \omega_M}{\omega_o^2 - \omega^2} \tag{13.4.2}$$

$$X_{yx} = -X_{xy} = j\frac{\omega \omega_M}{\omega_o^2 - \omega^2} \tag{13.4.3}$$

As seen from Eq. 13.4.2 and Eq. 13.3.15, the diagonal permeability is a real quantity and, as seen from Eq. 13.4.3 and Eq. 13.3.15, the nonzero off-diagonal permeability is an imaginary quantity. By custom, the diagonal permeability is represented by μ and the off-diagonal permeability is represented by jk. For the lossless ferrites,

$$\mu \equiv \mu_o(1 + X_{xx}) = \mu_o\left(1 + \frac{\omega_o \omega_M}{\omega_o^2 - \omega^2}\right) \tag{13.4.4}$$

$$jk \equiv \mu_o X_{xy} = j\mu_o \frac{\omega \omega_M}{\omega_o^2 - \omega^2} \tag{13.4.5}$$

$$\overleftrightarrow{\mu} = \begin{pmatrix} \mu & -jk & 0 \\ jk & \mu & 0 \\ 0 & 0 & \mu_o \end{pmatrix} \tag{13.4.6}$$

This assumes the dc magnetic field is applied in the z-direction as illustrated in Fig. 13.1. If the transverse magnetic (TM) mode of microwaves is fed along the z-direction parallel to the dc magnetic field, the microwave field lacks the z-component by definition of the TM mode. Then only b_x, b_y, h_x, and h_y exist. Using Eq. 13.4.6,

$$\begin{pmatrix} b_x \\ b_y \\ 0 \end{pmatrix} = \begin{pmatrix} \mu & -jk & 0 \\ jk & \mu & 0 \\ 0 & 0 & \mu_o \end{pmatrix} \begin{pmatrix} h_x \\ h_y \\ 0 \end{pmatrix} \tag{13.4.7}$$

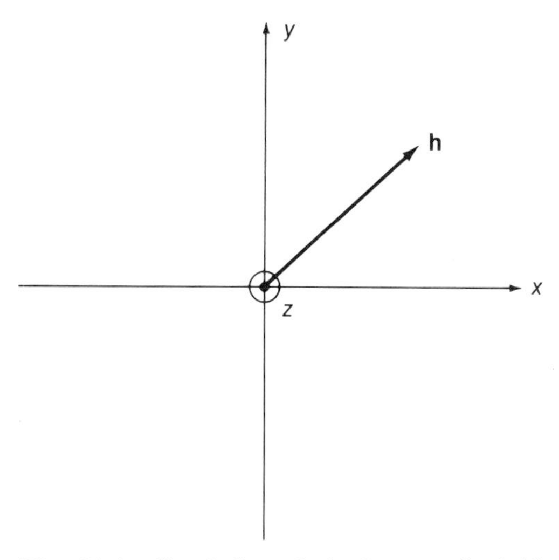

Fig. 13.1 Circularly polarized magnetic field rotating in the counterclockwise direction.

In Fig. 13.1, if **h** is a circularly polarized microwave frequency magnetic field rotating in the mathematically positive direction (counterclockwise direction), then h_x is 90° ahead of h_y. Therefore

$$h_x = jh_y \tag{13.4.8}$$

From Eq. 13.4.7,

$$b_x = \mu h_x - jkh_y \tag{13.4.9}$$

Substituting Eq. 13.4.8 in Eq. 13.4.9,

$$b_x = \mu h_x - jk\frac{h_x}{j} = (\mu - k)h_x \tag{13.4.10}$$

Further, from Eq. 13.4.7

$$b_y = jkh_x + \mu h_y = jk(jh_y) + \mu h_y = (\mu - k)h_y \tag{13.4.11}$$

The effective permeability of the ferrite for the positive circular polarization is therefore

$$\mu_{\text{eff}}^{+} \equiv \mu - k \tag{13.4.12}$$

For negative circular polarization, or when the magnetic field vector **b** is rotating in a clockwise direction or in the mathematically negative direction, h_y is 90° ahead of h_x as seen from Fig. 13.1.

$$h_y = jh_x \tag{13.4.13}$$

From Eq. 13.4.7,

$$b_x = \mu h_x - jkh_y = \mu h_x - jk(jh_x) = (\mu + k)h_x \tag{13.4.14}$$

$$b_y = jkh_x + \mu h_y = jk\frac{h_y}{j} + \mu h_y = (\mu + k)h_y \tag{13.4.15}$$

For negative circular polarization, the effective permeability is therefore

$$\mu_{\text{eff}}^{-} \equiv \mu + k \tag{13.4.16}$$

It is interesting to note that the same ferrite presents different permeabilities depending on the polarization. For waves of positive circular polarization, the permeability of the ferrite is μ_{eff}^{+}. For waves of negative circular polarization, the permeability of the same ferrite is μ_{eff}^{-}.

A trigonometric relation

$$\cos \omega t = \tfrac{1}{2}(e^{j\omega t} + e^{-j\omega t}) \tag{13.4.17}$$

states that a linear polarization $\cos \omega t$ consists of a positive circular polarization $\frac{1}{2}e^{j\omega t}$ and a negative circular polarization $\frac{1}{2}e^{-j\omega t}$. When a linearly polarized magnetic field $h \cos \omega t$ is propagating in the direction of the dc magnetic field in the ferrites, the linearly polarized magnetic field consists of a positive circularly polarized magnetic field $\frac{1}{2}he^{j\omega t}$ and a negative circularly polarized magnetic field $\frac{1}{2}he^{-j\omega t}$. The same ferrite presents different effective permeabilities to different circular polarizations. For $\frac{1}{2}he^{j\omega t}$, μ_{eff}^{+} and for $\frac{1}{2}he^{-j\omega t}$, μ_{eff}^{-}. If the phase constant of the positive circular polarization waves is β^{+} and the phase constant of the negative circular polarization waves is β^{-}, respectively, the propagating circularly polarized waves in the ferrite can be represented by the positive polarization based on the definition of the phase constant [4].

$$F^{+} \equiv \tfrac{1}{2}he^{j\omega t}e^{-j\beta_z^{+}} \tag{13.4.18}$$

The negative polarization is given by

$$F^{-} \equiv \tfrac{1}{2}he^{-j\omega t}e^{-j\beta_z^{-}} \tag{13.4.19}$$

All electromagnetic waves must follow the following differential equation (called the wave equation) which is derived from Maxwell's equations. For a lossless medium and uniform plane waves

$$\frac{\partial^2 F}{\partial z^2} + \omega^2 \varepsilon \mu_{\text{eff}} F = 0 \tag{13.4.20}$$

where ε is the permittivity of the ferrite [4].

For the positive circular polarization,

$$\frac{\partial^2 F^{+}}{\partial z^2} + \omega^2 \varepsilon \mu_{\text{eff}}^{+} F^{+} = 0 \tag{13.4.21}$$

For the negative circular polarization,

$$\frac{\partial^2 F^{-}}{\partial z^2} + \omega^2 \varepsilon \mu_{\text{eff}}^{-} F^{-} = 0 \tag{13.4.22}$$

Substituting Eq. 13.4.18 in Eq. 13.4.21,

$$-(\beta^{+})^2 F^{+} + \omega^2 \varepsilon \mu_{\text{eff}}^{+} F^{+} = 0$$

or

$$\beta^{+} = \pm \omega \sqrt{\varepsilon \mu_{\text{eff}}^{+}} \tag{13.4.23}$$

The positive sign is for the waves propagating in the z-direction.

Substituting Eq. 13.4.19 in Eq. 13.4.22,

$$\beta^{-} = \pm \omega \sqrt{\varepsilon \mu_{\text{eff}}^{-}} \tag{13.4.24}$$

The phase constants β^{+}, and β^{-} are by definition the phase shift per unit distance of propagation [4]. If the distance of propagation in the magnetized ferrite along the magnetizing dc magnetic field is L, then the phase shift of the waves is, for the positive circular polarization,

$$\theta^{+} = \beta^{+} L = \omega \sqrt{\varepsilon \mu_{\text{eff}}^{+}}\, L \tag{13.4.25}$$

and for negative circular polarization,

$$\theta^{-} = \beta^{-} L = \omega \sqrt{\varepsilon \mu_{\text{eff}}^{-}}\, L \tag{13.4.26}$$

As seen from Eqs. 13.4.18 and 13.4.19, if the linearly polarized

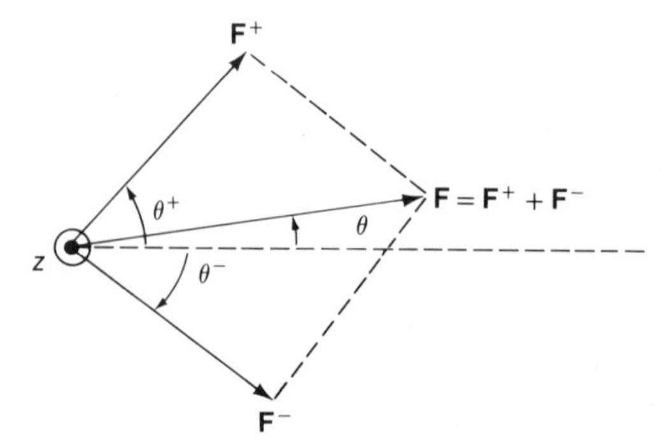

Fig. 13.2 Faraday rotation of a polarized field.

transverse magnetic field **h** enters the ferrite at $z = 0$,

$$\theta^{+}|_{z=0} = \theta^{-}|_{z=0} = 0 \tag{13.4.27}$$

This gives a reference to the amount of the phase shift. After propagating the distance L, the phase angles shift as shown in Eqs. 13.4.25 and 13.4.26, respectively. Since the field vectors $\mathbf{F}^{+}$ and $\mathbf{F}^{-}$ are rotating in space, the reference phase angle shifts θ^{+} at the distance L in space for the positive circular polarization and the reference phase angle shifts θ^{-} at the distance L in space for the negative circular polarization. The resultant at the distance L is then

$$\mathbf{F}|_{z=L} = \mathbf{F}^{+}|_{z=L} + \mathbf{F}^{-}|_{z=L} \tag{13.4.28}$$

The vector relationship is illustrated in Fig. 13.2 In this figure, the angle θ is the Faraday rotation angle for the propagation distance L. By inspection of Fig. 13.2,

$$\theta^{+} - \theta = \frac{\theta^{+} + \theta^{-}}{2} \tag{13.4.29}$$

Then,

$$\theta = \frac{\theta^{+} - \theta^{-}}{2} \tag{13.4.30}$$

Substituting Eqs. 13.4.25 and 13.4.26 in Eq. 13.4.30,

$$\theta = \frac{\omega L\sqrt{\varepsilon}}{2}\left(\sqrt{\mu^{+}_{\text{eff}}} - \sqrt{\mu^{-}_{\text{eff}}}\right) \tag{13.4.31}$$

Substituting Eqs. 13.4.12 and 13.4.16 in Eq. 13.4.31,

$$\theta = \frac{\omega L\sqrt{\varepsilon}}{2}\left(\sqrt{\mu - \text{k}} - \sqrt{\mu + \text{k}}\right) \tag{13.4.32}$$

Substituting Eqs. 13.4.4 and 13.4.5 in Eq. 13.4.32,

$$\theta = \frac{\omega L\sqrt{\varepsilon\mu_o}}{2}\left(\sqrt{1 + \frac{\omega_M(\omega_o - \omega)}{\omega_o^2 - \omega^2}} - \sqrt{1 + \frac{\omega_M(\omega_o + \omega)}{\omega_o^2 - \omega^2}}\right) \tag{13.4.33}$$

Substituting Eqs. 13.3.8 and 13.3.9 in Eq. 13.4.33,

$$\theta = \frac{\omega L\sqrt{\varepsilon\mu_o}}{2}\left(\sqrt{1 + \frac{|\gamma|M_o(|\gamma|H_o - \omega)}{(\gamma H_o)^2 - \omega^2}} - \sqrt{1 + \frac{|\gamma|M_o(|\gamma|H_o + \omega)}{(\gamma H_o)^2 - \omega^2}}\right) \tag{13.4.34}$$

For a given geometry L and material ε, γ, the angle of Faraday rotation varies significantly with the applied dc magnetic field $\mathbf{H}_o$ and the operating frequency ω_o. By selecting the material, operating frequency sample length, and the dc biasing magnetic field, a wide range of Faraday rotation angles can be obtained.

In the sample ferrite presented in Section 13.3, from Eq. 13.3.27 and Eq. 13.4.12,

$$\begin{aligned}\mu_{\text{eff}}^{+} &= (4\pi \times 10^{-7}\ \text{H/m})(13.1 - 11.4) \\ &= 2.14 \times 10^{-6}\ \text{H/m}\end{aligned} \tag{13.4.35}$$

From Eq. 13.3.27 and Eq. 13.4.16,

$$\begin{aligned}\mu_{\text{eff}}^{-} &= (4\pi \times 10^{-7}\ \text{H/m})(13.1 + 11.4) \\ &= 30.79 \times 10^{-6}\ \text{H/m}\end{aligned} \tag{13.4.36}$$

Since $\mathbf{H}_o$ and $\boldsymbol{\omega}_o$ are both parallel to the z-direction, the rotational direction of ω_o or precession is equal to the positive rotation of $\mathbf{h}$. As seen from 13.4.35 and 13.4.36, the ferrite shows low permeability for the positive rotation and high permeability for the negative rotation in the coordinate system shown in Fig. 13.1.

The permittivity of a ferrite $\varepsilon = \varepsilon_o\varepsilon_r$ varies widely depending on the composition of the ferrite. Some ferrites are metallic and some are dielectric. So if

$$\begin{aligned}\varepsilon &= 2.5\varepsilon_o = 2.5 \times 8.854 \times 10^{-12}\ \text{F/m} \\ &= 22.1 \times 10^{-12}\ \text{F/m}\end{aligned} \tag{13.4.37}$$

For example, for $f = 8$ GHz, and $L = 2$ cm, from Eq. 13.4.25,

$$\theta^{+} = 2\pi \times 8 \times 10^{9}\ \text{Hz}\ \sqrt{22.1 \times 10^{-12}\ \text{F/m} \times 2.14 \times 10^{-6}\ \text{H/m}} \times 2 \times 10^{-2}\ \text{m} = 6.91\ \text{rad} \qquad (13.4.38)$$

From Eq. 13.4.26,

$$\theta^{-} = 2\pi \times 8 \times 10^{9}\ \text{Hz}\ \sqrt{22.1 \times 10^{-12}\ \text{F/m} \times 30.79 \times 10^{-6}\ \text{H/m}} \times 2 \times 10^{-2}\ \text{m} = 26.23\ \text{rad} \qquad (13.4.39)$$

Then the angle of rotation of the total magnetic field **h**, is from Eq. 13.4.30,

$$\theta = \frac{\theta^{+} - \theta^{-}}{2} = \frac{6.91\ \text{rad} - 26.23\ \text{rad}}{2} = -9.66\ \text{rad} = -(2\pi + 3.38)\ \text{rad} \qquad (13.4.40)$$

This means that vector **h** rotates clockwise one turn plus 3.38 radians for 2 cm propagation in the z-direction. If the propagation direction is reversed, keeping all other directions the same, then the direction of rotation remains the same. The direction of propagation does not change or determine the direction of rotation of a vector **h**. It always rotates clockwise in the geometry of Fig. 13.1. If the rotation direction must be reversed, the only way to do it is to reverse the direction of $\mathbf{H}_o$. That is, if $\mathbf{H}_o$ is in the $-z$-direction, vector **h** rotates counterclockwise. The angle of 3.38 radians is 193.66 degrees.

13.5 Faraday Rotation Type Isolators

A microwave circuit component which permits microwave transmission in one direction and absorbs the microwave power for reverse transmission is called an isolator. Isolators are used for stabilizing transmitters and oscillators against load variations. Some isolators are built based on Faraday rotation.

Equation 13.4.34, is derived assuming that the microwaves are propagating in the positive z-direction is seen from Eqs. 13.4.23 and 13.4.24. Equations 13.4.8 and 13.4.13 hold regardless of the direction of propagation, as do Eqs. 13.4.12 and 13.4.16. These equations are basically rotation dependent but not propagation direction dependent. For reverse propagation, both Eqs. 13.4.25 and 13.4.26 hold. Consequently, for reverse propagation, the Faraday rotation angle θ is the same as for forward propagation and θ is given by Eq. 13.4.34.

An important Faraday rotation type isolator is the 45° rotator. The material, geometry, applied dc magnetic bias, and the operating frequency range are carefully chosen to make the angle of Faraday rotation $\theta = 45°$ according to Eq. 13.4.34. If $\theta = 45°$ for propagation in the positive z-direction, it should also be $\theta = 45°$ for reverse propagation. This situation is illustrated in Fig. 13.3. The situation of the microwave electromagnetic field for forward propagation is shown in Fig. 13.3 A. The input and the situation for reverse propagation are shown in Fig. 13.3 B. Microwaves are fed so that their electric field vector **e** is perpendicular to the resistive film to avoid dissipation in the film. Maxwell's equation and Poynting's theory require that the vector **e** should be perpendicular to the vector **h** and the direction of propagation. The waves pass through the resistive film with small loss and encounter the ferrite rod which produces a 45° Faraday rotation. The waves pass through the ferrite rod with little loss. For reverse propagation, the microwave fields are fed with their electric field vector **e** in the same direction as cited before. According to the Poynting theory, the direction of the **e** vector, **h** vector, and direction of propagation follow the right-hand rule. Passing through the 45° ferrite Faraday rotator, the magnetic vector is now perpendicular to the resistive film and the electric field vector **e** is parallel to the resistive film. The microwave energy is therefore dissipated in the resistive film and does not exit. The forward insertion loss of an isolator is therefore mainly due to the loss in the ferrite and the reverse insertion loss is mainly due to the resistive film. Thus the reverse propagation experiences a large amount of loss. This is the principle of the Faraday rotation type isolator. Isolators are used for reverse isolation between the oscillator and the load as well as interstage reverse isolation for multistage microwave amplifiers.

13.6 Faraday Rotation Type Circulators

Assuming that there is an n-port junction, where n is an integer, and p is an integer less than n, if the microwave power is fed to the $(p - 1)^{th}$ port, the output appears only at p^{th} port. When the microwave power is fed at the nth port, the output appears only at the first port, the n-port junction functions as a circulator.

An example of a 3-port circulator is illustrated in Fig. 13.4. A microwave field is fed to Port-1 of a rectangular waveguide with the electric field vector as illustrated in Fig. 13.4. The microwave

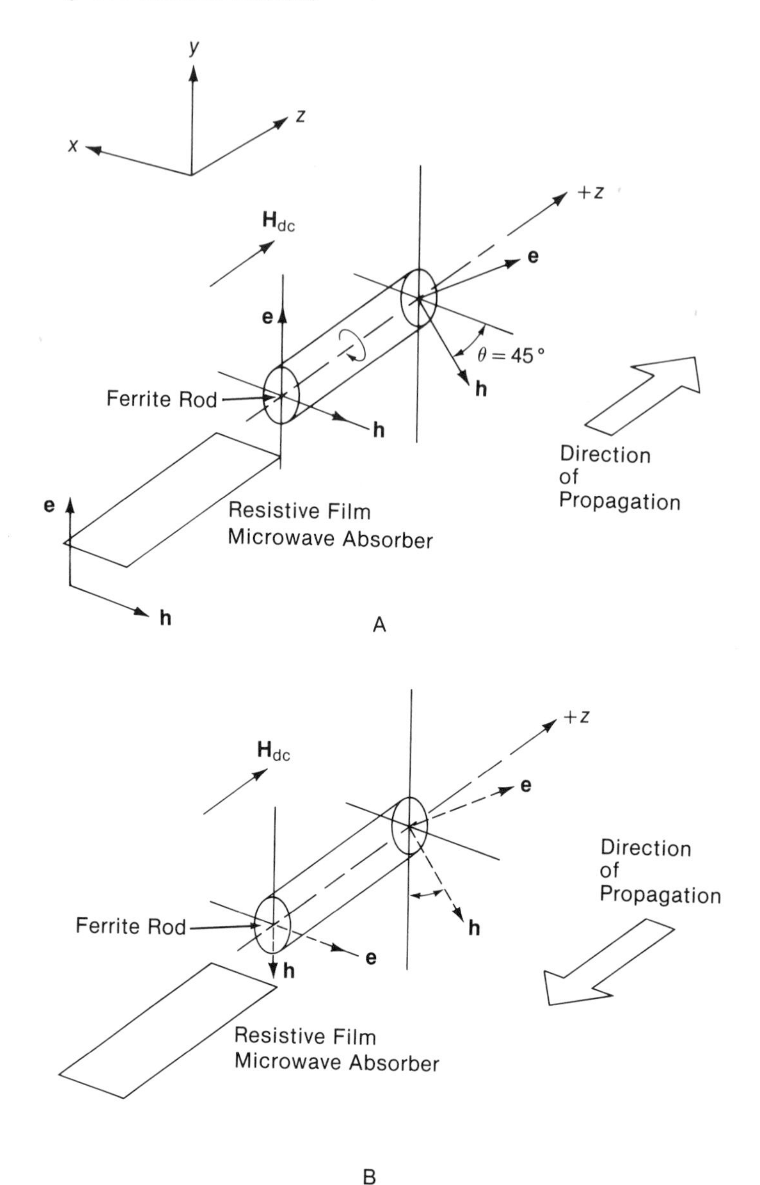

Fig. 13.3 Faraday rotation type isolator. A. Forward transmission. B. Reverse transmission.

field exits from Port-2 through a 45° Faraday rotator. The microwave field does not exit from Port-3 because Port-3 is the cutoff to the E-mode (Appendix 5). The *E*-mode exists when the electric field vector is parallel to the waveguide axis. When microwaves are fed into Port-2 with the electric field vector as illustrated, the output appears only at Port-3 through the 45° Faraday

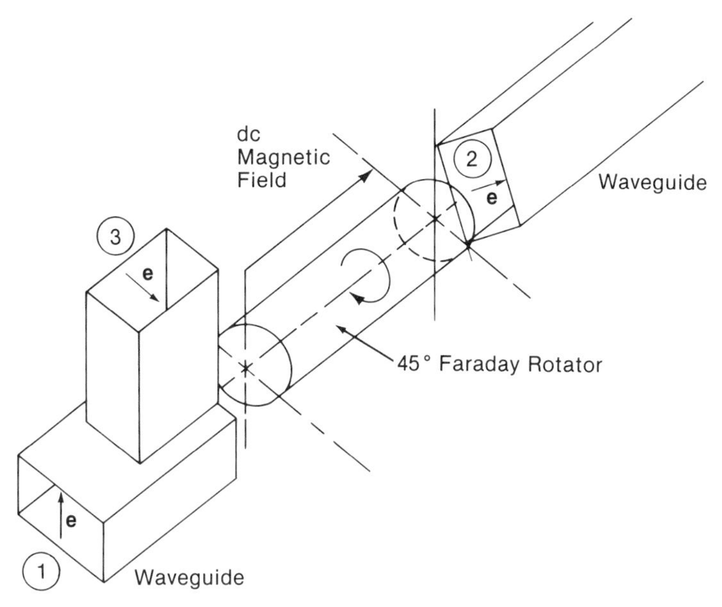

Fig. 13.4 Principle of Faraday rotation type three-port circulator.

rotator. The output will not appear at Port-1 due to the horizontal field cutoff (Appendix 5). When the microwaves are fed into Port-3 with the electric field vector as illustrated, the output appears only at Port-1 after traveling through the 45° Faraday rotator twice. The waveguide of Port-1 will not accept the horizontal field. The horizontal fields go to the Faraday rotator and appear at Port-2 with 45° orientation. The 45° oriented electric field vector is horizontal to the tilted waveguide at Port-2 and the field is rejected. The rejected field reflects back through the 45° Faraday rotator and arrives as a vertical field at the waveguide of Port-1. The vertical field can come out of Port-1 only. It cannot come out of Port-3 due to the *E*-mode cutoff. The *E*-mode is a propagation mode which has a longitudinal *E*-component in the direction of propagation. This is the principle of the Faraday rotation type circulator. The circulators are used for microwave multiplexing and interstage reverse direction isolation for cascade microwave amplifiers.

13.7 Faraday Rotation Type Switches

A Faraday rotation type distributor switch is illustrated in Fig. 13.5. The orientation of the input electric field vector can be rotated in desired different directions by properly adjusting the dc

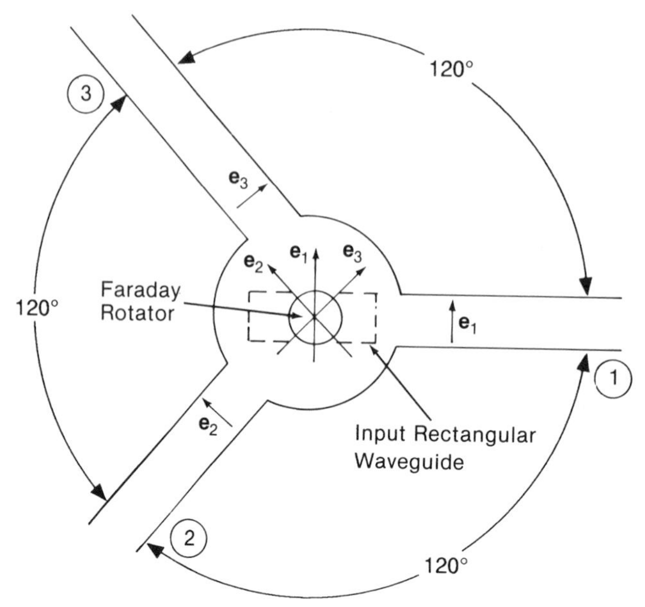

Fig. 13.5 Distributor switch.

bias magnetic field by an electromagnet. When the electric field vector is oriented in the direction of $\mathbf{e}_1$, the input waveguide is coupled mainly with the output waveguide-1. Faraday rotation type shut-off switches are schematically illustrated in Fig. 13.6. The dc magnetic bias is supplied by an external electromagnet. When there is no dc magnetic bias, microwaves transmit. When the dc magnetic bias is on, the rotator is adjusted to be a 90° rotator. The electric field vector then rotates to the horizontal orientation. For the case of a shutter switch, as in Fig. 13.6 A, the waveguide is cut off to the horizontal field, so the waves reflect back instead of transmitting. In the case of the reflectionless shut-off switch as illustrated in Fig. 13.6 B, the horizontally oriented electric fields are dissipated in the resistive film and do not reach the output waveguide. In this scheme, there is no reflection in the input waveguide if the switch is well designed and fabricated.

13.8 Field Displacement

The microwave magnetic field pattern in a rectangular waveguide is illustrated in Fig. 13.7 A [4]. Careful inspection of the microwave magnetic field pattern reveals that, for forward propagation as shown in Fig. 13.7 A, there is a clockwise rotating magnetic

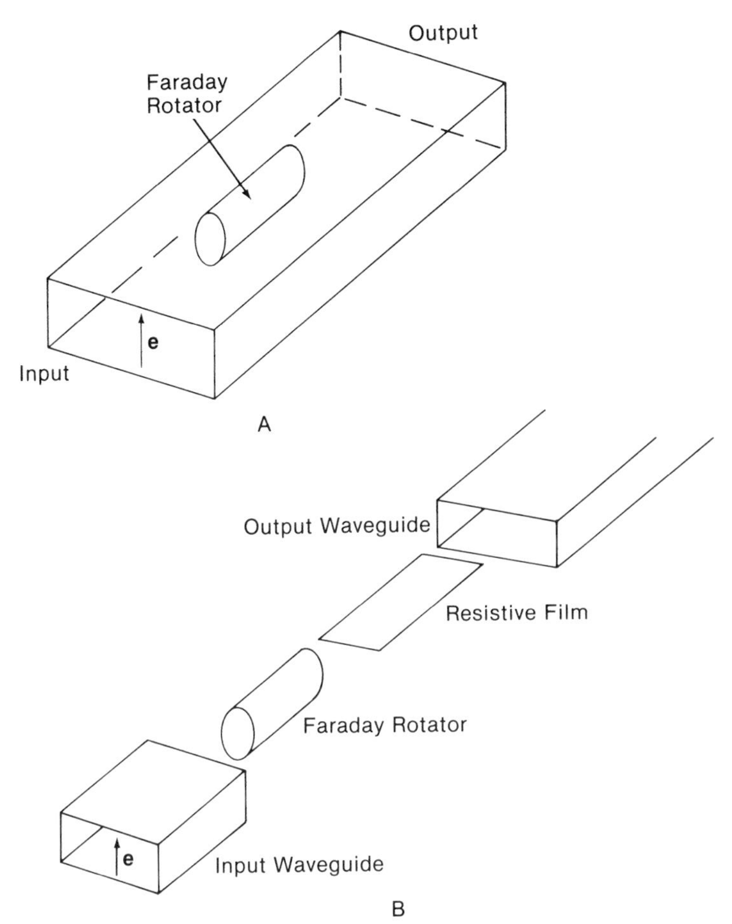

Fig. 13.6 Shut-off switches. A. Shutter switch. B. Reflectionless shut-off switch.

field at the left-hand side of the waveguide. For reverse propagation, there is a counterclockwise rotating magnetic field at the same location. If a magnetically biased ferrite slab is placed in the waveguide, as shown in Fig. 13.7 B and C, the ferrite slab presents μ^+_{eff} to the forward propagation and the same ferrite slab presents μ^-_{eff} to the reverse propagation. As seen from Eqs. 13.4.12, 13.4.16, 13.4.4, and 13.4.5,

$$\mu^-_{\text{eff}} = \mu_o\left(1 + \frac{|\gamma|M_o}{|\gamma|H_o - \omega}\right) \tag{13.8.1}$$

$$\mu^+_{\text{eff}} = \mu_o\left(1 + \frac{|\gamma|M_o}{|\gamma|H_o + \omega}\right) \tag{13.8.2}$$

If the operating frequency ω is less than the gyromagnetic resonance frequency $|\gamma|H_o \equiv \omega_o$, then

$$\mu_{eff}^{-} > \mu_{eff}^{+} \tag{13.8.3}$$

This is the reason that the magnetic flux is attracted by the ferrite slab for reverse propagation and the magnetic flux is not attracted as much by the ferrite slab for forward propagation, as illustrated in Fig. 13.7 B and C, respectively. As seen from these figures, the microwave field pattern of a magnetically biased ferrite loaded waveguide changes depending on the direction of propagation. This phenomenon is called field displacement.

13.9 Field Displacement Type Isolators

As seen from Eqs. 13.8.1 and 13.8.2, the effect of the field displacement is great if operated near $\omega = \omega_o = |\gamma|H_o$. Microwaves are attracted to the ferrite slab in reverse propagation. If the ferrite is a lossy material, it can produce appreciable reverse attenuation while maintaining a reasonably small forward loss. This type of isolator is called the resonance isolator. The resonance isolator is considered to be a special type of isolator based on the field displacement principle, and it is appropriately called the field displacement type isolator. The resonance isolator is a field displacement type isolator operated at or near the gyromagnetic resonance frequency.

If a low loss ferrite slab is employed or operated far below the resonance frequency to gain a wide bandwidth, the high reverse attenuation is not obtainable by the ferrite slab alone. A piece of resistive film is usually attached to the ferrite slab to absorb microwave fields attracted by the ferrite slab and thus produce sufficient reverse attenuation for practical applications. Without the resistive film, the ferrite loaded waveguide is basically a nonreciprocal phase shifter. The nonreciprocal phase shifter is called a gyrator if the forward phase shift is $2n\pi$ and the reverse is $(2n + 1)\pi$, where n is zero or an integer.

13.10 Field Displacement Type Circulators

If a magnetically biased piece of ferrite is placed at a waveguide junction as shown in Fig. 13.8, this is a circulator. When the dc magnetic field is in the direction which is out of the paper, the left-hand side looking into the direction of propagation is always

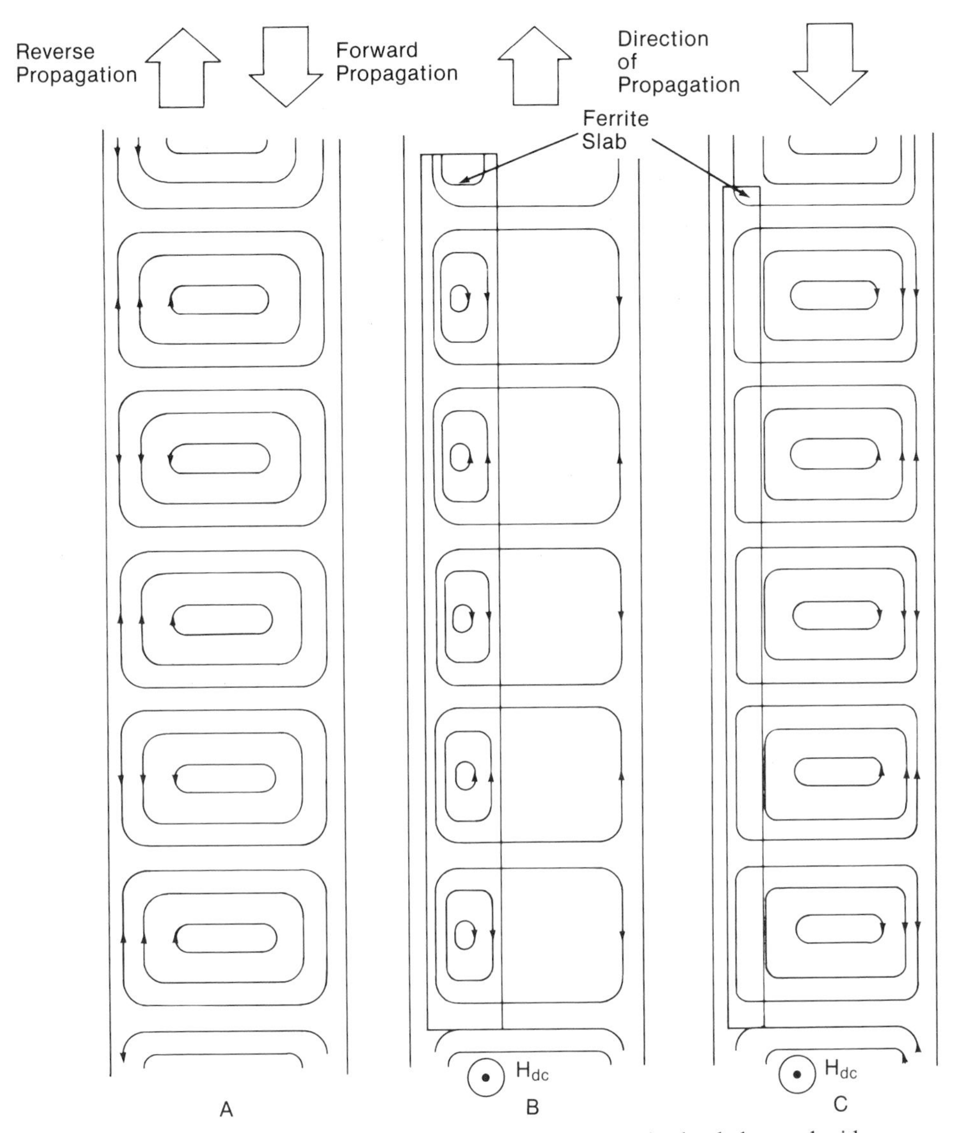

Fig. 13.7 Field displacement. A. Empty waveguide. B. Ferrite loaded wavedguide, reverse propagation. C. Ferrite loaded waveguide, forward propagation.

μ^{+}_{eff} and the right-hand side is μ^{-}_{eff}, and $\mu^{+}_{\text{eff}} < \mu^{-}_{\text{eff}}$. The waves usually couple toward the right-hand side waveguide as shown in Fig. 13.8. Thus the microwaves fed into Port-1 couple to Port-2; the microwaves fed into Port-2 couple to Port-3; and the microwaves fed into Port-3 couple to Port-1.

Similar explanations can be presented for the case of a microstripline circulator shown in Fig. 13.9. A microstripline junction is placed across the magnetically biased ferrite disk. The direction of

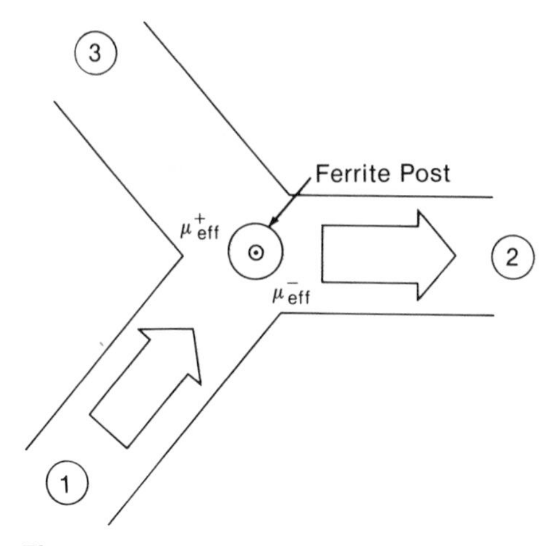

Fig. 13.8 Waveguide junction circulator.

a dc magnetic field is in the direction out of the paper. If microwaves are fed into Branch-1, the waves tend to couple to both Branches-2 and -3. Observe carefully the time sequence of the microwave magnetic field vector **h**. When microwaves couple from Branch-1 to Branch-2, the **h** vector rotates clockwise. When microwaves couple from Branch-1 to Branch-3, the **h** vector rotates counterclockwise. At the left-hand corner of the junction, the ferrite disk presents μ^{+}_{eff} and at the same time, at the right-hand corner of the junction, the same ferrite disk presents μ^{-}_{eff}. As studied before, if operated under the condition $\mu^{+}_{eff} < \mu^{-}_{eff}$, the

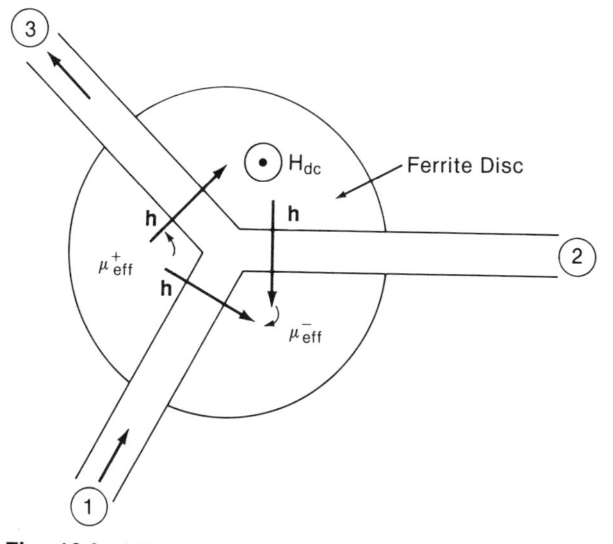

Fig. 13.9 Microstripline circulator.

microwave field is attracted to Branch-2 rather than Branch-3. So the majority of microwaves couple to Branch-2 rather than to Branch-3. Microwaves usually couple to the right-hand side branches when looking into the direction of propagation if the ferrite disk is dc magnetically biased in the direction out of the paper. So the microwaves fed into Branch-1 couple to Branch-2; the microwaves fed into Branch-2 couple to Branch-3; and the microwaves fed into Branch-3 couple to Branch-1. These circulators are biased on the principle of field displacement. They are therefore called field displacement type circulators.

13.11 Field Displacement Type Switches

In field displacement type circulators, the direction of the dc magnetic bias on a piece of ferrite determines the direction of coupling. If the magnetic bias is given by an external electromagnetic, by simply reversing the excitation current of the electromagnet, the coupling is switched from one waveguide to another. This explains the principle of a field displacement type microwave switch.

13.12 Ferrimagnetic Electron Devices

Isolators, circulators, gyrators, ferrimagnetic phase shifters, ferrimagnetic attenuators, and ferrimagnetic switches are all microwave ferrimagnetic devices [5–8]. Their function is based on either Faraday rotation principles or field displacement principles. Both Faraday rotation principles and field displacement principles stem from precessing and spinning unpaired bound electrons of Fe^{3+} atoms under a dc magnetic field [8]. The directional properties of these devices are determined by the direction of dc magnetic bias field. By controlling the direction of dc magnetic bias field, the directional properties of the ferrimagnetic electron device are controlled.

For devices such as ferrite isolators, circulators, and switches, the descriptive parameters are isolation, insertion loss, power handling capacity, VSWR, operating frequency, and the frequency bandwidth. These characterizing parameters are defined in ways similar to those described in Chapter 11 for microwave switching diodes, and will therefore not be repeated here. In commercial devices, isolation more than 15 dB, VSWR less than 1.5, and insertion loss less than 1.2 dB are often considered acceptable. Some devices have wide bandwidths over many GHz yet the isolation is not as high. Some have a narrow frequency bandwidth and yet have high isolation and low insertion loss. Circulators and isolators are used for reverse isolation, multistage amplifiers, and microwave multiplexing.

Problems

13.1 Define ferrimagnetic materials.

13.2 List various applications for ferrite materials in microwave electronics.

13.3 State the reason that **b** is not parallel to **h**.

13.4 Derive a gyromagnetic equation of magnetization in lossless ferrite.

13.5 Explain the Landau-Lifschitz gyromagnetic equation.

13.6 Point out the cause of ferrimagnetic phenomena.

13.7 Write down an equation of motion for a small magnet in a uniform magnetic field.

13.8 Define the gyromagentic ratio.

13.9 Write down the susceptibility tensor of a magnetically biased ferrite and explain its significance.

13.10 Write down the susceptibility tensor of a magnetically biased ferrite operating at the gyromagentic resonance frequency.

13.11 Explain why the real term of the susceptibility represents the dispersive susceptibility and the imaginary term represents the dissipative susceptibility.

13.12 Write down a permeability tensor matrix for a ferrite with a small dc magnetic bias.

13.13 Define Faraday rotation.

13.14 Explain why the direction of the Faraday rotation does not change regardless of the direction of propagation.

13.15 Explain why a linearly polarized field consists of two circularly polarized fields rotating opposite to each other.

13.16 Explain analytically why a dc magnetically biased ferrite presents different permeabilities to rotating magnetic fields of different rotational directions.

13.17 Write down the wave equation for a magnetized ferrite.

13.18 Derive a formula for the phase constant of microwaves in a magnetized ferrite medium.

13.19 Geometrically, relate the Faraday rotation angle θ to θ^+ and θ^-.

13.20 The angle $\theta = \beta L$ is, in reality, in phasor space. It coincidentally is equal to the rotation angle of **F** in physical space. Explain why this is so.
13.21 List the parameters which determine the angle of Faraday rotation.
13.22 Define an isolator.
13.23 Explain the principle of Faraday rotation type isolators.
13.24 Explain the Poynting theory.
13.25 Explain the principle of microwave dissipation in a resistive film.
13.26 List some possible applications of microwave isolators and state the advantages and objectives of using isolators for each example cited.
13.27 Define a circulator.
13.28 Explain the principle of a Faraday rotation type circulator.
13.29 List some possible applications of microwave circulators.
13.30 Explain how the microwave Faraday rotation type switch works.
13.31 Ascertain that there is a rotating magnetic field inside a rectangular waveguide.
13.32 Explain the reason for the nonreciprocal transmission characteristic of a field displacement type isolator.
13.33 Explain a resonance isolator.
13.34 Explain gyromagnetic resonance and its effect on the characteristics of isolators and circulators.
13.35 Explain gyromagnetic resonance physically.
13.36 Compare the advantages and disadvantages of resonance and nonresonance isolators.
13.37 Explain the principle of a field displacement type circulator.
13.38 Explain the principle of a microstripline circulator.
13.39 Explain the principle of a field displacement type microwave switch.
13.40 In principle, construct an isolator using a circulator.

References

1 B. Lax and K. J. Button, "Microwave Ferrites and Ferrimagnetics." McGraw-Hill, New York, 1962.
2 M. G. Mathew and T. K. Ishii, Microwave properties of magnetically unbiased ferrites. *IEEE Trans. Magn.* **MAG-13**(5), 1230–1235 (1977).
3 L. Landau and E. Lifshitz, On the theory of the dispersion of magnetic permeability in ferromagnetic bodies. *Physikz* (*U.S.S.R.*) **8**, 153 (1935) (in English).
4 T. K. Ishii, "Microwave Engineering." Ronald Press, New York, 1966.

5 D. Zhang and W. Lin, H-plane waveguide circulator design method. *Proc. Int. Conf. Infrared Millimeter Waves* pp. 342–343 (1987).
6 G. F. Dionne, Nonreciprocal 45° Faraday rotator for quasioptical beams at 35 GHz. *IEEE MTT-S Int. Microwave Symp. Dig.* **1**, 127–130 (1988).
7 D. Kother, Frequency dependence of wavedguide modes in premagnetized ferrite near resonance. *IEEE MTT-S Int. Microwave Symp. Dig.* **2**, 761–764 (1988).
8 M. Mrozowski and J. Mazur, Lower bound on the Eigenvalues of the characteristic equations for an arbitrary multilayered gyromagnetic structure with perpendicular magnetization. *IEEE Trans. Microwave Theory Tech.* **MTT-37** (3), 640–643 (1989).

14

Velocity Modulation Devices

14.1 Velocity Modulation Devices and Kinetic Energy Transfer

The process of producing a variation in the velocity of moving electrons by applied microwave electromagnetic fields is called velocity modulation. As seen from the definition, this phenomenon and its underlying principles appear widely in published techniques of microwave electronics for both solid-state and thermionic devices. For simplicity, velocity modulation in thermionic devices will be considered. By slight modification, the same concept applies to solid-state devices.

Suppose that an electron of charge q coulombs and mass m kg is moving with the velocity u m/s in a vacuum and is accelerated by an electric field which is the gradient of the electric potential V volts. By the law of the conservation of energy, the kinetic energy $1/2mu^2$ (J) must be equal to the potential energy qV (J) or

$$\frac{1}{2}mu^2 = qV \qquad \text{(J)} \tag{14.1.1}$$

Then velocity of the electron is

$$u = \sqrt{\frac{2qV}{m}} \qquad \text{(m/s)} \tag{14.1.2}$$

If the electric potential is a combination of a dc acceleration potential V_o and a microwave acceleration potential $v \sin \omega t$ of amplitude v and angular frequency ω, then

$$V = V_o + v \sin \omega t \tag{14.1.3}$$

Substituting Eq. 14.1.3 in Eq. 14.1.2,

$$u = \sqrt{\frac{2q}{m}(V_o + v \sin \omega t)} \tag{14.1.4}$$

The velocity of the electron, u, is modulated by microwave signals (velocity modulation). Eq. 14.1.4 is

$$u = \sqrt{\frac{2qV_o}{m}}\left(1 + \frac{v}{V_o}\sin \omega t\right)^{1/2} \tag{14.1.5}$$

This equation can be expanded by a binomial expansion if

$$\frac{v}{V_o} < 1 \tag{14.1.6}$$

The binomial expansion of Eq. 14.1.5 is

$$u = \sqrt{\frac{2qV_o}{m}}\left[1 + \frac{1}{2}\frac{v}{V_o}\sin \omega t + \frac{1}{8}\left(\frac{v}{V_o}\sin \omega t\right)^2 + \ldots\right] \tag{14.1.7}$$

In most practical microwave devices, both solid-state and thermionic, truncation at the second term is sufficient. Then,

$$u = \sqrt{\frac{2qV_o}{m}}\left(1 + \frac{1}{2}\frac{v}{V_o}\sin \omega t\right) \tag{14.1.8}$$

If the microwave potential is zero, Eq. 14.1.8 degenerates into

$$u_o = \sqrt{\frac{2qV_o}{m}} \tag{14.1.9}$$

This is the unmodulated dc velocity. Substituting Eq. 14.1.9 in Eq. 14.1.8,

$$u = u_o\left(1 + \frac{1}{2}\frac{v}{V_o}\sin \omega t\right) \tag{14.1.10}$$

The velocity modulation term, or the microwave velocity, is then

$$u_m = \frac{u_o}{2}\frac{v}{V_o}\sin \omega t \tag{14.1.11}$$

Note the sign of the microwave velocity. The velocity may be forward or backward depending on the phase angle ωt.

The acceleration of the velocity modulated electron can be obtained by differentiating Eq. 14.1.10.

$$a = \frac{\partial u}{\partial t} = \frac{1}{2}\frac{u_o \omega v}{V_o}\cos \omega t \tag{14.1.12}$$

As seen from this equation, there is no acceleration from the dc potential V_o at this point, if V_o is uniform in space near the point of consideration. It is of interest to note that there is a 90° phase difference between the velocity and the acceleration. The acceleration is more effective at a higher angular frequency ω. Many microwave thermionic devices—such as klystrons, magnetrons, and traveling wave tubes—are made so that V_o is uniform in the space where the electron beam interacts with microwaves. Equation 14.1.12 is therefore applicable in most cases. On the other hand, in microwave solid-state devices such as microwave FETs and active microwave diodes, the dc acceleration and velocity modulation are applied at the same time and space. Then both V_o and v are functions of the position of the interaction. If a one dimensional variation in the x-direction is considered, both V_o and v are functions of x. The velocity u is therefore a function of both x in space and t in time. Then the acceleration is

$$a = \frac{\mathrm{d}u}{\mathrm{d}t} = \frac{\partial u}{\partial x}\frac{\partial x}{\partial t} + \frac{\partial u}{\partial t} = \frac{\partial u}{\partial x} u + \frac{\partial u}{\partial t} \tag{14.1.13}$$

It is of interest to note that the higher the velocity, the more effective the acceleration. From Eq. 14.1.8,

$$u = \sqrt{\frac{2qV_o}{m}} + \frac{\sqrt{q}}{\sqrt{2mV_o}} v \sin \omega t \tag{14.1.14}$$

Substituting Eq. 14.1.14 in Eq. 14.1.13,

$$a = \sqrt{\frac{q}{2mV_o}}\frac{\partial V_o}{\partial x} u + \frac{\sqrt{q}}{\sqrt{2m}}\left(\frac{\partial v}{\partial x} \cdot V_o^{-1/2} - \frac{v}{2} V_o^{-3/2}\right)$$

$$\times u \sin \omega t + \frac{u_o}{2}\frac{\omega v}{V_o}\cos \omega t$$

$$= \sqrt{\frac{q}{2mV_o}}\left[E_o u + \left(E_m - \frac{v}{2V_o}\right) u \sin \omega t\right]$$

$$+ \frac{u_o}{2}\frac{\omega v}{V_o}\cos \omega t \tag{14.1.15}$$

where the dc acceleration field is

$$E_o \equiv \frac{\partial V_o}{\partial x} \tag{14.1.16}$$

and the microwave acceleration field is

$$E_m \equiv \frac{\partial v}{\partial x} \tag{14.1.17}$$

In practical microwave devices, the microwave potential v is produced by an externally applied voltage v_m across a pair of electrodes. The microwave potential v is proportional to the applied voltage v_i. If the proportionality constant is k,

$$v = \mathrm{k} v_i \tag{14.1.18}$$

This proportionality constant k is called the beam coupling coefficient. Substituting Eq. 14.1.18 in Eq. 14.1.10,

$$u = u_o\left(1 + \frac{\mathrm{k} v_i}{2V_o} \sin \omega t\right) \tag{14.1.19}$$

The term $(\mathrm{k} v_i/2V_o)$ is called the depth of velocity modulation.

If the density of the electrons is N (m^{-3}) and the cross section of the electron beam current is S, the electron beam current is then

$$I = qNuS \qquad (\mathrm{A}) \tag{14.1.20}$$

Substituting Eq. 14.1.19 in Eq. 14.1.20,

$$I = qNu_o\left(1 + \frac{\mathrm{k} v_i}{2V_o} \sin \omega t\right) S \tag{14.1.21}$$

The term qNu_oS is the dc beam current and can be expressed in the form of

$$I_o \equiv qNu_oS \tag{14.1.22}$$

Combining both Eqs. 14.1.21 and 14.1.22,

$$I = I_o\left(1 + \frac{\mathrm{k} v_i}{2V_o} \sin \omega t\right) \tag{14.1.23}$$

Now the depth of velocity modulation $(\mathrm{k} v_i/2V_o)$ is the depth of the beam current modulation.

In Eq. 14.1.23, the microwave beam current is

$$I_{\mathrm{m}} = \frac{\mathrm{k} v_{\mathrm{i}} I_{\mathrm{o}}}{2V_{\mathrm{o}}} \sin \omega t \tag{14.1.24}$$

The microwave current in the beam of this amount is associated with a power of

$$P_{\mathrm{m}} = \frac{1}{2} v_{\mathrm{i}} |I_{\mathrm{m}}| = \frac{\mathrm{k} v_{\mathrm{i}}^2 I_{\mathrm{o}}}{4V_{\mathrm{o}}} \quad (\mathrm{W}) \tag{14.1.25}$$

The energy for the velocity modulation can be analyzed from Eq. 14.1.11. The necessary energy for velocity modulation per electron is

$$W = \frac{1}{2} m u_{\mathrm{m}}^2 = \frac{1}{8} m \left(\frac{u_{\mathrm{o}} v}{V_{\mathrm{o}}} \right)^2 \sin^2 \omega t \tag{14.1.26}$$

If the electron beam is uniform and occupies a volume τ (m^3) of interaction space, the average energy necessary to produce the velocity modulation is, for first approximation,

$$\begin{aligned} W_{\mathrm{ave}} &= \frac{1}{8} m \tau N \left(\frac{u_{\mathrm{o}} v}{V_{\mathrm{o}}} \right)^2 \cdot \frac{1}{\pi} \int_0^{\pi} \sin^2 (\omega t) \, \mathrm{d}(\omega t) \\ &= \frac{1}{16} m \tau N \left(\frac{u_{\mathrm{o}} v}{V_{\mathrm{o}}} \right)^2 \quad (\mathrm{J}) \end{aligned} \tag{14.1.27}$$

This much energy must be maintained at the interaction region all the time.

Another aspect of the modulation power requirement can be analyzed from Eq. 14.1.26. The time rate of change of the kinetic energy W is the kinetic power P. Therefore, from Eq. 14.1.26,

$$P = \frac{\partial W}{\partial t} = \frac{1}{4} m \omega \left(\frac{u_{\mathrm{o}} v}{V_{\mathrm{o}}} \right)^2 (\sin \omega t)(\cos \omega t) \tag{14.1.28}$$

The average kinetic power per electron is

$$\begin{aligned} P_{\mathrm{ave}} &= \frac{1}{4} m \omega \left(\frac{u_{\mathrm{o}} v}{V_{\mathrm{o}}} \right)^2 \frac{1}{\Delta\theta} \int_0^{\Delta\theta} (\sin \omega t)(\cos \omega t) \, \mathrm{d}(\omega t) \\ &= \frac{1}{8} m \omega \left(\frac{u_{\mathrm{o}} v}{V_{\mathrm{o}}} \right)^2 \frac{1}{\Delta\theta} \sin^2 \Delta\theta \end{aligned} \tag{14.1.29}$$

where $\Delta\theta$ is the electron transit angle in the interaction region.

The electron transit angle $\Delta\theta$ is given by

$$\Delta\theta \equiv \omega\,\Delta t \tag{14.1.30}$$

where Δt is the electron transit time through the interaction region for velocity modulation.

For example, in practical velocity modulation thermionic devices, the electron acceleration voltage is on the order of 10^3 V. The electron velocity is calculated from Eq. 14.1.2

$$u_o = \sqrt{\frac{2 \times 1.602 \times 10^{-19}\ \text{C} \times 10^3}{9.109 \times 10^{-31}\ \text{kg}}} = 3.52 \times 10^7\ \text{m/s}$$

This is approximately 1/10 the velocity of light in a vacuum. This is a typical value of an electron velocity for a microwave thermionic device.

If $P = 10$ mW microwave power is fed through a coaxial cable of $Z_o = 50\ \Omega$, then the peak voltage across the coaxial cable is $P = v_i^2/2Z_o$ or

$$v_i = \sqrt{2Z_oP} = \sqrt{2 \times 50\ \Omega \times 10 \times 10^{-3}\ \text{W}} = 1\ \text{V}$$

If this microwave voltage is stepped up using a cavity resonator of $Q = 200$, then the available cavity voltage is

$$v = v_i\sqrt{Q} = (1\ \text{V})\sqrt{200} = 14.14\ \text{V}$$

this compares with the dc acceleration voltage $V_o = 1000$ V. The inequality 14.1.6 is therefore verified as

$$\frac{v}{V_o} = \frac{14.14\ \text{V}}{1000\ \text{V}} = 0.01414 < 1$$

According to Eq. 14.1.11, the amplitude of the microwave velocity u_m is

$$u_m = \frac{3.52 \times 10^7\ \text{m/s}}{2} \times 0.01414 = 0.249 \times 10^7\ \text{m/s}$$

This compares with the dc velocity $u_o = 3.52 \times 10^7$ m/s. If the device is operated at frequency $f = 9.5$ GHz, then the amplitude of acceleration a_o is obtained by Eq. 14.1.12 as

$$a_o = \frac{3.52 \times 10^7\ \text{m/s}}{2} \times 2\pi \times 9.5 \times 10^9\ \text{s}^{-1} \times 0.01414$$

$$= 1.49 \times 10^{16}\ \text{m/s}^2$$

This is an impressive amount of acceleration considering $(v/V_o) = 0.01414$ and $v = 14.14$ V.

Most velocity modulated thermionic devices are designed so that where microwave fields exist, there are no dc fields. So, in Eq. 14.1.16, $E_o = 0$. As for E_m in Eq. 14.1.17, this depends on how microwave fields interact with the electron beam. If the microwave voltage of $v = 14.14$ V (peak value) exists across the cavity resonator and is directly applied to the electron beam longitudinally across an interaction gap of $g = 1$ mm, then the microwave field strength at peak value is

$$E_m = \frac{v}{g} = \frac{14.14 \text{ V}}{10^{-3} \text{ m}} = 14.14 \text{ kV/m}$$

This is significant field strength. Longitudinal means that the direction of the electron beam and the direction of the applied microwave electric field are parallel to each other. In practical velocity modulated thermionic devices, the beam coupling coefficient k is 0.75 or higher and never exceeds 1 by definition. If k = 0.8, the depth of velocity modulation is, from Eq. 14.1.19,

$$\frac{\text{k}v}{2V_o} = \frac{0.8}{2} \times 0.01414 = 0.005656$$

It is interesting to note that in spite of the impressively strong E_m and large acceleration, the depth of velocity modulation is very shallow. This is due to short interaction time.

In medium power devices, the dc beam current can be as high as $I = 500$ mA. The beam radius can be as thick as $r = 1$ mm. Then from Eq. 14.1.20, the electron density is

$$N = \frac{I}{qu_o\pi r^2}$$

$$= \frac{500 \times 10^{-3} \text{ A}}{1.602 \times 10^{-19} \text{ C} \times 3.52 \times 10^7 \text{ m/s} \times \pi \times (10^{-3})^2 \text{ m}^2}$$

$$= 28.22 \times 10^{15} \text{ m}^{-3}$$

It is interesting to note that the electron concentration of thermionic devices is much less, a factor of 10^{-6} of the carrier concentration for solid-state devices.

According to Eq. 14.1.24, the amplitude of the microwave beam current is only

$$I_m = \frac{kv}{2V_o} I = 0.005656 \times 500 \text{ mA} = 2.83 \text{ mA}$$

This compares with the dc beam current $I = 500$ mA. In the initial stage of velocity modulation, not much microwave beam current is induced in the dc beam. According to Eq. 14.1.25, modulation of the beam consumes microwave power of

$$P_m = \tfrac{1}{2} v I_m = \tfrac{1}{2} \times 14.14 \text{ V} \times 2.83 \text{ mA} = 19.99 \text{ mW}$$

As seen from Eq. 14.1.25, this power is transferred in part from the dc beam current I_o. If I_o is less, P_m is less. If I_o is zero, P_m is zero no matter how much the input power P is.

The average microwave energy involved in the interaction space of the electron beam and microwave electric field is calculated using Eq. 14.1.27.

$$W_{ave} = \frac{1}{16} \times 9.109 \times 10^{-31} \text{ kg} \times \pi \times (10^{-3} \text{ m})^2 \times 10^{-3} \text{ m}$$

$$\times 28.22 \times 10^{15} \text{ m}^{-3} \left(\frac{3.52 \times 10^7 \text{ m/s} \times 14.14 \text{ V}}{10^3 \text{ V}} \right)^2$$

$$= 1.25 \times 10^{-12} \text{ J}$$

Not much microwave energy is transferred into the beam at the interaction space. However, this energy transfer is done very rapidly. In this case, the electron velocity is $u_o = 3.52 \times 10^7$ m/s and the interaction space length $g = 10^{-3}$ m. Then the electron transit time across the space, Δt, is

$$\Delta t = \frac{g}{u_o} = \frac{10^{-3} \text{ m}}{3.52 \times 10^7 \text{ m/s}} = 0.28 \times 10^{-10} \text{ s}$$

$$= 0.028 \text{ ns}$$

The electron transit angle across the interaction space $g = 10^{-3}$ m is

$$\Delta\theta = \omega \Delta t = 2\pi \times 9.5 \times 10^9 \text{ s}^{-1} \times 0.028 \times 10^{-9} \text{ s}$$

$$= 1.67 \text{ rad}$$

From Eq. 14.1.29, multiplying the total number of electrons in the interaction region SgN,

$$P_{\text{ave}} = \frac{1}{8} \times 9.109 \times 10^{-31}\ \text{kg} \times 2\pi \times 9.5 \times 10^{9}\ \text{s}^{-1}$$

$$\times \left(\frac{3.52 \times 10^{7}\ \text{m/s} \times 14.14\ \text{V}}{1000\ \text{V}} \right)^{2} \frac{\sin^{2}(1.67\ \text{rad})}{1.67\ \text{rad}}$$

$$= 0.498 \times 10^{-9}\ \text{W}$$

Very small power levels are needed for velocity modulation.

14.2 Bunching

If the velocity modulation shown in Eq. 14.1.4 is applied to a smoothly flowing electron beam as shown in Eq. 14.1.2, a modulated beam current as shown in Eq. 14.1.23 is produced. This modulated beam current can be modified by dividing both sides of Eq. 14.1.23 or Eq. 14.1.21 by Squ_{o},

$$\frac{I}{qu_{\text{o}}S} = N\left(1 + \frac{\text{k}v_{\text{i}}}{2V_{\text{o}}} \sin \omega t\right) \tag{14.2.1}$$

Actually, $I/qu_{\text{o}}S$ is the electron density of the velocity modulated beam, represented by n:

$$n = \frac{I}{qu_{\text{o}}S} \tag{14.2.2}$$

Combining Eqs. 14.2.1 and 14.2.2,

$$n = N\left(1 + \frac{\text{k}v_{\text{i}}}{2V_{\text{o}}} \sin \omega t\right) \tag{14.2.3}$$

This equation shows the density modulation explicitly. As soon as the electrons are velocity modulated at a particular location, density modulation is produced. This density modulation can be increased by the use of the electron bunching technique.

The electron beam with the velocity modulated beam is introduced to a drift space. The drift space is a space of equipotential. In this space there is neither acceleration nor deceleration of the electrons. The electrons therefore maintain their initial velocity when they enter into the drift space.

If an electron leaves the drift space at $x = x_1$ at $t = t_1$, having arrived at $x = x_1 + l_d$ at $t = t_2$, then

$$t_2 = t_1 + \frac{l_d}{u} \tag{14.2.4}$$

Applying Eq. 14.1.19 to Eq. 14.2.4,

$$t_2 \approx t_1 + \frac{l_d}{u_o}\left(1 + \frac{kv_i}{2V_o}\sin\omega t_1\right)^{-1}$$

$$\approx t_1 + \frac{l_d}{u_o}\left(1 - \frac{kv_i}{2V_o}\sin\omega t_1\right) \tag{14.2.5}$$

The principle of binomial expansion has been employed in the above derivation.

If the beam current at $x = x_1$ is I_1 and at $x = x_2$ the current is I_2, then the law of conservation of electric charge applied to the drift space between $x = x_1$ and $x = x_2$ requires that the amount of charge entering at $x = x_1$ during Δt_1 at $t = t_1$ must be equal to the amount of the charge exiting at $x = x_2$ during the time interval Δt_2 at $t = t_2$, if no charge accumulation or depletion is permitted in the drift space. Then

$$I_1\,dt_1 = I_2\,dt_2 \tag{14.2.6}$$

The current at $x = x_2$ is therefore

$$I_2 = I_1\frac{dt_1}{dt_2} \tag{14.2.7}$$

Differentiating Eq. 14.2.5,

$$dt_2 = dt_1\left(1 - \frac{kv_i}{2V_o}\frac{l_d}{u_o}\omega\cos\omega t_1\right) \tag{14.2.8}$$

$$\frac{dt_1}{dt_2} = \frac{1}{1 - \dfrac{kv_i}{2V_o}\dfrac{l_d}{u_o}\omega\cos\omega t_1} \tag{14.2.9}$$

Substituting Eq. 14.2.9 in Eq. 14.2.7,

$$I_2 = \frac{I_1}{1 - \dfrac{kv_i}{2V_o}\dfrac{l_d}{u_o}\omega\cos\omega t_1} \tag{14.2.10}$$

If

$$\frac{kv_i}{2V_o}\frac{l_d}{u_o}\omega \cos \omega t_1 \ll 1 \tag{14.2.11}$$

then

$$I_2 \approx I_1\left(1 + \frac{kv_i}{2V_o}\frac{l_d}{u_o}\omega \cos \omega t_1\right) \tag{14.2.12}$$

Now the term $kv_i/2V_o$ is the depth of velocity modulation, the term $(l_d/u_o)\omega$ is the electron transit angle between $x = x_1$ and $x = x_2$ in the drift space. If $x = x_1$ is chosen at the beginning of the velocity modulation, then

$$I_1 = I_o \tag{14.2.13}$$

which is the dc beam current. Substituting Eq. 14.2.13 in Eq. 14.2.12,

$$I_2 \approx I_o\left(1 + \frac{kv_i}{2V_o}\frac{l_d}{u_o}\omega \cos \omega t_1\right) \tag{14.2.14}$$

The term $(kv_i/2V_o)(l_d/u_o)\omega$ is the measure of density modulation. This is called the bunching parameter and is represented by X_b:

$$X_b \equiv \frac{kv_i}{2V_o}\frac{l_d}{u_o}\omega \tag{14.2.15}$$

Under a fixed system structure with constant dc voltage and microwave voltage, the electron transit angle

$$\theta_b = \omega\frac{l_d}{u_o} \tag{14.2.16}$$

will determine the density modulation. Within the restriction of Eq. 14.2.11, the density modulation gets greater with a larger electron transit angle θ. Under constant system structure and constant voltage, the electron transit time l_d/u_o is fixed, so the higher the operating frequency, the greater the electron bunching. Under constant frequency operation, by adjusting u_o, the bunching parameter can be increased or, by selecting $x = x_2$ to make l_d greater, the bunching parameter is increased.

It is important to compare Eq. 14.2.12 with Eq. 14.1.23. The modulation term at $x = x_2$ is θ times greater than at $x = x_1$, and this is the effect of electron bunching. Slow electrons that leave

$x = x_1$ early are caught by fast electrons that leave $x = x_1$ later. Thus, at $x = x_2$, electron bunching takes place.

In practice, the condition 14.2.11 is not always met. In this case, Eq. 14.2.10 can be expanded into a Fourier series and the definition of the Bessel function can be applied to it. When this is done,

$$I_2 = I_1\left[1 + 2\sum_{n=1}^{\infty} J_n(nX_b)\cos n(\omega t - \theta)\right] \qquad (14.2.17)$$

where n is an integer, $J_n(nX_b)$ is a Bessel function of order n and argument nX_b, and θ is a part of the delayed phase angle from the modulation voltage to the microwave beam current at $x = x_2$ [1–5]. Equation 14.2.17 contains the Bessel function $J_n(nX_b)$. The Bessel function is an oscillatory function with respect to its argument. The function reaches its maxima at certain values of the argument nX_b. This means that at a certain value of l_d, there will be a maximum beam current, or there is electron bunching at that location. Beyond the value of the optimum l_d, the current will decrease, or debunching takes place. This relationship will become clear if the fundamental frequency component of Eq. 14.2.17 is observed. For the fundamental frequency,

$$n = 1 \qquad (14.2.18)$$

Then, from Eq. 14.2.17,

$$I_2 = I_1[1 + 2J_1(X_b)\cos(\omega t - \theta)] \qquad (14.2.19)$$

where θ is again a part of the delayed phase angle of the microwave beam current at $x = x_2$ from the modulating voltage at $x = x_1$. According to the table of Bessel functions, $J_1(X_b)$ is an increasing function from zero to $X_b \approx 2$ for the first maximum. Beyond that point, the function decreases. It is also known from Eq. 14.2.17 that the velocity modulated and bunched electron current contains harmonics. The bunched electron beam can be used as a harmonic generator [6, 7].

Continuing the example presented in Section 14.1, to maximize the microwave modulated current, according to Eq. 14.2.19, $J_1(X_b)$ must be maximized. According to the table of Bessel functions, $J_1(X_b)$ is maximum when $X_b = 1.8$. This means that, from Eq. 14.2.15,

$$X_b = \frac{k v_i}{2V_o}\frac{l_d}{u}\omega = 1.8$$

or

$$l_d = (uX_b/\omega)\frac{kv_1}{2V_o}$$

$$= (3.52 \times 10^7 \text{ m/s} \times 1.8/2\pi \times 9.5 \times 10^9 \text{ s}^{-1})(0.005656)$$

$$= 1.78 \text{ m}$$

This is a long device. At $X_b = 1.8$, $J_1(1.8) = 0.5815$. Though this is not impractical, it is inconveniently long. If l_d is made a manageable size of 15 cm, as in the case of a practical device, then $X_b = 1.8 \times 15$ cm/178 cm $= 0.15$ and $J_1(0.15) = 0.08$. This means that by making the drift space 1.78 m/0.15 m = 12 times longer, the modulated current gets $J_1(1.8)/J_1(0.15) = 0.5815/0.08 = 7.3$ times stronger.

If $l_d = 1.78$ m is chosen then, according to Eq. 14.2.18, the magnitude of microwave beam current is

$$I_m = 2I_1J_1(X_b) \approx 2I_oJ_1(X_b) = 2 \times 500 \times 10^{-3} \text{ A} \times 0.585$$

$$= 0.585 \text{ A}$$

If $l_d = 15$ cm is chosen, then

$$I_m = 2 \times 500 \times 10^{-3} \text{ A} \times 0.08 = 0.08 \text{ A}$$

As shown in Section 14.1, when the beam exits the velocity modulation region and enters the drift space, the microwave beam current is only $I_m = 2.83$ mA. At the end of the drift space, for long devices, the microwave beam current grows to $I_m = 585$ mA, which is 585 mA/2.83 mA = 206 times greater or $20 \log_{10} 206 =$ 46.3 dB microwave beam current amplification. For short devices, the microwave beam current grows by a factor of 80 mA/ 2.83 mA = 28.3, which is $20 \log_{10} 28.3 = 29$ dB microwave beam current amplification.

14.3 Dynamic Induction Current

If an electron beam is velocity modulated and bunched, the electron beam current is no longer smooth. The well established modulated beam current can be approximately described by

$$I_b = I_o + I \sin \omega\left(t - \frac{x}{u}\right) \tag{14.3.1}$$

where I_o is the dc beam current and the term $I \sin \omega(t - x/u)$

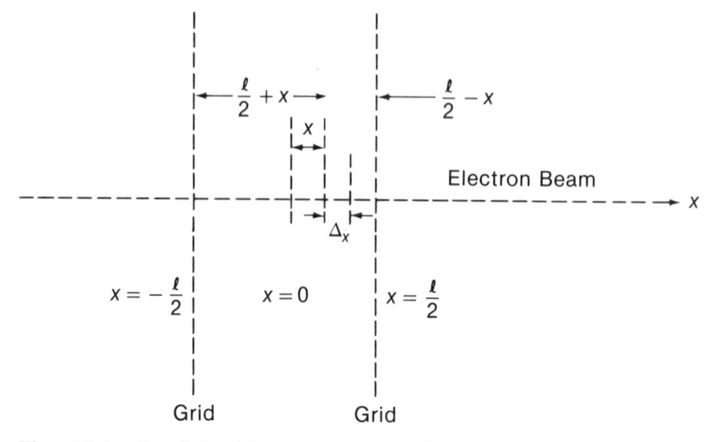

Fig. 14.1 Parallel grids on an electron beam.

represents the modulated beam current of the magnitude I fluctuating with the microwave frequency ω if observed at a constant location x and the variation is moving with the electron beam velocity u m/s. If a pair of parallel grids with a very small gap between them is introduced perpendicular to the modulated beam current as shown in Fig. 14.1, the electron bunch in the beam will induce an electric charge on th grid by simple electrostatic induction. In Fig. 14.1, the amount of electron charge in the electron beam between x and $x + \Delta x$ is, by mathematical manipulation, ΔQ:

$$\Delta Q = \frac{\Delta Q}{\Delta t}\frac{\Delta t}{\Delta x}\Delta x \tag{14.3.2}$$

In Eq. 14.3.2,

$$\frac{\Delta Q}{\Delta t} = I_{\mathrm{b}} \tag{14.3.3}$$

$$\frac{\Delta x}{\Delta t} = u \tag{14.3.4}$$

Then

$$\Delta Q = I_{\mathrm{b}}\frac{\Delta x}{u} \tag{14.3.5}$$

This electron charge, which is negatively charged, induces positive charge on the grid at $(l/2 + x)$ distance from x

$$\Delta Q_{\mathrm{i}} = \Delta Q\frac{\frac{l}{2} + x}{l} \tag{14.3.6}$$

where l is the separation distance of the grids as shown in Fig. 14.1. The amount of the induced charge on the grids is inversely proportional to the distance between the inducing charge and the grids. The amount of inducing charge ΔQ is therefore distributed inversely proportionally to the distance to the grids in Eq. 14.3.6. If the two grids are electrically connected to an external circuit, an induced current of $\Delta Q_i/\Delta t$ A would flow.

Actually, the electron charge packet ΔQ exists across the gap and induces the current $\Delta Q_i/\Delta t$ in the external circuit. The total induced current in the external circuit is therefore

$$i = \int_{-l/2}^{l/2} \frac{\Delta Q_i}{\Delta t} \, dx. \tag{14.3.7}$$

Substituting Eq. 14.3.6 in Eq. 14.3.7,

$$i = \int_{-l/2}^{l/2} \frac{\left(\frac{l}{2} + x\right)}{l} \frac{\Delta Q}{\Delta t} \, dx \tag{14.3.8}$$

Substituting Eq. 14.3.5 in Eq. 14.3.8,

$$i = \int_{-l/2}^{l/2} \left(\frac{1}{2} + \frac{x}{l}\right) \frac{I_b}{u} \frac{\Delta x}{\Delta t} \, dx = \int_{-l/2}^{l/2} \left(\frac{1}{2} + \frac{x}{l}\right) I_b \, dx \tag{14.3.9}$$

Substituting Eq. 14.3.1 in Eq. 14.3.9,

$$i = \int_{-l/2}^{l/2} \left(\frac{1}{2} + \frac{x}{l}\right)\left(I_o + I \sin \omega\left(t - \frac{x}{u}\right)\right) dx \tag{14.3.10}$$

After tremendous and laborious integration, the result is [2,3]

$$i = I_o + \frac{\sin\left(\frac{\omega l}{2u}\right)}{\frac{\omega l}{2u}} I \sin \omega t \tag{14.3.11}$$

It should be emphasized that this is the induction current in the circuit which is connected externally to the pair of grids. I_o is the dc current in the circuit which is the same amount of dc current in the beam. In the circuit, there is the same amount of dc current induced but this does not contribute to microwave current generation directly. In Eq. 14.3.11, l/u is the electron transit time across the pair of grids and $\omega l/u$ is the electron transit angle across the pair of grids. Let k_o be

$$k_o \equiv \frac{\sin\left(\frac{\omega l}{2u}\right)}{\frac{\omega l}{2u}} \tag{14.3.12}$$

Then, from Eq. 14.3.11 and 14.3.12,

$$i = I_o + k_o I \sin \omega t \tag{14.3.13}$$

In this equation, $k_o I$ is the magnitude of the microwave current in the circuit, and I is the magnitude of the microwave current in the electron beam. Therefore, k_o represents the ratio of the microwave current in the circuit to the microwave current in the electron beam. This is the reason that k_o is called the beam coupling coefficient.

The microwave beam current I induces the microwave circuit current $k_o I$. By reciprocity, the same beam coupling coefficient k_o takes effect for the velocity modulation as shown in Eq. 14.1.18. In a phasor form, the microwave circuit current is represented by

$$\dot{i}_m = k_o \dot{i} \tag{14.3.14}$$

If the external circuit impedance across the pair of grids is $\dot{Z}_L$, then the induced voltage across the gap is

$$\dot{v}_m = \dot{i}_m \dot{Z}_L = \dot{k}_o \dot{I} \dot{Z}_L \tag{14.3.15}$$

Continuing the example of a velocity modulated thermionic device presented in Section 14.2, if $l = 10^{-3}$ m, then the beam coupling coefficient of the parallel grids is, from Eq. 14.3.12,

$$k_o \equiv \frac{\sin\left(\frac{\omega l}{2u}\right)}{\frac{\omega l}{2u}} = \frac{\sin\left(\frac{2\pi \times 9.5 \times 10^9\ \mathrm{s}^{-1} \times 10^{-3}\ \mathrm{m}}{2 \times 3.52 \times 10^7\ \mathrm{m/s}}\right)}{\left(\frac{2\pi \times 9.5 \times 10^9\ \mathrm{s}^{-1} \times 10^{-3}\ \mathrm{m}}{2 \times 3.52 \times 10^7\ \mathrm{m/s}}\right)}$$

$$= \frac{\sin 0.848}{0.848} = 0.75.$$

According to Section 14.1, the microwave beam current in this device is $I_m = 2.83$ mA when the dc electron beam is velocity modulated at the beginning. After drifting for a while, for the long device, I_m increases to 585 mA. The magnitude of the induced microwave current in the circuit is therefore, from Eq. 14.3.14,

$$i_m = k_o I = k_o I_m = 0.75 \times 585 \text{ mA} = 438.8 \text{ mA}$$

If the load impedance across the grids is $\dot{Z}_L = 200 + j0\ \Omega$ at resonance, then the microwave voltage across the grid is, from Eq. 14.3.15,

$$v_m = i_m \dot{Z}_L = 438.8 \times 10^{-3} \text{ A} \times 200\ \Omega$$
$$= 87.75 \text{ V}$$

As stated in Section 14.1, the velocity modulating voltage was $v = 14.14$ V. The voltage amplification is therefore $v_m/v = 87.75 \text{ V}/14.14 \text{ V} = 6.2$ times, or the voltage gain is $20 \log_{10} 6.2 = 15.9$ dB.

For the short device,

$$v_m = i_m \dot{Z}_L = 80 \times 10^{-3} \text{ A} \times 200\ \Omega = 16 \text{ V}$$

In this case, the voltage amplification is

$$v_m/v = 16 \text{ V}/14.14 \text{ V} = 1.13$$

and the voltage gain is

$$20 \log_{10} 1.13 = 1.06 \text{ dB}$$

14.4 Klystrons

A klystron is a microwave amplifying or generating vacuum tube invented by the Varian brothers [1]. This tube is based on velocity modulation, bunching, and dynamic induction [8–10]. The tube is capable of operating at high power with high microwave frequencies beyond the solid-state microwave device range.

A schematic diagram of one of the most simple two-cavity klystrons is shown in Fig. 14.2. As seen from this figure, microwave input signals are fed into a cavity resonator which is called the rhumbatron to produce the velocity modulation voltage v_i across the pair of grids [1]. The velocity modulation voltage v_i gives the dc beam current velocity modulation as shown in Eq. 14.1.23. Before the dc beam current hits the first rhumbatron, the electron beam is smooth and there is no modulation in it. The electrons are accelerated by the dc acceleration voltage V_o. After passing through the grids of the first rhumbatron, the electron beam enters the drift space. As seen in Fig. 14.2, the first rhumbatron and the second rhumbatron are directly connected to each

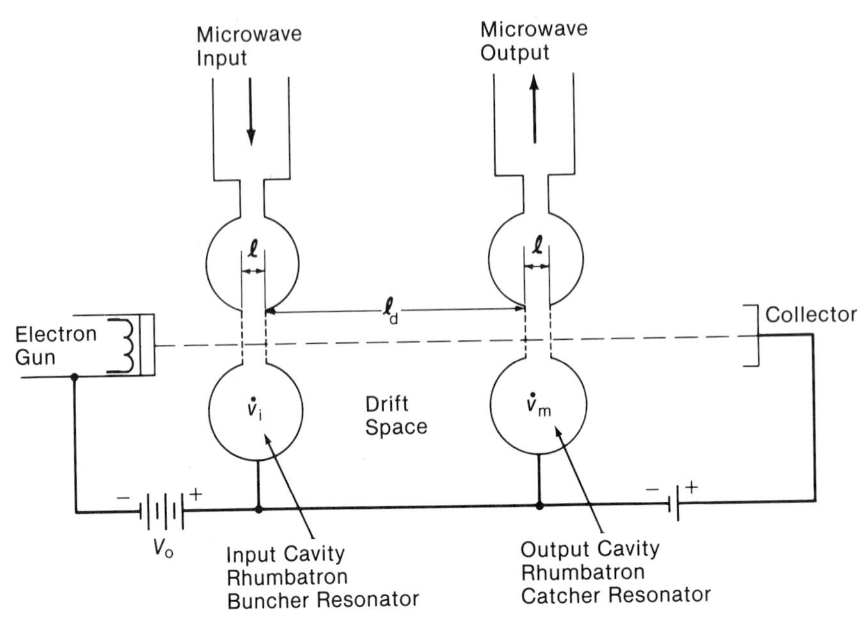

Fig. 14.2 Schematic diagram of a two-cavity klystron.

other, with no electric field in this space. The electrons drift in this space and there will be no significant change in the velocity during the flight in this drift space. The drift velocity is equal to the initial velocity of the electrons when they leave the first rhumbatron. After drifting some time, the electrons begin to bunch, forming the bunched beam current as seen from Eq. 14.2.17. This is the reason that the first rhumbatron is called the buncher. At the location where the bunched beam current is at its maximum value, a second rhumbatron is placed. By the principle of dynamic induction, the dynamic induction current, as seen from Fig. 14.3.14, will flow in the rhumbatron circuit. The microwave voltage $\dot{v}_m$ as seen from Eq. 14.3.15 is induced across the grids of the second rhumbatron and radiates out through the output waveguide. The second rhumbatron is called the catcher. Electron beams passed through the catcher rhumbatron are collected by a collector as shown in Fig. 14.2. If the klystron system is properly adjusted, it is possible to make $v_m > v_i$, a klystron amplifier. If the catcher and the buncher are connected by a positive feedback link, the klystron is an oscillator.

If the buncher admittance is $\dot{Y}_g = G_g + jB_g$, then the input power is

$$P_i = \tfrac{1}{2} G_g \dot{v}_i \dot{v}_i^* \qquad \text{(W)} \tag{14.4.1}$$

where $\dot{v}_i^*$ is the complex conjugate of $\dot{v}_i$. This microwave input power produces the velocity modulation and the bunched beam current as shown in Eqs. 14.2.19 and 14.3.1. Comparing Eqs. 14.2.19 and 14.3.1,

$$I_o = I_1 \tag{14.4.2}$$

$$I = 2I_1 J_1(X_b) \tag{14.4.3}$$

$$\sin \omega\left(t - \frac{x}{u}\right) = \cos(\omega t - \theta) = \sin\left(\omega t - \theta + \frac{\pi}{2}\right) \tag{14.4.4}$$

Eq. 14.4.4 suggests that

$$\frac{\omega x}{u} = \left(\theta - \frac{\pi}{2}\right) \tag{14.4.5}$$

$$x = \frac{u}{\omega}\left(\theta - \frac{\pi}{2}\right) \tag{14.4.6}$$

This is the location of the catcher rhumbatron.

In a phasor form, the input voltage across the buncher grids is represented by

$$\dot{v}_i = v_i e^{j\omega t} \tag{14.4.7}$$

Comparing Eq. 14.1.3 and Eq. 14.2.19, the bunched beam current is lagging the phase angle $(\theta - \pi/2)$ from the input voltage. Then, from Eq. 14.4.3,

$$\dot{I} = \mathrm{I} e^{j[\omega t - (\theta - \pi/2)]} \tag{14.4.8}$$

or

$$\dot{I} = 2I_o J_1(X_b) e^{j[\omega t - (\theta - \pi/2)]} \tag{14.4.9}$$

From Eq. 14.3.15, this bunched beam produces microwave power at the catcher circuit

$$\dot{P}_o = \tfrac{1}{2}\dot{v}_m i_m^* = \tfrac{1}{2}\mathrm{k}_o \dot{Z}_L \dot{I}\dot{I}^* \tag{14.4.10}$$

The actual output is

$$P_o = Re\,\dot{P}_o = \tfrac{1}{2}\mathrm{k}_o R_L \dot{I}\dot{I}^* = 2\mathrm{k}_o R_L I_o^2 J_1^2(X_b) \tag{14.4.11}$$

The power gain is then, using Eq. 14.4.1,

$$A = \frac{P_o}{P_i} = \frac{2\mathrm{k}_o R_L I_o^2 J_1^2(X_b)}{\dfrac{1}{2} G_g v_i^2} = \frac{4\mathrm{k}_o R_L I_o^2}{G_g}\left(\frac{J_1(X_b)}{v_i}\right)^2 \tag{14.4.12}$$

Remembering Eq. 14.2.14 for the bunching parameter, newly interpreted specifically for this case,

$$X_b \equiv \frac{k v_i}{2V_o} \frac{l_d}{u_o} \omega \tag{14.4.13}$$

If

$$X_b \ll 1 \tag{14.4.14}$$

$$J(X_b) \approx \frac{X_b}{2} \tag{14.4.15}$$

Substituting Eq. 14.4.15 in Eq. 14.4.12, the gain of this klystron amplifier is

$$A = \frac{\omega^2 k_o k^2 R_L I_o^2 l_d^2}{4 G_g V_o^2 u_o^2} \tag{14.4.16}$$

The voltage gain is then

$$A_v = \sqrt{A} \tag{14.4.17}$$

The voltage gain including the phase shift can also be obtained from Eq. 14.4.9 and Eq. 14.3.15:

$$\dot{v}_m = k_o \dot{Z}_L \dot{I} = 2k_o \dot{Z}_L I_o J_1(X_b) e^{j[\omega t - (\theta - \pi/2)]} \tag{14.4.18}$$

The phasor voltage gain is then

$$\dot{A}_v \equiv \frac{\dot{v}_m}{\dot{v}_i} = \frac{2k_o \dot{Z}_L I_o J_1(X_b) e^{j\omega t} \cdot e^{-j(\theta - \pi/2)}}{v_i e^{j\omega t}}$$

$$= 2\frac{k_o \dot{Z}_L I_o J_1(X_b)}{v_i} e^{-j(\theta - \pi/2)} \tag{14.4.19}$$

Gain equations 14.4.12, 14.4.17, and 14.4.19 show that the gain can be optimized electronically by adjusting V_o, I_o, ω, and u or by optimizing the tube structure and circuit by adjusting l_d, l, g, G_g, and $\dot{Z}_L$. As seen from Eqs. 14.4.12 and 14.4.19, the gain is inherently a function of the operating signal level. The klystron is therefore a nonlinear amplifier. When the input signal is small enough to allow the approximation 14.4.14, the gain is approximately linear as seen from Eq. 14.4.16.

A power oscillator can be built by connecting the catcher rhumbatron to the buncher rhumbatron externally. If the transfer admittance from the buncher to the catcher is $\dot{Y}_m$, then the input voltage $\dot{v}_i$ at the buncher produces the catcher current $\dot{Y}_m \dot{v}_i$. The output voltage is $\dot{v}_m$, and the output power available at the catcher is $\frac{1}{2}(\dot{Y}_m\dot{v}_i)^*\dot{v}_m$. A portion of this is consumed by the load impedance $\dot{v}_m\dot{v}_m^*/2R_L$ while another portion of the power at the catcher is fed back to the buncher. If the transfer admittance from the catcher to the buncher is $\dot{Y}_t$, then the microwave catcher voltage $\dot{v}_m$ at the catcher induces the buncher current $\dot{Y}_t\dot{v}_m$. The power fed back to the buncher from the catcher is then $\frac{1}{2}(\dot{Y}_t\dot{v}_m)^*\dot{v}_i$. Then

$$\frac{1}{2}\left(\dot{Y}_m\dot{v}_i\right)^*\dot{v}_m = \frac{1}{2}\frac{\dot{v}_m^*}{R_L}\dot{v}_m + \frac{1}{2}\left(\dot{Y}_t\dot{v}_m\right)^*\dot{v}_i$$

$$\left(\dot{Y}_m\dot{v}_i\right)^* = \frac{v_m^*}{R_L} + \left(\dot{Y}_t\dot{v}_m\right)^*\frac{\dot{v}_i}{\dot{v}_m} \tag{14.4.20}$$

Substituting Eq. 14.4.19,

$$\left(\dot{Y}_m\dot{v}_i\right)^* = \frac{\dot{v}_m^*}{R_L} + \frac{\left(\dot{Y}_t\dot{v}_m\right)^*}{\dot{A}_v} \tag{14.4.21}$$

Then

$$\dot{Y}_m\dot{v}_i = \frac{\dot{v}_m}{R_L} + \left(\frac{1}{\dot{A}_v}\right)^*\dot{Y}_t\dot{v}_m \tag{14.4.22}$$

Dividing through $\dot{v}_i$,

$$\dot{Y}_m = \frac{1}{R_L}\frac{\dot{v}_m}{\dot{v}_i} + \left(\frac{1}{\dot{A}_v}\right)^*\dot{Y}_t\frac{\dot{v}_m}{\dot{v}_i} \tag{14.4.23}$$

Substituting Eq. 14.4.19 in Eq. 14.4.23,

$$\dot{Y}_m = \frac{\dot{A}_v}{R_L} + \dot{Y}_t\dot{A}_v\left(\frac{1}{\dot{A}_v}\right)^* \tag{14.4.24}$$

If the phase angle of $\dot{A}_v$ is ϕ,

$$\dot{A}_v = A_v e^{j\phi} \tag{14.4.25}$$

$$\frac{1}{\dot{A}_v} = \frac{1}{A_v}e^{-j\phi} \tag{14.4.26}$$

$$\left(\frac{1}{\dot{A}_v}\right)^* = \frac{1}{A_v}e^{j\phi} \tag{14.4.27}$$

$$\dot{A}_v\left(\frac{1}{\dot{A}_v}\right)^* = A_v e^{j\phi}\frac{1}{\dot{A}_v}e^{j\phi} = e^{2j\phi} \tag{14.4.28}$$

Substituting Eq. 14.4.28 in Eq. 14.4.24,

$$\dot{Y}_{\mathrm{m}} = \frac{\dot{A}_{\mathrm{v}}}{R_{\mathrm{L}}} + \dot{Y}_{\mathrm{t}} e^{2j\phi} \quad \text{or} \quad \dot{Y}_{\mathrm{m}} - \dot{Y}_{\mathrm{t}} e^{2j\phi} = \frac{\dot{A}_{\mathrm{v}}}{R_{\mathrm{L}}} \qquad (14.4.29)$$

This is the steady state condition for klystron oscillation. The oscillator output power is obtained from Eq. 14.4.20 by noting that the output power to the load is

$$\begin{aligned} P_{\mathrm{o}} &= \frac{v_{\mathrm{m}} v_{\mathrm{m}}^*}{2R_{\mathrm{L}}} = \frac{1}{2}\left(\dot{Y}_{\mathrm{m}} \dot{v}_{\mathrm{i}}\right)^* \dot{v}_{\mathrm{m}} - \frac{1}{2}\left(\dot{Y}_{\mathrm{t}} \dot{v}_{\mathrm{m}}\right)^* \dot{v}_{\mathrm{i}} \\ &= \frac{1}{2}\left(\dot{Y}_{\mathrm{m}} \dot{v}_{\mathrm{i}}\right)^* \dot{A}_{\mathrm{v}} \dot{v}_{\mathrm{i}} - \frac{1}{2}\left(\dot{Y}_{\mathrm{t}} \dot{A}_{\mathrm{v}} \dot{v}_{\mathrm{i}}\right)^* \dot{v}_{\mathrm{i}} \\ &= \frac{1}{2}\left[\left(\dot{Y}_{\mathrm{m}} \dot{v}_{\mathrm{i}}\right)^* A_{\mathrm{v}} - \left(\dot{Y}_{\mathrm{t}} \dot{A}_{\mathrm{v}} \dot{v}_{\mathrm{i}}\right)^*\right] \dot{v}_{\mathrm{i}} \end{aligned} \qquad (14.4.30)$$

The klystron is thus an electron tube based on velocity modulation, bunching, and dynamic induction.

Continuing the examples presented in Sections 14.1, 14.2, and 14.3, after tuning and matching the impedance of the buncher and catcher at $f = 9.5$ GHz, the buncher conductance G_{g} can be calculated using Eq. 14.4.1:

$$G_{\mathrm{g}} = \frac{2P_{\mathrm{i}}}{|v_{\mathrm{i}}|^2} = \frac{2 \times 10 \times 10^{-3}\ \mathrm{W}}{14.14^2\ \mathrm{V}^2} = 0.1 \times 10^{-3}\ \mathrm{S}$$

If the catcher has the same structure as the buncher, then $R_{\mathrm{L}} = 1/G_{\mathrm{g}} = 10 \times 10^3\ \Omega$.

From Eq. 14.4.11, the output power of the klystron with the long drift space is

$$\begin{aligned} P_{\mathrm{o}} &= 2\mathrm{k}_{\mathrm{o}} R_{\mathrm{L}} I_{\mathrm{o}}^2 J_1^2(X_{\mathrm{b}}) \\ &= 2 \times 0.75 \times (10^4\ \Omega)(0.5\ \mathrm{A})^2 (0.5815)^2 \\ &= 126.8\ \mathrm{W} \end{aligned}$$

The power gain is, from Eq. 14.4.12,

$$A = \frac{P_{\mathrm{o}}}{P_{\mathrm{i}}} = \frac{126.8\ \mathrm{W}}{10^{-2}\ \mathrm{W}} = 12{,}680$$

or $10 \log_{10} 12{,}680 = 41$ dB. The voltage gain is, from Eq. 14.4.17, $A_{\mathrm{v}} = \sqrt{A} = \sqrt{12{,}680} = 112.6$ or $20 \log_{10} 112.6 = 41$ dB. For the short drift space klystron,

$$\begin{aligned} P_o &= 2 \times 0.75 \times (10^4\ \Omega)(0.5\ \text{A})^2 \times (0.08)^2 \\ &= 24\ \text{W} \end{aligned}$$

Then the power gain is

$$A = \frac{P_o}{P_i} = \frac{24\ \text{W}}{10^{-2}\ \text{W}} = 2400$$

or $10 \log_{10} 2400 = 33.8$ dB. The voltage gain is $A_v = \sqrt{2400} = 49$ or $20 \log_{10} 49 = 33.8$ dB.

When the electronic tuning, resonator tuning, and impedance matching are not perfect, the voltage gain will be a complex quantity or phasor as shown in Eq. 14.4.19.

The electron transit angle θ_b across the drift space l_d is, from Eq. 14.2.16, $\omega l_d/u$. For the long drift space device,

$$\begin{aligned} \left(\theta - \frac{\pi}{2}\right) &= \theta_b \\ &= \omega l_d/u = 2\pi \times 9.5 \times 10^9\ \text{s}^{-1} \times 1.78\ \text{m}/3.52 \times 10^7\ \text{m/s} \\ &= 3018.435\ \text{rad} = 480 \times 2\pi + 0.3977 \times 2\pi \\ &= 480 \times 2\pi + 2.4988\ \text{rad} \\ &= 480 \times 2\pi\ \text{rad} + 143.17^\circ \end{aligned}$$

This is the phase angle of the phasor voltage gain in Eq. 14.4.19,

$$\begin{aligned} \theta &= 480 \times 2\pi\ \text{rad} + 143.17^\circ + 90^\circ \\ &= 480 \times 2\pi\ \text{rad} + 233.172^\circ \\ &= 480 \times 2\pi\ \text{rad} + 4.07\ \text{rad} \end{aligned}$$

This is the phase angle which first appeared in Eq. 14.2.16. For the short drift space klystron,

$$\begin{aligned} \left(\theta - \frac{\pi}{2}\right) &= 2\pi \times 9.5 \times 10^9\ \text{s}^{-1} \times 0.15\ \text{m}/3.52 \times 10^7\ \text{m/s} \\ &= 254.36\ \text{rad} = 40 \times 2\pi\ \text{rad} + 0.483 \times 2\pi\ \text{rad} \\ &= 40 \times 2\pi\ \text{rad} + 2.544\ \text{rad} \\ &= 40 \times 2\pi\ \text{rad} + 145.74^\circ \end{aligned}$$

Then

$$\begin{aligned} \theta &= 40 \times 2\pi\ \text{rad} + 145.74^\circ + 90^\circ \\ &= 40 \times 2\pi\ \text{rad} + 235.74^\circ \\ &= 40 \times 2\pi\ \text{rad} + 4.11\ \text{rad} \end{aligned}$$

From these calculations and Eq. 14.4.19, the actual value of the voltage gain is maximum when

$$\theta - \frac{\pi}{2} = 2N\pi \qquad (14.4.31)$$

where N is an integer representing the number of electron transit cycles across the drift space.

$$\theta - \frac{\pi}{2} = \theta_b \approx \frac{\omega_o l_d}{u_o} = 2N\pi \qquad (14.4.32)$$

This is at the center frequency at $\omega = \omega_o$. Assuming that $|\dot{A}_v|$ in Eq. 14.4.19 is a slower varying function of ω than $e^{-j(\theta - \pi/2)}$ then, at the edge of the frequency bandwidth ω_e, $Re\, e^{j(\theta - \pi/2)} = 1/\sqrt{2}$ or

$$\theta - \frac{\pi}{2} = \frac{\omega_e l_d}{u_o} = 2N\pi \pm \frac{\pi}{4} \qquad (14.4.33)$$

This means that the high side of the frequency band edge is

$$\omega_e^+ = \left(2N\pi + \frac{\pi}{4}\right)\Big/(l_d/u_o) \qquad (14.4.34)$$

The low side of the frequency band edge is

$$\omega_e^- = \left(2N\pi - \frac{\pi}{4}\right)\Big/(l_d/u_o) \qquad (14.4.35)$$

The frequency bandwidth is then

$$\Delta f = \frac{\omega_e^+ - \omega_e^-}{2\pi} = \frac{\pi/2}{2\pi(l_d/u_o)} = \frac{u_o}{4l_d} \qquad (14.4.36)$$

The center frequency is then, from Eq. 14.4.32,

$$f_o = \frac{\omega_o}{2\pi} = \frac{2N\pi u_o}{2\pi l_d} = N\frac{u_o}{l_d} \qquad (14.4.37)$$

The above analysis assumes that the applied voltages, currents, and geometries are kept constant and only the operating frequency is changed.

For the example of the short drift space klystron, the center frequency can be at $N = 41$ by the previous calculation for $(\theta - \pi/2) = 40 \times 2\pi$ rad $+ 145.74°$. Then, from Eq. 14.4.37, the center frequency should be

$$f_o = 41 \times \frac{3.52 \times 10^7 \text{ m/s}}{0.15 \text{ m}} = 9.62 \text{ GHz}$$

and the frequency bandwidth is, from Eq. 14.4.36,

$$\Delta f = \frac{3.52 \times 10^7 \text{ m/s}}{4 \times 0.15 \text{ m}} = 58.67 \text{ MHz}$$

If the buncher and the catcher are connected by a feedback loop transfer admittance of $\dot{Y}_t$, the oscillation condition of this oscillator is given by Eq. 14.4.29. This equation can be rewritten as

$$\frac{1}{R_L} \gtreqless \frac{\dot{Y}_m - \dot{Y}_t e^{2j\phi}}{\dot{A}_v} \tag{14.4.38}$$

If $1/R_L$ is greater than the right-hand side of Eq. 14.4.38, then the demand of the load is greater than the klystron can supply. The oscillation then dies down. If $1/R_L$ is less than the right-hand side of Eq. 14.4.38, then the load is too light and the oscillation builds up. When $1/R_L$ is equal to the right-hand side of Eq. 14.4.38, the oscillation is in its steady state.

In Eq. 14.4.38, A_v is maximum when $\phi = 2\pi N$, where N is an integer and represents the number of electron transit cycles across the drift space.

According to Eq. 14.3.14, the microwave circuit current at the catcher circuit is

$$\dot{i}_m = k_o \dot{I} = \dot{Y}_m \dot{v}_i \tag{14.4.39}$$

Then

$$\dot{Y}_m = \frac{k_o \dot{I}}{\dot{v}_i} = \frac{k_o}{\dot{v}_i} 2 I_o J_1(X_b) e^{-j(\theta - \pi/2)} \tag{14.4.40}$$

For the short drift space klystron of this example, at the operating frequency of $f_o = 9.62$ GHz, the beam coupling coefficient of the catcher gap $k_o \approx 0.75$ $\dot{I} \approx 0.08 e^{j(2\pi \times 41)}$ A for $\dot{v}_i = 14.14 e^{j(0)}$ V. Then

$$\dot{Y}_m = 0.75 \frac{(0.8 \text{ A}) e^{-j(2\pi \times 41)}}{(14.14 \text{ V}) e^{j(0)}} = 4.24 e^{-j(2\pi \times 41)} \text{ mS}$$

$$= 4.24 \text{ mS}$$

According to previous calculations on the short drift space

klystron, $\phi = 2\pi \times 41$, $\dot{A}_{\text{v}} \approx 49$ and $R_{\text{L}} = 10^4\ \Omega$. Then, from Eq. 14.4.29,

$$\begin{aligned}
\dot{Y}_{\text{t}} &= \left(\dot{Y}_{\text{m}} - \frac{\dot{A}_{\text{v}}}{R_{\text{L}}}\right) e^{-2j\phi} \\
&= \left(4.24 \times 10^{-3}\ \text{S} - \frac{49}{10^4\ \Omega}\right) e^{-2j(41 \times 2\pi)} \\
&= (-0.66 \times 10^{-3}\ \text{S}) e^{-2j(41 \times 2\pi)} \\
&= \left[(0.66 \times 10^{-3}) e^{j\pi}\ \text{S}\right] e^{-2j(41 \times 2\pi)} \\
&= (0.66 \times 10^{-3}) e^{-j(4 \times 41 - 1)\pi}\ \text{S}
\end{aligned}$$

The output power of the short drift space klystron is then, from Eq. 14.4.30,

$$P_{\text{o}} = \frac{v_{\text{m}} v_{\text{m}}^*}{2R_{\text{L}}} = \frac{(14.14\ \text{V} \times 49)^2}{2 \times 10^4\ \Omega} = 24\ \text{W}$$

If the drift space is inconveniently long, the entire length of the klystron can be made short by increasing the number of bunchers as illustrated in Fig. 14.3 [8–10]. This multistage velocity modulation produces a large magnitude of the depth of velocity modulation and also produces tight bunching with a short distance in drift space.

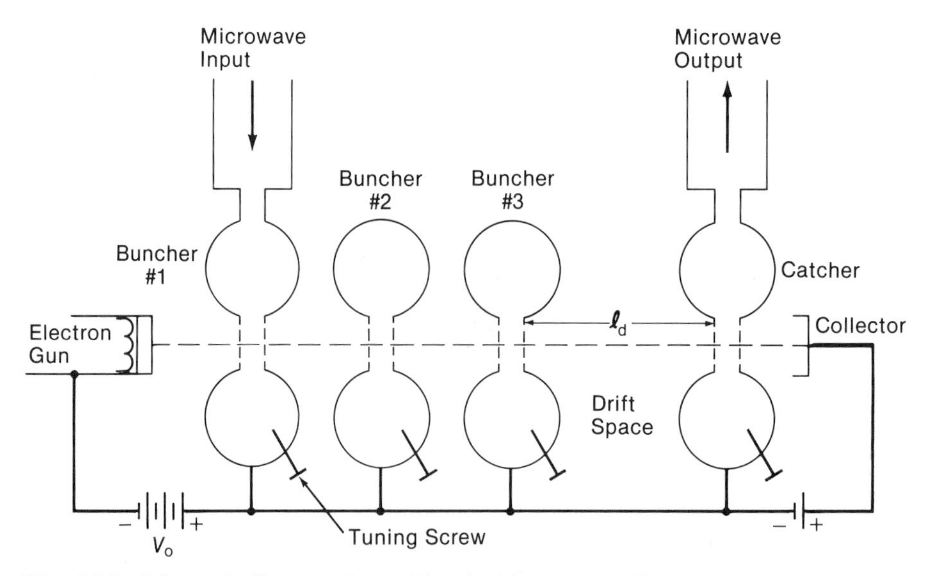

Fig. 14.3 Schematic diagram of a multicavity klystron amplifier.

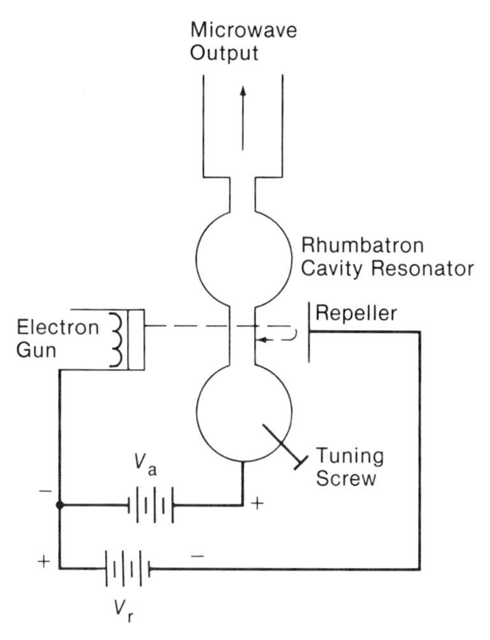

Fig. 14.4 Reflex klystron oscillator.

A positive feedback loop two-cavity klystron can be converted to a single cavity klystron as illustrated in Fig. 14.4. In this structure, the rhumbatron is a buncher for departing electrons. The departing electrons from the rhumbatron enter a decelerating field space since the reflector or repeller is biased negatively. The velocity modulated electrons enter the repeller space and, after reaching maximum penetration, these electrons are repelled back to the rhumbatron. By this time, the electrons are bunched and induce a microwave current as in the catcher rhumbatron of a klystron. Thus, the rhumbatron in Fig. 14.4 plays the role of the buncher for departing electrons and also plays the role of the catcher for bunched returning electrons. The velocity modulation is initially done by the microwave voltage induced in the rhumbatron at the resonance frequency of the rhumbatron cavity resonator by the electrical noise in the cavity. This noise is predominantly shot noise in the beam, partition noise created by the electron beam hitting the grids, and thermal noise in the cavity. This type of single cavity klystron is called a reflex klystron.

In reflex klystrons, positive feedback is achieved electronically by the bunched electrons returning to the rhumbatron. Reflex klystrons are used for low power or midlevel power applications. Reflex klystrons can be replaced by solid-state devices such as active diodes and FETs. Reflex klystrons today are employed only for special cases such as high or low temperature use or when there is a need for wide tunability or extremely low phase noise. On the contrary, multicavity klystrons are used for high power and high frequencies. Solid-state devices are expensive and inconvenient for these applications.

14.5 Velocity Modulation Devices

Most microwave electron devices, whether solid-state devices or thermionic devices, rely on the velocity modulation principle of the charge carriers in various forms. A velocity modulation thermionic device with a lienar electron beam and buncher and catcher rhumbatron cavities is called a klystron. A klystron is a microwave amplifier. Small input microwave power fed to the buncher rhumbatron cavity resonator produces velocity modulation in the electron beam. The velocity modulated electrons drift in the drift space. After drifting for a while, the electrons bunch. A catcher rhumbatron cavity resonator is located at that point. By electrostatic induction from the bunched electrons passing through the catcher rhumbatron cavity, the resonator induces a large microwave current in the catcher cavity. The induced microwave current is coupled to the output waveguide. Thus, microwave amplification is complete.

The performance of a klystron amplifier is characterized not only by gain and frequency bandwidth, but also by the bunching parameter, the depth of velocity modulation, the electron transit cycles in the buncher and catcher gaps, the beam coupling coefficient of the buncher and catcher gaps, and the number of electron transit cycles across the drift space.

Multicavity klystrons produce tight bunching over a short drift. They are therefore suited for high power klystrons with short drift space. Klystrons are primarily high microwave frequency and high power amplifiers. Output powers on the order of kW and MW are common. By adding an external positive feedback loop, the klystron can be made into a power oscillator. A single cavity klystron is a reflex klystron. A reflex klystron is an oscillator of low or medium power. For this type of application, more inexpensive and compact solid-state devices are preferred in most cases. However, for special applications of high frequency, tunability, low phase noise, or when extremely high or low temperature operating conditions are required, reflex klystrons are preferred over solid-state devices. Reflex klystron oscillators are negative resistance oscillators like tunnel diodes, Gunn diodes, and IMPATT diodes. Any negative resistance oscillator can be made into a negative resistance amplifier if the load resistance is properly adjusted. In a harsh environment such as extremely high or low temperatures, high or low pressures, or high humidity, where most solid-state devices fail to operate, the reflex klystron amplifier may be a good alternative.

Problems

14.1 Define velocity modulation.

14.2 Study the possibility of velocity modulation by a microwave magnetic field. Compare the efficiency or effectiveness of magnetic field velocity modulation and electric field velocity modulation.

14.3 Point out some examples of velocity modulation in solid-state microwave devices.

14.4 In microwave power tubes, the operating acceleration voltage is on the order of 10^3 V. Find the electron velocity and kinetic energy.

14.5 In a microwave power klystron, if the dc acceleration voltage is 10^3 V and the magnitude of the input microwave voltage is 10^0 V, find the maximum velocity and the minimum velocity of the electron. List all simplified assumptions.

14.6 Compute the energy, power and acceleration necessary to produce velocity modulation with a microwave voltage of 10^0 V and an acceleration voltage of 10^3 V.

14.7 When the electron velocity is a function of both space and time, find the acceleration. Determine whether or not zero acceleration is possible under this condition.

14.8 Define the beam coupling coefficient.

14.9 An electron beam current of 30 mA is focused to a diameter of 2 mm. The beam acceleration voltage is 10^3 V. Find the electron density.

14.10 Define electron bunching.

14.11 If a velocity modulated beam is introduced to a space of uniform acceleration, is it possible for electron bunching to take place? If so, formulate the process analytically.

14.12 Define the bunching parameter.

14.13 List parameters to optimize electron bunching.

14.14 Relate the electron transit angle in the drift space to the effect of electron bunching.

14.15 Does velocity modulation change in the drift space? If so, formulate the process analytically.

14.16 Show that the bunched beam current contains harmonics.

14.17 A bunched electron beam of $(30 + 5 \sin \omega t)$ mA is passing through an induction grid pair. The beam coupling coefficient is 0.75. Find the circuit current if the induction grids are connected to each other by an external circuit. The operating microwave frequency is 10 GHz. Also, find the velocity of the electrons.

14.18 The beam coupling coefficient can be defined based on the process of velocity modulation or the process of dynamic induction. Reconcile these two approaches.

14.19 Describe a klystron.

14.20 Find the optimum location for the catcher rhumbatron resonator.

14.21 Show that the voltage gain is the square root of the power gain.

14.22 The voltage gain of a klystron is obtained via the power gain or directly by the ratio of the output voltage to the input voltage. Show that the voltage gain of the klystron obtained by the two different approaches are the same.

14.23 In a steady state of klystron oscillation, show analytically that a part of the induced output current flows into the load and the rest of the current is fed back into the input.

References

1 R. H. Varian and S. F. Varian, A high frequency oscillator amplifier. *J. Appl. Phys.*, **10,** 321 (May 1939).

2 D. R. Hamilton, J. K. Knipp, and J. B. H. Kuper, "Klystron and Microwave Triodes." McGraw-Hill, New York, 1948.

3 H. J. Reich, P. F. Ordrung, H. L. Krauss, and J. G. Skalnik, "Microwave Theory and Techniques," Van Nostrand, Princeton, New Jersey, 1953.

4 T. K. Ishii, "Microwave Engineering." Ronald Press, New York, 1966.

5 R. F. Soohoo, "Microwave Electronics." Addison-Wesley, Reading, Massachusetts, 1971.

6 L. G. Dillon, R. A. Koenen, and T. K. Ishii, Conventional reflect klystron generates millimeter waves. *Microwaves* **3**(2), 38–43 (1964).

7 F. J. Tischer, "Microwellenmesztechnik." Springer-Verlag, Berlin, 1958.

8 J. T. Coleman, "Microwave Devices." Reston Publ., Reston, Virginia, 1982.

9 S. Y. Liao, "Microwave Electron-Tube Devices." Prentice-Hall, Englewood Cliffs, New Jersey, 1988.

10 E. A. Wolff and R. Kaul, "Microwave Engineering and Systems Applications." Wiley, New York, 1988.

15

Magnetrons

15.1 Formation of a Re-entry Beam

In a klystron, the electron beam is generated at the electron gun, goes through the process of velocity modulation, bunching, and dynamic induction, and finally hits the collector, ending the electron beam. If the end of the beam can be brought to the beginning and repeatedly recirculated, the output power and efficiency will be greatly increased. This re-entry of the electron beam is conceptually desirable for high power and high efficiency microwave electron devices. This concept has been utilized in the magnetron since 1927 in thermionic technology [1–7], but to date it is not utilized in solid-state devices technology.

In a magnetron, the re-entry electron beam is formed by a thermionic diode of cylindrical structure as illustrated in Fig. 15.1 [1–8]. In this figure, the radius of the cathode is represented by r_c and the radius of the anode is represented by r_a. A cylindrical coordinate is attached by coinciding the z-axis with the cylinder axis. In a magnetron, a magnetic flux density B_z (Tesla) is applied parallel to the z-axis of the cylindrical coordinate. If the cathode potential is zero, the anode potential is V_a, and the charge of an electron is q, then the electrons are initially accelerated in the radial direction by the radial electric field, but because of the

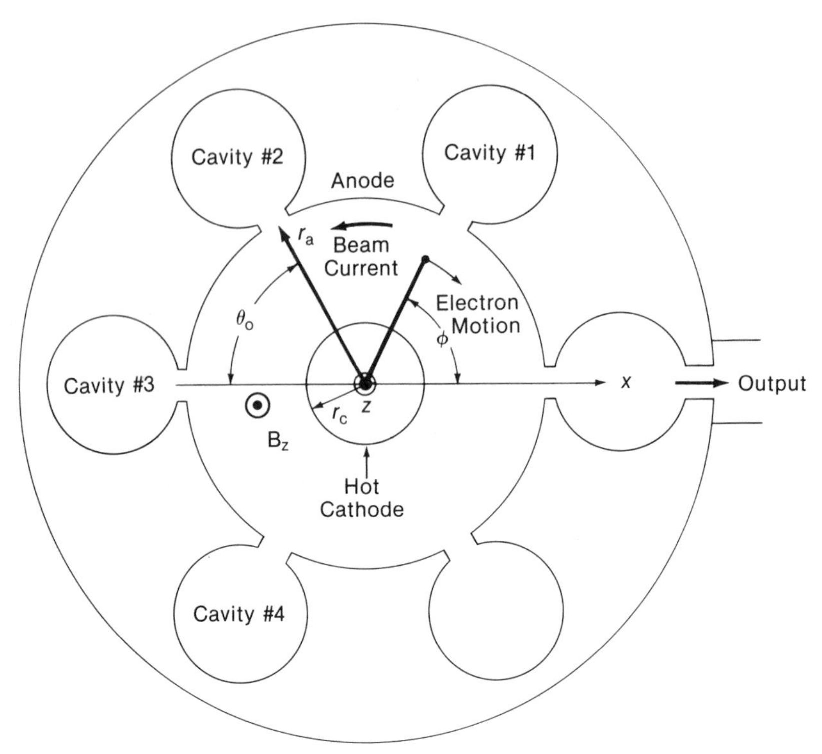

Fig. 15.1 Formation of a re-entrant electron beam.

presence of the magnetic flux B_z, the electrons are pushed in the tangential direction by the Lorentz force. The magnitude of the tangential Lorentz force is $qB\dot{r}$ (Newton). The torque on the electron with respect to the z-axis will therefore be $r(qB\dot{r})$ $(m - N)$. However, according to Newtonian mechanics, the torque is equal to the time rate of change of the angular momentum. Therefore

$$r(qB\dot{r}) = \frac{d}{dt}(rmr\dot{\phi})$$

or

$$\frac{d}{dt}(mr^2\dot{\phi}) = qBr\dot{r} \tag{15.1.1}$$

$$mr^2\dot{\phi} = qB\int_{t=0}^{t=t} r\dot{r}\,dt = qB\frac{r^2}{2}\Big|_{r_c}^{r} = qB\frac{r^2 - r_c^2}{2} \tag{15.1.2}$$

where it is assumed that the electron leaves the cathode at $t = 0$.

Then

$$\dot{\phi} = \frac{qB}{2m}\left(1 - \frac{r_c^2}{r^2}\right) \tag{15.1.3}$$

The electrons revolve around the cathode with the angular velocity given in Eq. (15.1.3). As seen from this equation, at $r = r_c$, $\dot{\phi} = 0$. This means that the electrons are not moving in a tangential direction on the surface of the cathode. The angular velocity reaches its greatest value at $r = r_a$, or close to the anode. Fortunately, the magnetron's microwave interaction is at or near the anode most of the time. Near the anode, the tangential velocity of electrons is

$$u_\phi = r_a\dot{\phi} \tag{15.1.4}$$

Combining Eqs. 15.1.3 and 15.1.4,

$$u_\phi = \frac{r_a qB}{2m}\left(1 - \frac{r_c^2}{r_a^2}\right) \tag{15.1.5}$$

If the height of the cylindrical structure is h, the thickness of the beam interacting with the microwave structure on the anode is δ, and the electron density near the anode is N, then the tangential electron beam current which is interacting with the microwave circuit on the anode cylinder is,

$$I_{o\phi} = h\,\delta N\,qu_\phi = \frac{r_a q^2 Bh\,\delta N}{2m}\left(1 - \frac{r_c^2}{r_a^2}\right) \tag{15.1.6}$$

This is the tangential beam current. It flows near the anode and the beginning and end are undefined. This is the re-entry electron beam current.

It is of interest to note that the anode voltage V_a does not enter in Eq. 15.1.6 explicitly, yet V_a is important to generate the radial velocity $\dot{r}$ as in Eq. 15.1.1. For example, a magnetron of $r_a = 7$ mm, $r_c = 2$ mm, and $B = 0.2$ T, with sufficient filament power and anode voltage should produce an angular velocity of Eq. 15.1.3 on electrons near the anode at $r = r_a$

$$\dot{\phi} = \frac{1.602 \times 10^{-19}\ \text{C} \times 0.2\ \text{T}}{2 \times 9.109 \times 10^{-31}\ \text{kg}}\left(1 - \frac{(2 \times 10^{-3}\ \text{m})^2}{(7 \times 10^{-3}\ \text{m})^2}\right)$$

$$= 1.62 \times 10^{10}\ \text{rad / s}$$

By Eq. 15.1.4, the tangential speed of electrons near the anode is

$$u_\phi = 7 \times 10^{-3}\ \text{m} \times 1.62 \times 10^{10}\ \text{rad}/\text{s} = 1.131 \times 10^{8}\ \text{m/s}$$

This is about one third of the velocity of light in a vacuum. If the tangential electron beam in the magnetron near the anode has an equivalent thickness of $\delta = 2$ mm and width of $h = 15$ mm, and the hot filament with sufficient anode voltage on the order of 10^3 V produces an electron density of $N = 10^{16}\ \text{m}^{-3}$ then, from Eq. 15.1.6, the tangential beam current is

$$\begin{aligned} I_{o\phi} &= h\,\delta N\,qu_\phi \\ &= 15 \times 10^{-3}\ \text{m} \times 2 \times 10^{-3}\ \text{m} \times 10^{16}\ \text{m}^{-3} \times 1.602 \\ &\quad \times 10^{-19}\ \text{C} \times 1.131 \times 10^{8}\ \text{m/s} \\ &= 5.44\ \text{A} \end{aligned}$$

This does not mean that the anode current of the magnetron is 5.44 A. This is only a tangential beam current. This large amount of current is created by recirculating electrons many times. The anode current will be created by the radial beam current I_{or}. This radial current is suppressed by the magnetic Lorentz force. Therefore, when the magnetron is oscillating, $I_{or} \ll I_{o\phi}$. This aspect will be investigated further.

15.2 Formation of Electron Poles

In Fig. 15.1, suppose that Cavity-1 is analogous to a buncher and Cavity-2 is analogous to a catcher in a klystron. Then Cavity-1 and Cavity-2 are spaced by the drift space. Cavity-1 and Cavity-2 then form a klystron with respect to the circulating beam current near the anode, as was studied in Section 15.1. Cavity-2 can be considered a buncher for Cavity-3. Though Cavity-3 is a catcher for Cavity-2, it is also a buncher for Cavity-4 and this process continues repeatedly around the anode. Each cavity receives an induction current by beam coupling and in turn gives velocity modulation and bunching in the circulating re-entry electron beam. The rotating bunched electrons take on a form of rotating electron clouds or electron poles. The cavity resonators pick their own resonance frequency signals while filtering out electromagnetic noise. By velocity modulation, drifting, bunching, catching, and proper synchronization, the electrons tend to bunch at the positive maxima of the microwave electric potential which rotates around the anode as the cavity resonators change their polarity sequentially. Borrowing the concept of the bunched electron beam in a klystron, from Eq. 14.3.1,

$$I_{m\phi} = I_{o\phi} + I_{\phi} \sin\left(\omega t - \frac{r_a \dot{\phi}}{u_{\phi}} \right) \tag{15.2.1}$$

where $I_{o\phi}$ is given by Eq. 15.1.6 and u_{ϕ} is given by Eq. 15.1.5. The microwave beam current I_{ϕ} is, from Eq. 14.2.14,

$$I_{\phi} = I_{o\phi} \frac{kv}{2V_o} \frac{l_d \omega}{u_{\phi}} \tag{15.2.2}$$

In this equation, V_o is an equivalent tangential acceleration potential. This is given by the law of conservation of energy:

$$qV_o = \tfrac{1}{2} m u_{\phi}^2 \tag{15.2.3}$$

$$V_o = \frac{m u_{\phi}^2}{2q} \tag{15.2.4}$$

This is called the equivalent kinetic potential. Substituting Eq. 15.2.4 in Eq. 15.2.2,

$$I_{\phi} = I_{o\phi} \frac{q k v l_d \omega}{m u_{\phi}^3} \tag{15.2.5}$$

Suppose that the particular magnetron presented in Section 15.1 is operated at the microwave output power at $P_o = 10^3$ W, with the output impedance across the interaction gap $Z_L = 10^2 + j0\ \Omega$, then the microwave voltage across the interaction gap is

$$v = \sqrt{2P_o R} = \sqrt{2 \times 10^3 \text{ W} \times 10^2\ \Omega} = 447 \text{ V}$$

The kinetic potential of this electron is, from Eq. 15.2.4,

$$V_o = \frac{m u_{\phi}^2}{2q} = \frac{9.109 \times 10^{-31} \text{ kg} \times \left[1.131 \times 10^8 \text{ m/s}\right]^2}{2 \times 1.602 \times 10^{-19} \text{ C}}$$

$$= 3.64 \times 10^4 \text{ V}$$

The equivalent drift distance l_d can be calculated by

$$l_d = \frac{2\pi r_a - ng}{n} \tag{15.2.6}$$

where n is the number of cavities in the magnetron and g is the interaction gap distance. In this particular example, $n = 6$, $g = 10^{-3}$ m, $r_a = 7 \times 10^{-3}$ m, and

$$l_d = \frac{2\pi \times 7 \times 10^{-3}\ \text{m} - 6 \times 10^{-3}\ \text{m}}{6} = 6.33 \times 10^{-3}\ \text{m}$$

The electron transit time from cavity to cavity is then

$$\tau_c = \frac{l_d}{u_\phi} = \frac{6.33 \times 10^{-3}\ \text{m}}{1.131 \times 10^8\ \text{m/s}} = 0.597 \times 10^{-12}\ \text{s}$$

The electron transit time across the interaction gap is

$$\tau_g = \frac{g}{u_\phi} = \frac{10^{-3}\ \text{m}}{1.131 \times 10^8\ \text{m/s}} = 0.0884 \times 10^{-12}\ \text{s}$$

The electron transit angle from cavity to cavity for the operating frequency $f = 2.45$ GHz is

$$\theta_c = \omega\tau_c = 2\pi f\tau_c = 2\pi \times 2.45 \times 10^9\ \text{Hz} \times 0.597 \times 10^{-12}\ \text{s} = 9.2 \times 10^{-3}\ \text{rad}$$

For the interaction gap,

$$\theta_g = \omega\tau_g = 2\pi \times 2.45 \times 10^9\ \text{Hz} \times 0.0884 \times 10^{-12}\ \text{s} = 1.36 \times 10^{-3}\ \text{rad}$$

With the beam coupling coefficient of the interaction gap k = 0.75, the magnitude of microwave beam current I_ϕ is calculated from Eq. 15.2.2 as

$$I_\phi = 5.44\ \text{A} \times \frac{0.75 \times 477\ \text{V} \times 9.2 \times 10^{-3}\ \text{rad}}{2 \times 3.64 \times 10^4\ \text{V}}$$

$$= 5.44\ \text{A} \times 4.52 \times 10^{-5} = 246 \times 10^{-6}\ \text{A}$$

This is not much current, but this current at Gap-6 is due to modulation at Gap-5. In reality, each of the six gaps contributes to the velocity modulation at each gap including itself, due to the re-entry structure of the magnetron. As seen from Eq. 15.2.2, the microwave beam current I_ϕ is proportional to $\theta_c = \omega l_d/u_\phi = \omega\tau_c$. If the magnetron reaches a steady state of oscillation within $\tau_c = 10^{-9}$ s of operation then, taking the effect of six gaps,

$$I_\phi = 6 \times 246 \times 10^{-6}\ \text{A} \times \frac{10^{-9}\ \text{s}}{0.597 \times 10^{-12}\ \text{s}} = 2.472\ \text{A}$$

During one nanosecond, or $\tau_c = 10^{-9}$ s, electrons rotate approximately 10^{-9} s/6 × ($\tau_c + \tau_g$) = 10^{-9} s/(6 × (0.597 + 0.0884) × 10^{-12} s) = 243 times in the magnetron. By this time, the rotating or circulating tangential beam current has 5.441 A dc and 2.47 A microwave frequency current.

15.3 Magnetron Oscillation

If the circulating bunched electron beam current in Eq. 15.2.1 is combined with the anode configuration shown in Fig. 15.1, the magnetron is considered to be a re-entry configuration of a klystron. In the case of Fig. 15.1, there are six klystrons connected in cascade, each cavity playing the role of a catcher for the preceding cavity as well as the role of a buncher for the succeeding cavity. The circulating beam current goes through six equivalent two-cavity klystrons per revolution. However the beam current re-enters. After one revolution, the second revolution will begin, and this process continues as long as the magnetron is energized. Practically, a magnetron can be considered a cascade of infinite numbers of two-cavity klystrons. According to Eq. 14.4.12, the power gain between the nth cavity and the $n + 1^{th}$ cavity is

$$A_n = \frac{\mathrm{k}_{n+1} R_{\mathrm{L},n+1} I_{\mathrm{o}\phi}^2}{G_{gn}} \left(\frac{J_1(X_n)}{v_n} \right)^2 \tag{15.3.1}$$

where the tangential bunching parameter

$$X_n = \frac{\mathrm{k}_n \mathrm{v}_n}{2V_\mathrm{o}} \frac{l_\mathrm{d}\omega}{u_\phi} \tag{15.3.2}$$

and V_o is given by Eq. 15.2.4.

Assuming proper phase, when the n increases, v_n increases and $J_1(X_n)$ decreases. As a result, A_n decreases and saturation or steady state will set in.

$$\text{At } n \to \infty,\ A_n \to 1 \tag{15.3.3}$$

If oscillation starts at the noise power level

$$P_1 = \mathrm{k}T\Delta f \qquad (\mathrm{W}) \tag{15.3.4}$$

where k is the Boltzmann constant 1.389054×10^{-23} J/K, T is the absolute temperature of the Cavity-1, and Δf is the frequency bandwidth of the cavity resonator. The power at the second cavity is

$$P_2 = A_1 P_1 \tag{15.3.5}$$

The power at the third cavity is

$$P_3 = A_2 P_2 = A_2 A_1 P_1 \tag{15.3.6}$$

Repeating the same procedure n times,

$$P_n = A_{n-1} A_{n-2} A_{n-3} \cdots A_3 A_2 A_1 P_1 = \left(\prod_{m=1}^{m=n-1} A_m \right) P_1 \tag{15.3.7}$$

After the magnetron has reached steady state or saturation, the nth cavity can be any cavity. If an external circuit is connected as illustrated in Fig. 15.1, the output can be taken from the magnetron.

Substituting Eqs. 15.3.1 and 15.3.4 in Eq. 15.3.7,

$$P_n = \mathrm{k}T\Delta f \prod_{m=1}^{m=n-1} \frac{\mathrm{k}_{m+1} R_{\mathrm{L},\, m+1} I_{\mathrm{o}\phi}^2}{G_{gm}} \left(\frac{J_1(X_m)}{v_m} \right)^2 \tag{15.3.8}$$

In most magnetrons, all cavities are identical except the one with the external circuit connected. If the load coupling is light, Eq. 15.3.8 can be slightly simplified as

$$P_n = \mathrm{k}T\Delta f \frac{\mathrm{k}^{n-1} R_{\mathrm{L}}^{n-1} I_{\mathrm{o}\phi}^{2(n-1)}}{G_{\mathrm{g}}^{n-1}} \prod_{m=1}^{m=n-1} \left(\frac{J_1(X_m)}{v_m} \right)^2 \tag{15.3.9}$$

Proper phasing can be accomplished by controlling the drift electrical phase angle

$$\theta = \omega \frac{l_{\mathrm{d}}}{u_\phi} \tag{15.3.10}$$

The electron transit time between two adjacent cavities is l_{d}/u_ϕ. Therefore, $\omega(l_{\mathrm{d}}/u_\phi)$ is the electron transit angle. If the angle of the arc l_{d} with respect to the z-axis is θ_{o}, then

$$l_{\mathrm{d}} = r_{\mathrm{a}} \theta_{\mathrm{o}} \tag{15.3.11}$$

and

$$u_\phi = r_{\mathrm{a}} \dot{\phi} \tag{15.3.12}$$

Substituting Eqs. 15.3.11 and 15.3.12 in Eq. 15.3.10,

$$\theta = \omega \frac{\theta_{\mathrm{o}}}{\dot{\phi}} \tag{15.3.13}$$

As seen from Eq. 15.1.3, $\dot{\phi}$ is controlled by the applied magnetic flux density B.

The steady state cavity gap voltage can be obtained from the conditions in Eq. 15.3.3. Combining Eqs. 15.3.3 and 15.3.1,

$$\frac{J_1(X_n)^2}{v_n} = \frac{G_{gn}}{\mathrm{k}_{n+1} R_{\mathrm{L},\,n+1} I_{\mathrm{o}\phi}^2} \qquad (15.3.14)$$

This is a transcendental equation of v_n, since $J_1(X_n)$ contains v_n. The solution of Eq. 15.3.14 with respect to v_n is the steady state cavity gap voltage of the magnetron. If the gap load resistance is R_L, the output power is

$$P_\mathrm{o} = \frac{v_n^2}{2R_\mathrm{L}} \qquad (15.3.15)$$

In the example of a 2.45 GHz, 1 kW magnetron presented in Sections 15.1 and 15.2, the bunching parameter between two adjacent cavities under the presaturation condition is, using Eq. 15.3.2 with $v_n = 477 \times 10^{-6}$ V,

$$X_n = \frac{\mathrm{k}_n v_n}{2V_\mathrm{o}} \frac{l_\mathrm{d}}{u_\phi} \omega = \frac{0.75 \times 477 \times 10^{-6}\ \mathrm{V} \times 9.2 \times 10^{-3}\ \mathrm{rad}}{2 \times 3.64 \times 10^4\ \mathrm{V}}$$

$$= 4.52 \times 10^{-11}$$

This is very small. Then

$$J_1(X_n) \approx \frac{X_n}{2}$$

$$\frac{J_1(X_n)}{v_n} \approx \frac{X_n}{2v_n} = \frac{\mathrm{k}_n}{4\mathrm{V}_\mathrm{o}} \theta_\mathrm{c} = \frac{0.75}{4 \times 3.64 \times 10^4\ \mathrm{V}} \times 9.2 \times 10^{-3}\ \mathrm{rad}$$

$$= 0.474 \times 10^{-7}\ \mathrm{V}^{-1}$$

Before saturation $A_n \geq 1$. Then, from Eq. 15.3.1,

$$\frac{R_{\mathrm{L},\,n+1}}{G_{gn}} = \frac{A_n}{\mathrm{k}_{n+1} I_{\mathrm{o}\phi}^2 (J_1(X_n)/v_n)^2}$$

$$\geq \frac{1}{0.75 \times (5.44\ \mathrm{A})^2 (0.474 \times 10^{-7}\ \mathrm{V}^{-1})^2}$$

$$= 0.2 \times 10^{14} = 20 \times 10^{12}\ \Omega^2$$

This means that the interaction gaps must be lightly loaded, on the order of 10^6 Ω each, except output Cavity-6 in Fig. 15.1. As a matter of fact, five cavities are not externally loaded at all. So R_L

or $1/G_g$ can be as high as 10^6 Ω. At any rate, the cavity to cavity gain is not much more than 1 or

$$A_n \approx 1 + \varepsilon$$

where ε is a small quantity $\varepsilon \ll 1$.

However, there are six cavities in this case and the electron beam re-enters. The total gain will therefore be

$$\prod_{m=1}^{m=n-1} A_m \approx (1+\varepsilon)^N \times \frac{\theta}{2\pi} \times T_s$$

$$= (1+\varepsilon)^6 \times \frac{1.63 \times 10^{10} \text{ rad / s}}{2\pi} \times 0.25 \times 10^{-6} \text{ s}$$

In this calculation, T_s is the output stabilizing time and assumed to be $T_s = 0.25 \times 10^{-6}$ s. Then the equivalent gain is

$$\prod_{m=1}^{m=n-1} A_m = (1+\varepsilon)^{1.55 \times 10^4}$$

If $\varepsilon = 0.01$, then

$$10 \log\left(\prod_{m=1}^{m=n-1} A_m\right) = (0.388 \times 10^5) \log_{10}(1+0.01) \text{ dB}$$

$$= 0.388 \times 10^5 \times 4.32 \times 10^{-3} \text{ dB}$$

$$= 167.4 \text{ dB}$$

If the Q of the cavity resonator is $Q = 200$ at operating frequency $f = 2.45$ GHz, then the frequency bandwidth Δf is

$$Q = \frac{f}{\Delta f}$$

$$\Delta f = f/Q = 2.45 \times 10^9 \text{ Hz}/200 = 12.25 \times 10^6 \text{ Hz}$$

From Eq. 15.3.4, the noise power is then

$$P_1 = kT\Delta f = 1.38054 \times 10^{-23} \text{ J/K} \times 500 \text{ K} \times 12.25 \times 10^6 \text{ Hz}$$

$$= 8.456 \times 10^{-14} \text{ W}$$

$$= -130.7 \text{ dBW} = -100.7 \text{ dBm}$$

Here the electron temperature is assumed to be $T = 500$ K. Then the amplified output of the magnetron at the output cavity gap is from Eq. 15.3.7,

$$P_n \approx \prod_{m=1}^{m=n-1} A_m P_1 = 167.4 \text{ dBW} - 130.7 \text{ dBW} = 36.7 \text{ dBW}$$

$$= 30 \text{ dBW} + 6.7 \text{ dBW} = 10^3 \text{ W} + 4.7 \text{ W}$$

$$= 1004.7 \text{ W}$$

The oscillation starts at the noise level of 8.456×10^{-14} W. After only one quarter microsecond, the power grows to 1004.7 W, with a cavity-to-cavity gain of 1.01 = 0.043 dB. By this time, the magnetron is saturated and reaches the equilibrium condition. Since this microwave power is available at the interaction gap of the output cavity resonator, if the impedance matching is perfect then, from Eq. 15.3.15,

$$v_n = \sqrt{2R_L P_o}$$

If the microwave power is launched into a waveguide of R_L = 400 Ω, then the waveguide microwave voltage will be

$$v_n = \sqrt{2 \times 400\ \Omega \times 1004.7\ \text{W}} = 896.5 \text{ V}$$

In the magnetron, the circulating bunched beam is used over and over again. It consumes relatively small amounts of anode current due to the strong dc magnetic flux density B_z. The axial magnetic fields tend to prevent electrons from reaching the anode. Strong magnetron oscillation occurs near the anode currents, cut off by the axial magnetic field. Some magnetrons operate under dc-to-microwave power conversion efficiencies greater than 90%. One of the reasons that the magnetron is popular for use as a power oscillator is its high efficiency over other microwave electron devices.

In a magnetron at a location (r, ϕ, z), if the electric potential at the point is V volts and the potential energy of the electron is qV joules, then the kinetic energy of the electron associated with the radial velocity $\dot{r}$ is $\frac{1}{2}m\dot{r}^2$ joules. The kinetic energy of the electron associated with the tangential velocity $(r\dot{\phi})$ is $\frac{1}{2}m(r\dot{\phi})^2$. Therefore, by the law of conservation of energy,

$$\tfrac{1}{2}m\dot{r}^2 + \tfrac{1}{2}m\left(r\dot{\phi}\right)^2 = qV \qquad (15.3.16)$$

Substituting Eq. 15.1.3 in Eq. 15.3.16,

$$\tfrac{1}{2}m\dot{r}^2 + \tfrac{1}{2}mr^2\left(\frac{\bar{q}B}{2m}\right)^2\left(1 - \frac{r_c^2}{r^2}\right)^2 = qV \qquad (15.3.17)$$

At the anode,

$$\left.\begin{aligned} r &= r_a \\ V &= V_a \end{aligned}\right\} \tag{15.3.18}$$

Then

$$\tfrac{1}{2}m\dot{r}_a^2 + \tfrac{1}{2}mr_a^2\left(\frac{qB}{2m}\right)^2\left(1 - \frac{r_c^2}{r_a^2}\right)^2 = qV_a \tag{15.3.19}$$

Under a constant anode voltage, if B increases, the electron will not reach the anode at a certain value of the magnetic flux B_c. This flux density is called the cutoff magnetic flux density. If the cutoff of the anode current occurs at $r = r_a$, then $\dot{r}_a = 0$ in Eq. 15.3.19. Under this condition, if Eq. 15.3.19 is solved for B, the cut off flux density is obtained. From Eq. 15.3.19,

$$\frac{r_a^2 q^2 B_c^2}{8m}\left(1 - \frac{r_c^2}{r_a^2}\right)^2 = qV_a$$

thus,

$$B_c^2 = \frac{8mV_a r_a^2}{q\left(r_a^2 - r_c^2\right)^2} \tag{15.3.20}$$

This equation gives the cutoff magnetic flux density and is a measure of the proper amount of magnetic flux density to be given to the magnetron for oscillation. The optimum value of B can be obtained by trimming B around the value of B_c.

Solving Eq. 15.3.19 for $\dot{r}_a^2$,

$$\dot{r}_a^2 = \frac{2q}{m}V_a - \left(\frac{qB}{2m}\right)^2\left(\frac{r_a^2 - r_c^2}{r_a^2}\right)^2 \tag{15.3.21}$$

If the electron density near the anode is N (m^{-3}) and the height of the anode cylinder is h (m), then the dc anode current can be obtained by

$$I_a = qN\dot{r}_a(2\pi r_a h) \tag{15.3.22}$$

In this equation, $qN\dot{r}_a$ is the dc current density at the anode and $2\pi r_a h$ is the total surface area of the anode.

Substituting Eq. 15.3.21 in Eq. 15.3.22,

$$I_a = 2\pi r_a hqN\sqrt{\frac{2q}{m}V_a - \left(\frac{qB}{2m}\right)^2\left(\frac{r_a^2 - r_c^2}{r_a^2}\right)^2} \tag{15.3.23}$$

The dc anode loss is then

$$P_{dc} = V_a I_a \tag{15.3.24}$$

This loss may be small at the cutoff condition, but power will be needed to heat the hot cathode. Power will be needed to generate B if the magnetron is biased by an electromagnet and must be included to calculate the magnetron efficiency.

In the example presented in this section, near the anode at $r \approx r_a$, the tangential kinetic energy of an electron is

$$\begin{aligned}\tfrac{1}{2}m\left(r_a\dot{\phi}\right)^2 &= \tfrac{1}{2}(9.1085 \times 10^{-31}\ \text{kg}) \times \left(1.131 \times 10^8\ \text{m/s}\right)^2 \\ &= 5.83 \times 10^{-15}\ \text{J}\end{aligned}$$

Since the magnetron is oscillating, $\dot{r}_a = 0$. From Eq. 15.3.16,

$$\begin{aligned}V_a &= \frac{1}{q}\frac{1}{2}m\left(r_a\dot{\phi}\right)^2 = 5.83 \times 10^{-15}\ \text{J}/1.602 \times 10^{-19}\ \text{C} \\ &= 3.64 \times 10^4\ \text{V}\end{aligned}$$

The cutoff magnetic flux density B_c can be calculated from Eq. 15.3.20:

$$\begin{aligned}B_c &= \sqrt{\frac{8mV_a}{q}}\,\frac{r_a}{r_a^2 - r_c^2} \\ &= \sqrt{\frac{8 \times 9.1085 \times 10^{-31}\ \text{kg} \times 3.64 \times 10^4\ \text{V}}{1.602 \times 10^{-19}\ \text{C}}} \\ &\quad \cdot \frac{7 \times 10^{-3}\ \text{m}}{\left(7 \times 10^{-3}\ \text{m}^2 - (2 \times 10^{-3}\ \text{m}\right)^2} = 0.2\ \text{T}\end{aligned}$$

If $B = 0.2$ T, most electrons do not reach the anode, but some electrons reach due to fringing fields, debunching, and mutual repulsion. Thus, the anode current is extremely small.

If the efficiency of the magnetron is $\eta = 0.5$, then

$$\eta \approx \frac{P_o}{P_o + P_{dc}} = 0.5$$

or

$$P_{dc} = \left(\frac{1}{\eta} - 1\right)P_o = \left(\frac{1}{0.5} - 1\right)\text{kW} = 1\ \text{kW}$$

Then, from Eq. 15.3.24,

$$I_a = \frac{P_{dc}}{V_a} = \left(\frac{1}{\eta} - 1\right)\frac{P_o}{V_a}$$

$$= \left(\frac{1}{0.5} - 1\right)\frac{10^3\ \text{W}}{3.64 \times 10^4\ \text{V}} = 0.027\ \text{A}$$

If $\eta = 0.8$, then

$$I_a = \left(\frac{1}{0.8} - 1\right)\frac{10^3\ \text{W}}{3.64 \times 10^4\ \text{V}} = 0.00675\ \text{A}$$

Likewise, if $\eta = 0.95$, then $I_a = 0.001$ A. Thus, by monitoring the anode current, the approximate efficiency is known.

15.4 Magnetron Principles

The underlying principles for magnetron oscillation are the same as for klystrons. Velocity modulation, drifting, bunching, and dynamic induction generate microwaves in magnetrons. The mechanism of microwave power generation therefore resembles the mechanism of a multiple cavity klystron. However, in magnetrons the re-entry structure of the electron beam is employed. The electron beam which deposits microwave power at the catcher cavity re-enters the drift region leading to the first buncher cavity. The bunched electron beam is then recirculated. The tangentially moving bunched electrons are called electron poles. The electron poles induce microwave current in the resonating cavities.

The circulating electron beam is formed by a radial dc electric field and an axial dc magnetic field from a magnet. The dc anode current is reduced by controlling the axial dc magnetic field. The high efficiency of a magnetron is due to two reasons. The magnetron oscillates only when the dc axial magnetic flux is sufficiently large enough to cut off the anode current, reducing the dc power loss. The other reason is that the electron beam is recirculated, and the bunched tangentially moving electron beam current is used repeatedly.

Problems

15.1 Sketch the possible motion of electrons in a magnetron.

15.2 Write down the equations of motion for an electron in a magnetron.

15.3 Define a magnetron.
15.4 Sketch the general structure of a magnetron and explain how it works.
15.5 Explain the reason that the magnetic flux is applied in the axial direction of a magnetron.
15.6 Obtain analytically the radial Lorentz force and the tangential Lorentz force in a magnetron.
15.7 Analytically find the tangential velocity of an electron.
15.8 Propose various approaches to determine the thickness of an interacting electron beam current near the anode in a magnetron.
15.9 Derive a formula for the tangential electron beam current near the anode of a magnetron.
15.10 Express the circulating electron beam current as a function of the anode voltage V_a.
15.11 Define the electron pole.
15.12 Sketch the electron poles.
15.13 Drift space in a klystron and the drift space in a magnetron are quite different. Clearly differentiate between them.
15.14 Find an equivalent acceleration potential for the circulating electron beam in a magnetron.
15.15 Formulate analytically the output of a magnetron.
15.16 Obtain the steady state gap voltage for a magnetron cavity.
15.17 Derive an equation to calculate the efficiency of a magnetron.
15.18 Estimate the efficiency of a domestic cooking magnetron in a microwave oven.

References

1 G. B. Collins, "Microwave Magnetrons." McGraw-Hill, New York, 1948.
2 H. J. Reich, P. F. Ordung, H. L. Krauss, and J. G. Skalnik, "Microwave Theory and Techniques." Van Nostrand, Princeton, New Jersey, 1953.
3 R. F. Soohoo, "Microwave Electronics." Addison-Wesley, Reading, Massachusetts, 1971.
4 R. G. E. Hutter, "Beam and Wave Electronics in Microwave Tubes." Van Nostrand, Princeton, New Jersey, 1960.
5 T. K. Ishii, "Microwave Engineering." Ronald Press, New York, 1966.
6 S. Y. Liao, "Microwave Electron-Tube Devices." Prentice Hall, Englewood Cliffs, New Jersey 1988.
7 J. T. Coleman, "Microwave Devices." Reston Publ., Reston, Virginia, 1982.
8 G. E. Dombrowski, Numerical simulation of magnetron oscillator and crossed-field amplifiers. *IEEE Trans. Electron Devices* **ED-35**(11), 2060–2067 (1988)

16

Traveling Wave Devices

16.1 Velocity Modulation by Traveling Waves

The speed of electromagnetic waves in traveling wave devices is usually much greater than the velocity of electrons in an ordinary electron beam. The velocity of electrons accelerated by an accelerator potential V_a is given by Eq. 14.1.2 or

$$u = \sqrt{\frac{2qV_a}{m}} \tag{16.1.1}$$

If V_a is large so that u approaches the velocity of light c, then the electron mass m becomes relativistic or

$$m^* = \frac{m}{\sqrt{1 - \left(\frac{u}{c}\right)^2}} \tag{16.1.2}$$

In most thermionic devices,

$$u \ll c \tag{16.1.3}$$

This case is inconvenient if the electrons and electromagnetic waves are to interact. To reduce the relative speed of electromagnetic waves along the electron beam, a slow wave structure as illustrated in Fig. 16.1 has been invented [1,13]. In this illustration,

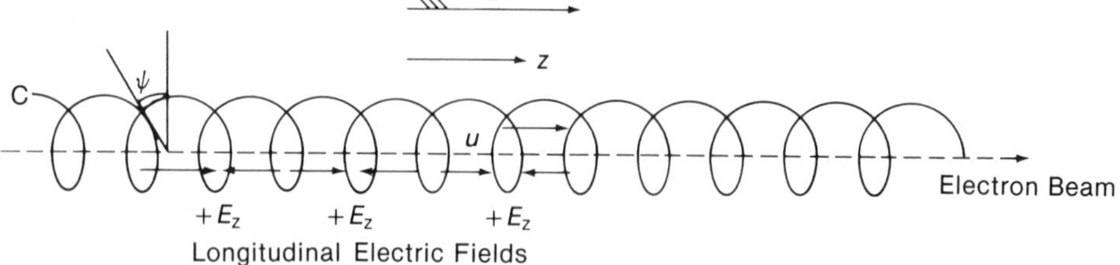

Fig. 16.1 Electron beam and slow wave structure.

a helix is shown. Assuming that the electromagnetic waves, or microwaves, propagate along the helix with the velocity of light c, and the angle of the helix is ψ, then the longitudinal component of the tangential velocity c is [1–3]

$$u_z = c \sin \psi \tag{16.1.4}$$

As seen from this equation, by adjusting the angle of the helix ψ, the axial propagation velocity of microwaves u_z can be adjusted to any value from 0 to c. It is therefore possible to adjust u_z to be equal to the electron velocity u which is in the direction along the helix axis as shown in Fig. 16.1. If

$$u_z \approx u \tag{16.1.5}$$

the longitudinal electric field component E_z, which is associated with the traveling waves, will travel with the velocity u_z. The microwave field E_z can then be expressed in the following traveling wave form:

$$E_z = E_o e^{-\alpha z} e^{-j(\beta z - \omega t)} \tag{16.1.6}$$

where E_o is the amplitude of the longitudinal electric field at $z = 0$, α is the attenuation constant in nepers per meter, and β is the phase constant in radians per meter.

Depending on the phase angle $(\beta z - \omega t)$, the longitudinal electric field E_z can be positive or negative. Electrons in the positive field will be decelerated and electrons in the negative field will be accelerated. Since the electron velocity u and the wave velocity u_z are made equal, the electrons and fields travel at the same speed. This means that electrons under the accelerating field are kept accelerating and electrons in the decelerating field are kept decelerating. This is called velocity modulation on the electron beam. In this traveling wave structure, the electrons travel at approximately equal speeds of accelerating or decelerating fields.

The duration of velocity modulation is therefore large in both space and time in comparison with the klystron. Due to the lengthy effective velocity modulation time, the electron beam is tightly bunched. After velocity modulation, the velocity of electrons may be expressed in the form of

$$u = u_o + u_m e^{-\alpha_e z} e^{-j(\beta_e z - \omega t)} \tag{16.1.7}$$

where u_o is the dc velocity of the electron in the beam and u_m is the amplitude of velocity modulation at $z = 0$. α_e is the attenuation constant of the velocity modulation. α_e is actually negative because, if the amplifier tube is properly adjusted, the velocity modulation grows as it propagates. β_e is the phase constant of the propagation for the velocity modulation in the electron beam.

If the electron beam is velocity modulated as shown in Eq. 16.1.7, then the density of the electrons in the electron beam is also modulated. The modulated electron density can be expressed by

$$\rho = \rho_o + \rho_m e^{-\alpha_e z} e^{-j(\beta_e z - \omega t)} \tag{16.1.8}$$

where ρ_o is the dc charge density and ρ_m is the amplitude of microwave electron charge density at $z = 0$.

The electron beam current density is therefore

$$\begin{aligned} i = \rho u = \rho_o u_o &+ (\rho_o u_m + \rho_m u_o) e^{-\alpha_e z} e^{-j(\beta_e z - \omega t)} \\ &+ \rho_m u_m e^{-2\alpha_e z} e^{-2j(\beta_e z - \omega t)} \end{aligned} \tag{16.1.9}$$

The bunched electron beam inherently contains the second harmonic. The dc beam current density is then

$$i_o \equiv \rho_o u_o \tag{16.1.10}$$

and the microwave beam current density at the fundamental frequency is

$$i_m = (\rho_o u_m + \rho_m u_o) e^{-\alpha_e z} e^{-j(\beta_e z - \omega t)} \tag{16.1.11}$$

If the cross section of the electron beam is S (m^2), the electron bunched beam current is

$$\begin{aligned} I = Si = S\rho_o u_o &+ S(\rho_o u_m + \rho_m u_o) e^{-\alpha_e z} e^{-j(\beta_e z - \omega t)} \\ &+ S\rho_m u_m e^{-2\alpha_e z} e^{-2j(\beta_e z - \omega t)} \end{aligned} \tag{16.1.12}$$

For example, the acceleration voltage for a typical traveling wave tube is on the order of 10^3 V. From Eq. 16.1.1, the dc velocity of the electron is

$$u_o = \sqrt{\frac{2qV_a}{m}} = \sqrt{\frac{2 \times 1.602 \times 10^{-19} \times 10^3}{9.1085 \times 10^{-31}}}$$

$$= 1.876 \times 10^7 \text{ m/s}$$

This is about one tenth the velocity of light c, which is typical for electrons in microwave thermionic devices. As far as the relativistic factor is concerned, from Eq. 16.1.2,

$$1\bigg/\sqrt{1 - \left(\frac{u_o}{c}\right)^2} = 1\bigg/\sqrt{1 - \left(\frac{1.876 \times 10^7 \text{ m/s}}{3 \times 10^8 \text{ m/s}}\right)^2} = 1.002$$

Thus, at $V_a = 1000$ V, the relativistic contribution is small but discernible. So $u \ll c$ in Eq. 16.1.3 means that, in reality, $u \approx c/10$.

If a helix is used to slow the wave velocity to $u_z = u_o = 1.876 \times 10^7$ m/s, then the angle of the helix ψ must be adjusted to, using Eq. 16.1.4,

$$\psi = \sin^{-1}\frac{u_z}{c} = \sin^{-1}\frac{1.876 \times 10^7 \text{ m/s}}{3 \times 10^8 \text{ m/s}} = 0.0626 \text{ rad}$$

$$= 3.59°$$

If the length of active region of the helix is $L = 30$ cm, the duration of velocity modulation τ is approximately

$$\tau \approx \frac{L}{u_o} = \frac{30 \times 10^{-2} \text{ m}}{1.876 \times 10^7 \text{ m/s}} = 15.99 \times 10^{-9} \text{ s}$$

On the other hand, in the case of a klystron with the gap distance $g = 0.5$ mm,

$$\tau = \frac{g}{u_o} = \frac{0.5 \times 10^{-3} \text{ m}}{1.876 \times 10^7 \text{ m/s}} = 0.0267 \times 10^{-9} \text{ s}$$

A traveling wave tube therefore has a longer period of velocity modulation than a klystron. Velocity modulation in traveling wave tubes is therefore more effective than in klystrons.

According to transmission line theory, the phase velocity u_z is related to phase constant β by [6,7]

$$u_z = \frac{\omega}{\beta} \tag{16.1.13}$$

Then

$$\beta = \frac{\omega}{u_z} \approx \frac{2\pi f}{u_o} \tag{16.1.14}$$

If $f = 3000$ MHz, then

$$\beta = \frac{2\pi \times 3 \times 10^9 \text{ s}^{-1}}{1.1876 \times 10^7 \text{ m/s}} = 1.005 \times 10^3 \text{ rad/m}$$

Then the wavelength in the traveling wave tube is

$$\lambda_l = \frac{2\pi}{\beta} = \frac{2\pi \text{ rad}}{1.005 \times 10^3 \text{ rad/m}} = 6.25 \times 10^{-3} \text{ m}$$

In free space, a microwave of $f = 3000$ MHz should have a wavelength of

$$\lambda_o = \frac{c}{f} = \frac{3 \times 10^8 \text{ m/s}}{3000 \times 10^6 \text{ s}^{-1}} = 0.1 \text{ m}$$

If the beam current is $I_o = 5$ mA, and the beam radius $r_o =$ 0.5 mm, then the cross-sectional area S is

$$S = \pi r_o^2 = \pi \times (0.5 \times 10^{-3} \text{ m})^2 = 0.7854 \times 10^{-6} \text{ m}^2$$

From Eq. 16.1.10, the electron charge density of this beam is

$$\rho_o = \frac{i_o}{u_o} = \frac{I_o/S}{u_o} = \frac{5 \times 10^{-3} \text{ A}}{0.7854 \times 10^{-6} \text{ m}^2 \times 1.876 \times 10^7 \text{ m/s}}$$

$$= 3.39 \times 10^{-4} \text{ C/m}^3$$

At this point in analysis, there is no way of knowing the values of u_m and ρ_m, but the values of u_o and ρ_o give the upper bounds of these modulated values.

16.2 Dynamic Induction in a Traveling Wave Structure

If the beam coupling coefficient between the electron beam and the slow wave structure such as a helix or meanderline or interdigital circuit is k, the induced microwave current in the circuit is, by the definition of the beam coupling coefficient,

$$I_1 = \mathrm{k} I_{b1} \tag{16.2.1}$$

where I_1 is the fundamental frequency component of the induced

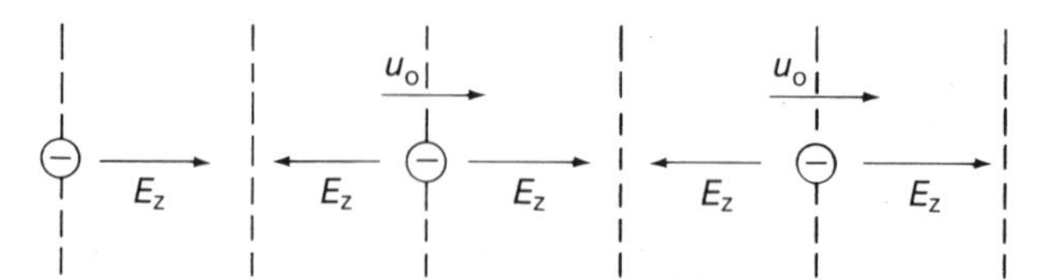

Fig. 16.2 Centers of the electron bunches and longitudinal electric fields in a traveling wave tube.

microwave current in the wave slowing circuit due to the bunched beam current shown by Eq. 16.1.12. Then, from Eq. 16.1.12,

$$I_{\mathrm{b1}} = S(\rho_{\mathrm{o}} u_{\mathrm{m}} + \rho_{\mathrm{m}} u_{\mathrm{o}}) e^{-\alpha_{\mathrm{e}} z} \tag{16.2.2}$$

According to Eq. 14.3.12, the beam coupling coefficient between the bunched electron beam current and the inducing interaction catcher gap is given by

$$\mathrm{k}_1 = \frac{\sin\left(\dfrac{\omega l}{2u_{\mathrm{o}}}\right)}{\dfrac{\omega l}{2u_{\mathrm{o}}}} \tag{16.2.3}$$

In this equation, l is the space length of the interaction region. Unlike the space length of the klystron, the traveling wave tube interaction space length l is not well defined physically. However, it is possible to define l electronically.

The relationship between the bunched electrons and the longitudinal electric fields due to the wave slowing structure is illustrated in Fig. 16.2. The center of the electron bunch is represented by the symbol ⊖. As explained in Section 14.1, to put energy in the beam, the electrons must be accelerated by the electric field. If the electrons are decelerated by the electric field, then the electric field is receiving energy from the electrons. In other words, kinetic energy of the electron is converted to the energy of microwave electric fields. By careful observation of Fig. 16.2, it appears that, as soon as the center of the bunch enters the space of the decelerating field, the microwave field receives energy from the bunched electrons. The length of the deceleration region z_l is one half-electronic wavelength or, in Eq. 16.1.8,

$$\beta_{\mathrm{e}} z_l = \pi \tag{16.2.4}$$

According to Eq. 16.1.8, if the peak microwave electron charge density of the bunched electrons is ρ_{m} and the cross-sectional area of the electron beam is S, then the microwave electron charge in the beam is, at a location z,

$$Sz_l\rho_m e^{-\alpha_e z} = \frac{S\pi\rho_m}{\beta_e} e^{-\alpha_e z} \tag{16.2.5}$$

By electrostatic induction, the same amount of positive charge is induced in the wave-slowing structure. The induced positive charge will move at approximately the same speed of the bunched electrons after all the induced positive charge is bound to the bunched electrons. The induced microwave longitudinal current in the wave-slowing circuit is then

$$I_{1l} \approx u_o \frac{S\pi\rho_m}{\beta_e} e^{-\alpha_e z} \tag{16.2.6}$$

If the wave-slowing circuit is a helix of angle ψ and the amplitude of the microwave current along the helix pitch is I_1, then

$$I_{1l} = I_1 \sin\psi \tag{16.2.7}$$

$$I_1 = \frac{I_{1l}}{\sin\psi} = \frac{u_o S\pi\rho_m}{\beta_e \sin\psi} e^{-\alpha_e z} \tag{16.2.8}$$

Since microwaves in a traveling wave structure are adjusted to travel at approximately equal speeds with the electrons, the phase constant β_e is, by definition,

$$\beta_e \equiv \frac{2\pi}{\lambda_e} = \frac{2\pi f}{\lambda_e f} \approx \frac{\omega}{u_o} \tag{16.2.9}$$

where λ_e is the wavelength of the electronic waves in the beam.

Substituting Eq. 16.2.9 in Eq. 16.2.8,

$$I_1 = \frac{u_o^2 S\pi\rho_m}{\omega \sin\psi} e^{-\alpha_e z} \tag{16.2.10}$$

In properly adjusted traveling wave structures, α_e is negative. I_1 increases as it travels in the wave-slowing circuit.

Combining Eqs. 16.2.1, 16.2.2, and 16.2.10,

$$\frac{u_o^2 S\pi\rho_m}{\omega \sin\psi} e^{-\alpha_e z} = \mathrm{k}S(\rho_o u_m + \rho_m u_o) e^{-\alpha_e z}$$

Thus,

$$\mathrm{k} = \frac{\pi u_o^2 \rho_m}{\omega(\rho_o u_m + \rho_m u_o)\sin\psi} \tag{16.2.11}$$

Combining Eqs. 16.2.11 and 16.2.3,

$$\frac{\sin\left(\dfrac{\omega l}{2u_o}\right)}{\dfrac{\omega l}{2u_o}} = \frac{\pi u_o^2 \rho_m}{\omega(\rho_o u_m + \rho_m u_o)\sin\psi} \tag{16.2.12}$$

This is a transcendental equation with respect to l. A numerical solution of Eq. 16.2.12 for l will give an equivalent interaction distance between the bunched electron beam and the wave-slowing circuit.

When a bunched electron beam current (16.1.9) is put through a periodical circuit structure, by dynamic induction, a longitudinal microwave current given by Eq. 16.2.6 will be generated. If the wave-slowing structure is a helix, the microwave helix current is given by Eq. 16.2.10. This current travels along the helix with a velocity approximately equal to the electronic wave velocity in the beam. The microwave current in the beam is therefore, in a phasor form,

$$I_1 = \frac{u_o^2 S \pi \rho_m}{\omega \sin\psi} e^{-\alpha_e z} e^{-j(\beta_e z - \omega t)} \tag{16.2.13}$$

It should be noted that ψ should be chosen to make Eq. 16.1.5 valid.

In the traveling wave tube of the example in Section 16.1, the wavelength of the helical line was calculated to be $\lambda_l = 6.25 \times 10^{-3}$ m. Since $u_z \approx u_o$,

$$\lambda_l = \frac{u_z}{f} \approx \frac{u_o}{f} = \lambda_e \tag{16.2.14}$$

where λ_e is the electronic wavelength. So, in this example $\lambda_e = 6.25 \times 10^{-3}$ m. Therefore, according to Eq. 16.2.4, $z_l = \lambda_e/2 = 6.25 \times 10^{-3}$ m$/2 = 3.125 \times 10^{-3}$ m is the length of one section of the traveling deceleration region. When a traveling wave tube is properly designed, fabricated, and operated with a beam acceleration voltage V_a, the starting region of the helix line is the velocity modulation region and the end region is the catching region. In the velocity modulation region, traveling $\lambda_e/2 = 3.13$ mm sections of acceleration regions and other traveling $\lambda_e/2 = 3.13$ mm sections of deceleration regions alternately exist in succession as sketched in Fig. 16.2. In the early stage of electron beam–microwave field interaction, these sections are occupied by electron beams. However, due to velocity modulation, sooner or later there will be bunching. By the time the electron beam reaches the catching region of the helix line, electrons are bunched and the centers of the bunches are all in the traveling $\lambda_e/2 = 3.13$ mm sections of deceleration regions and very few electrons are left in the traveling acceleration sections.

According to Eq. 16.2.11, the beam coupling coefficient depends on modulation. Equation 16.2.11 is rewritten as

$$k = \frac{\pi u_o^2}{\omega\left(\rho_o \dfrac{u_m}{\rho_m} + u_o\right)\sin\psi} \tag{16.2.15}$$

At the beginning of velocity modulation, $u_m \to 0$. Therefore

$$k \approx \frac{\pi u_o}{\omega \sin\psi} \tag{16.2.16}$$

In this example

$$k \approx \frac{\pi \times 1.876 \times 10^7 \text{ m/s}}{2\pi(3 \times 10^9 \text{ s}^{-1})\sin 0.1176} = 0.00156$$

This is a small quantity. Then

$$\frac{\sin \dfrac{\omega l}{2u_o}}{\dfrac{\omega l}{2u_o}} = 0.00156$$

From a table of $(\sin X)/X$, where

$$X = \frac{\omega l}{2u_o} = 3.125$$

$$l = 3.125\left(\frac{2u_o}{\omega}\right) = \frac{3.125 \times 2 \times 1.876 \times 10^7 \text{ m/s}}{2\pi \times 3 \times 10^9 \text{ s}^{-1}}$$

$$= 6.22 \times 10^{-3} \text{ m}$$

This is the equivalent length of the traveling interaction section in Eq. 16.2.3 or Eq. 16.2.12.

If the electron beam is 100% density modulated at the end of the traveling wave tube and the entire length of the interacting helix line is $L = 30$ cm, then

$$\rho_m e^{-\alpha_e L} = 2\rho_o$$

or

$$-\alpha_e L = \ln\left(\frac{2\rho_o}{\rho_m}\right)$$

In a typical traveling wave tube of 3 GHz, α_e is a negative quantity and $e^{-\alpha_e L}$ is at least 20 dB which is ten times the current ratio.
This means that

$$\frac{2\rho_o}{\rho_m} = e^{-\alpha_e L} = 10$$

Then

$$\rho_m = \frac{2\rho_o}{10} = \frac{2 \times 1.81 \times 10^{-4} \text{ C/m}^3}{10}$$

$$= 0.362 \times 10^{-4} \text{ C/m}^3$$

and

$$-\alpha_e L = \ln 10 = 2.3 \text{ nepers} = 2.3 \times 8.686 \text{ dB}$$

$$= 20 \text{ dB}$$

The gain parameter or the gain constant $-\alpha_e$ is

$$-\alpha_e = \frac{2.3 \text{ neper}}{0.3 \text{ m}} = 7.67 \text{ nepers/m}$$

or

$$-\alpha_e = \frac{20 \text{ dB}}{0.3 \text{ m}} = 66.67 \text{ dB/m}$$

According to Eq. 16.2.10, the microwave helix line current is, at the output end,

$$I_1 = \frac{u_o^2 S \pi \rho_m e^{-\alpha_e L}}{\omega \sin \psi}$$

$$= \frac{(1.876 \times 10^7 \text{ m/s})^2 \times 0.7854 \times 10^{-6} \text{ m}^2 \pi \times 0.362 \times 10^{-4} \text{ C/m}^3 \times 10}{2\pi(3 \times 10^9 \text{ s}^{-1})\sin(0.0626 \text{ rad})}$$

$$= 0.297 \times 10^{-3} \text{ A}$$

Since the current gain is $e^{-\alpha_e L} = 10$, the input current to the helix line must be I_1 in $= 0.0297 \times 10^{-3}$ A. If this input current is fed from a $Z_o = 50\ \Omega$ line, then the input power is

$$P_{in} = I_{1,\,in}^2 Z_o = (0.0297 \times 10^{-3}\ \text{A})^2 (50\ \Omega)$$
$$= 0.0441 \times 10^{-6}\ \text{W} = -43.5\ \text{dBm}$$

If the output current $I_1 = 0.297 \times 10^{-3}$ A is fed to $Z_o = 50\ \Omega$ output load, then the output power is

$$P_{out} = I_1^2 Z_o = (0.297 \times 10^{-3}\ \text{A})^2 (50\ \Omega)$$
$$= 4.41 \times 10^{-6}\ \text{W} = -29.36\ \text{dBm}$$

The microwave input voltage to this traveling wave amplifier is then

$$V_{in} = Z_o I_{1,\,in} = 50\ \Omega \times 0.0297 \times 10^{-3}\ \text{A}$$
$$= 1.485 \times 10^{-3}\ \text{V}$$

The microwave output voltage of this traveling wave amplifier is

$$V_{out} = Z_o I_1 = 50\ \Omega \times 0.5 \times 10^{-3}\ \text{A}$$
$$= 14.85 \times 10^{-3}\ \text{V}$$

16.3 Traveling Wave Amplification

As seen from Eq. 16.2.13, to obtain microwave amplification, the microwave attenuation constant α_e must be negative. The coupling of the bunched electron beam current and the wave-slowing circuit can be considered a transmission line that is continuously energized by the bunched electron beam current. A generalized equivalent circuit of a traveling wave amplification system is shown in Fig. 16.3 [4, 5].

Any transmission line can be represented by a series–parallel combination of a series impedance per unit linelength $\dot{Z}$, and a parallel admittance per unit linelength $\dot{Y}$. By Ohm's law, if the line voltage across the transmission line is $\dot{V}$ and the line current is $\dot{I}$, and assuming the transmission line extends in the z-direction and the voltage wave $\dot{V}$ and the current wave $\dot{I}$ are propagating in the z-direction with no reflection, then the voltage drop per unit length of the line is

$$\frac{d\dot{V}}{dz} = -\dot{Z}\dot{I} \tag{16.3.1}$$

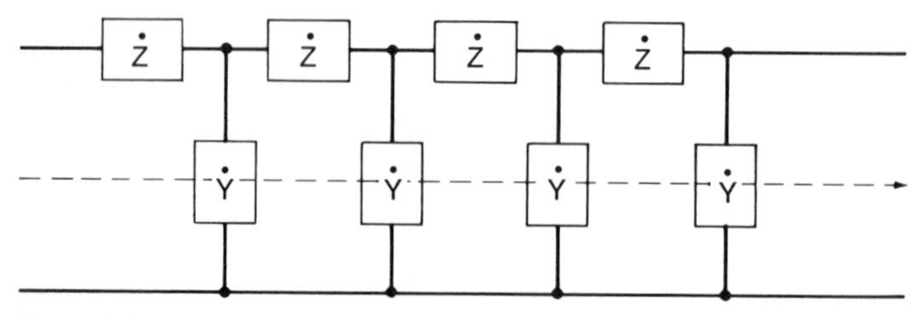

Fig. 16.3 Coupling of a transmission line and a bunched electron beam current. $\dot{I} = \dot{I}_o e^{-\dot{\gamma}z}$, $V = \dot{V}_o e^{-\dot{\gamma}z}$, and $\dot{i}_1 = \dot{i}_{o1} e^{-\dot{\gamma}z}$.

The current decrease per unit length of the line is

$$\frac{d(\dot{I} + k\dot{I}_{b1})}{dz} = -\dot{Y}\dot{V} \tag{16.3.2}$$

where $\dot{I}_{b1}$ is the microwave beam current and k is the beam coupling coefficient.

Differentiating Eq. 16.3.1 with respect to z once more,

$$\frac{d^2\dot{V}}{dz^2} = -\dot{Z}\frac{d\dot{I}}{dz} \tag{16.3.3}$$

Substituting Eq. 16.3.2 in Eq. 16.3.3,

$$\frac{d^2\dot{V}}{dz^2} = \dot{Z}\dot{Y}\dot{V} + k\dot{Z}\frac{d\dot{I}_{b1}}{dz} \tag{16.3.4}$$

When the electron beam is shut off

$$\dot{I}_{b1} = 0 \tag{16.3.5}$$

and Eq. 16.3.4 degenerates to

$$\frac{d^2\dot{V}}{dz^2} = \dot{Z}\dot{Y}\dot{V} \tag{16.3.6}$$

In this cold state with no electron beam, as long as the transmission line is fed by the input signal, there will be a traveling wave on the line. The traveling wave can be expressed in the phasor form

$$\dot{V} = \dot{V}_o e^{j\omega t - \dot{\gamma}_1 z} \tag{16.3.7}$$

where $\dot{V}_o$ is the amplitude of the microwave line voltage at $z = 0$ and $t = 0$. $\dot{\gamma}_1$ is the propagation constant of the circuit only, or the circuit under the "cold" state. Substituting Eq. 16.3.7 in Eq. 16.3.6,

$$\dot{\gamma}_1^2 \dot{V}_o e^{j\omega t - \dot{\gamma}_1 z} = \dot{Z}\dot{Y}\dot{V}_o e^{j\omega t - \dot{\gamma}_1 z}$$

or

$$\dot{\gamma}_1^2 = \dot{Z}\dot{Y} \tag{16.3.8}$$

When the bunched electron beam is turned on, the governing differential equation is Eq. 16.3.4. In a well adjusted traveling wave amplifier, the microwaves on the transmission line and the microwaves in the electron beam current form a combined wave propagating in unison in the same direction. Under this condition, the traveling voltage wave on the transmission line is expressed by

$$\dot{V} = \dot{V}_o e^{j\omega t - \dot{\gamma} z} \tag{16.3.9}$$

where $\dot{\gamma} = \alpha + j\beta$ is the propagation constant of the traveling wave, α is the attenuation constant, and β is the phase constant. The microwave beam current is then represented by

$$\dot{i}_1 = \dot{i}_{o1} e^{j\omega t - \dot{\gamma} z} \tag{16.3.10}$$

where $\dot{i}_{o1}$ is the amplitude of the microwave beam current at $z = 0$ and $t = 0$. Substituting Eqs. 16.3.9 and 16.3.10 in Eq. 16.3.4,

$$\dot{\gamma}^2 \dot{V} = \dot{Z}\dot{Y}\dot{V} - \mathrm{k}\dot{Z}\dot{\gamma}\dot{i}_1 \tag{16.3.11}$$

Substituting Eq. 16.3.8 in Eq. 16.3.11,

$$\dot{\gamma}^2 \dot{V} = \dot{\gamma}_1^2 \dot{V} - \mathrm{k}\dot{Z}\dot{\gamma}\dot{i}_1$$

or

$$\left(\dot{\gamma}^2 - \dot{\gamma}_1^2\right)\dot{V} = -\mathrm{k}\dot{Z}\dot{\gamma}\dot{i}_1 \tag{16.3.12}$$

Noting that $\dot{i}_1$ is the microwave beam current, by setting a cylindrical Gaussian surface S surrounding the beam current and applying Gauss' law,

$$\int_{S_1} \dot{J}_T \, dS_1 = \int_{S_1} \left(\frac{\dot{i}_1}{S_1} + \varepsilon_o \frac{\partial \dot{E}}{\partial t} \right) dS_1 = 0 \tag{16.3.13}$$

where S_1 is the Gaussian surface which surrounds the microwave total current density $\dot{J}_T$ which consists of the convection current density $\dot{i}_1/S$ and the displacement current density $\varepsilon_o \partial \dot{E}/\partial t$. S is the cross-sectional area of the electron beam. Since the time dependence of the electric field is $e^{j\omega t}$, the differential operator

$\partial/\partial t$ can be replaced by $j\omega$. Then, in Eq. 16.3.13,

$$\int_{S_1}\left(\frac{\dot{i}_1}{S} + j\omega\varepsilon_o\dot{E}\right)\mathrm{d}S_1 = 0 \qquad (16.3.14)$$

If the Gaussian surface S_1 is a coaxial cylinder of radius r_o and length Δz surrounding the electron beam, then the integral 16.3.14 is evaluated to be

$$\left(\frac{1}{S}\frac{\partial i_1}{\partial z} + j\omega\varepsilon_o\frac{\partial\dot{E}_z}{\partial z}\right)\Delta z\pi r_o^2 + j\omega\varepsilon_o\dot{E}_r 2\pi r_o\Delta z = 0 \quad (16.3.15)$$

The first part is the longitudinal flux and the second part is the radial flux. Since the z dependency of all functions is $e^{-\dot{\gamma}z}$, the operator $\partial/\partial z$ is replaced by $-\dot{\gamma}$. Then

$$\left(-\dot{\gamma}\frac{\dot{i}_1}{S} - j\omega\varepsilon_o\dot{\gamma}\dot{E}_z\right)r_o + j2\omega\varepsilon_o\dot{E}_r = 0$$

or

$$\dot{\gamma}r_o\frac{\dot{i}_1}{S} + j\omega\varepsilon_o r_o\dot{\gamma}\dot{E}_z - j2\omega\varepsilon_o\dot{E}_r = 0 \qquad (16.3.16)$$

According to Newton's law, force per unit mass is equal to acceleration. The longitudinal electric field $\dot{E}_z$ exerts the longitudinal force $q\dot{E}_z$ on an electron. The electron receives the longitudinal acceleration $(qE_z)/m$.

$$\frac{q\dot{E}_z}{m} = \frac{\mathrm{d}\dot{u}}{\mathrm{d}t} = \frac{\partial\dot{u}}{\partial z}\frac{\partial z}{\partial t} + \frac{\partial\dot{u}}{\partial t} = -\dot{\gamma}u_m u_o + j\omega u_m$$

$$= u_m u_o\left(\frac{j\omega}{u_o} - \dot{\gamma}\right) \qquad (16.3.17)$$

Using the concept of Eq. 16.2.9,

$$\frac{q\dot{E}_z}{m} = u_m u_o(j\beta_e - \dot{\gamma})$$

Thus,

$$\dot{E}_z = \frac{m}{q}u_m u_o(j\beta_e - \dot{\gamma}) = j\frac{m\varepsilon_o}{q\rho_o}u_m u_o\frac{\rho_o}{\varepsilon_o}(j\beta_e - \dot{\gamma}) \quad (16.3.18)$$

From Eq. 16.1.11,

$$\frac{i_1}{S} = \rho_o u_m + \rho_m u_o$$

so

$$\rho_o u_m = \frac{i_1}{S} - \rho_m u_o \tag{16.3.19}$$

The law of conservation of the charge requires that

$$\frac{1}{S}\nabla \cdot \dot{i}_1 = -\frac{\partial \rho_m}{\partial t}$$

$$\frac{1}{S}\frac{\partial i_1}{\partial z} = -j\omega\rho_m$$

$$-\frac{\dot{\gamma} i_1}{S} = -j\omega\rho_m \tag{16.3.20}$$

$$\rho_m = -j\frac{\dot{\gamma} i_1}{S\omega} \tag{16.3.21}$$

Substituting Eq. 16.3.21 in Eq. 16.3.19,

$$\rho_o u_m = \left(1 + j\frac{u_o}{\omega}\dot{\gamma}\right)\frac{i_1}{S} = \frac{1}{j\beta_e}(j\beta_e - \dot{\gamma})\frac{i_1}{S} \tag{16.3.22}$$

Substituting Eq. 16.3.22 in Eq. 16.3.18,

$$\dot{E}_z = j\frac{m\varepsilon_o}{q\rho_o}\frac{1}{j\beta_e}(j\beta_e - \dot{\gamma})^2\frac{u_o}{\varepsilon_o}\frac{i_1}{S} \tag{16.3.23}$$

The quantity $\sqrt{q\rho_o/m\varepsilon_o}$ is called the plasma oscillation frequency and is represented by

$$\omega_p^2 \equiv \frac{q\rho_o}{m\varepsilon_o} \tag{16.3.24}$$

Substituting Eq. 16.3.24 in Eq. 16.3.23, noting that $\beta_e \equiv \omega/u_o$,

$$\dot{E}_z = \frac{u_o^2}{\omega\varepsilon_o\omega_p^2}(j\beta_e - \dot{\gamma})^2\frac{i_1}{S} \tag{16.3.25}$$

Substituting Eq. 16.3.25 in Eq. 16.3.16,

$$\left[\dot{\gamma} r_o + j\omega\varepsilon_o r_o\dot{\gamma}\frac{u_o^2}{\omega\varepsilon_o\omega_p^2}(j\beta_e - \dot{\gamma})^2\right]\frac{i_1}{S} - j2\omega\varepsilon_o\dot{E}_r = 0$$

or

$$\dot{\gamma} r_o\left[1 + j\frac{u_o^2}{\omega_p^2}(j\beta_e - \dot{\gamma})^2\right]\frac{i_1}{S} - j2\omega\varepsilon_o\dot{E}_r = 0 \tag{16.3.26}$$

Let

$$\frac{u_o^2}{\omega_p^2} \equiv \frac{1}{\beta_p^2} \tag{16.3.27}$$

Combining Eq. 16.3.27 with Eq. 16.3.26,

$$\dot{\gamma} r_o \left[\beta_p^2 + j(j\beta_e - \dot{\gamma})^2 \right] \frac{i_1}{S} - j2\omega\varepsilon_o \beta_p^2 \dot{E}_r = 0$$

Therefore,

$$i_1 = \frac{2\omega\varepsilon_o \beta_p^2 S \dot{E}_r}{\dot{\gamma} r_o \left[(j\beta_e - \dot{\gamma})^2 - j\beta_p^2 \right]} \tag{16.3.28}$$

Substituting Eq. 16.3.28 in Eq. 16.3.12,

$$(\dot{\gamma}^2 - \dot{\gamma}_1^2)\dot{V} = -\frac{2k\dot{Z}\dot{\gamma}\omega\varepsilon_o \beta_p^2 S \dot{E}_r}{\dot{\gamma} r_o \left[(j\beta_e - \dot{\gamma})^2 - j\beta_p^2 \right]} \tag{16.3.29}$$

$$(\dot{\gamma}^2 - \dot{\gamma}_1^2)\left[(\dot{\gamma} - j\beta_e)^2 - j\beta_p^2 \right] = \frac{-2k\dot{Z}\omega\varepsilon_o \beta_p^2 S \dot{E}_r}{r_o \dot{V}} \tag{16.3.30}$$

In practical traveling wave amplifiers, the value of the right-hand side of Eq. 16.3.30 is a small quantity. For simplicity, if it is assumed to be negligible,

$$(\dot{\gamma}^2 - \dot{\gamma}_1^2)\left[(\dot{\gamma} - j\beta_e)^2 - j\beta_p^2 \right] \approx 0 \tag{16.3.30 A}$$

This means that either

$$\dot{\gamma}^2 - \dot{\gamma}_1^2 = 0 \tag{16.3.31}$$

or

$$(\dot{\gamma} - j\beta_e)^2 - j\beta_p^2 = 0 \tag{16.3.32}$$

If Eq. 16.3.31 stands,

$$(\dot{\gamma} + \dot{\gamma}_1)(\gamma - \dot{\gamma}_1) = 0 \tag{16.3.33}$$

Then

$$\dot{\gamma} = -\dot{\gamma}_1 \tag{16.3.34}$$

or

$$\dot{\gamma} = \dot{\gamma}_1 \tag{16.3.35}$$

The definition of $\dot{\gamma}$ as seen from Eq. 16.3.9 and Eq. 16.3.34 implies backward waves or waves propagating toward the $-z$-

direction. Equation 16.3.35 represents a forward wave. As seen from Eq. 16.3.8, this wave is the same wave propagating in the wave-slowing circuit. As seen from Eq. 16.3.8,

$$\dot{\gamma}_1 = \sqrt{\dot{Z}\dot{Y}} \equiv \alpha_c + j\beta_c \tag{16.3.36}$$

where

$$\alpha_c \equiv Re\sqrt{\dot{Z}\dot{Y}} \tag{16.3.37}$$

and

$$\beta_c \equiv Im\sqrt{\dot{Z}\dot{Y}} \tag{16.3.38}$$

Combining Eq. 16.3.35 and Eq. 16.3.36,

$$\dot{\gamma} = \alpha_c + j\beta_c \tag{16.3.39}$$

Substituting Eq. 16.3.39 in Eq. 16.3.9,

$$V = V_o e^{-\alpha_c z} e^{-j(\beta_c z - \omega t)} \tag{16.3.40}$$

This is a voltage wave propagating in the z-direction with attenuation constant α_c and phase constant β_c or the phase velocity

$$v_c = \frac{\omega}{\beta_c} \tag{16.3.41}$$

This forward wave is completely determined by the wave-slowing circuit impedance $\dot{Z}$ and admittance $\dot{Y}$. This mode of waves is attenuating and does not contribute to the amplification at all.

From Eq. 16.3.32,

$$\dot{\gamma} - j\beta_e = \pm\beta_p e^{j\pi/4} = \pm\beta_p\left(\frac{1}{\sqrt{2}} + j\frac{1}{\sqrt{2}}\right) \tag{16.3.42}$$

Knowing that the propagation constant $\dot{\gamma} = \alpha + j\beta$, then

$$\beta = \beta_e \pm \frac{\beta_p}{\sqrt{2}} \tag{16.3.43}$$

or

$$\frac{\omega}{u} = \frac{\omega}{u_o} \pm \frac{\omega_p}{u_o\sqrt{2}} \tag{16.3.44}$$

$$\frac{\omega}{u} = \frac{\omega}{u_o}\left(1 \pm \frac{\omega_p}{\omega}\frac{u_o}{\sqrt{2}\,u_o}\right) = \frac{\omega}{u_o}\frac{\sqrt{2}\,\omega u_o \pm \omega_p u_o}{\sqrt{2}\,\omega u_o}$$

$$= \frac{\omega}{u_o\left(\dfrac{\sqrt{2}\,\omega u_o}{\sqrt{2}\,\omega u_o \pm \omega_p u_o}\right)} \tag{16.3.45}$$

or

$$u = u_o \frac{\omega}{\omega \pm \dfrac{\omega_p}{\sqrt{2}}} \qquad (16.3.46)$$

The positive sign represents u as being slower than u_o and the negative sign represents u as being faster than u_o. Condition 16.3.32 is actually a resonance condition. If this happens to Eq. 16.3.28, then the microwave current $i_1 \to \infty$ and the traveling wave amplifier is oscillating instead of amplifying.

In actuality, the right-hand side of Eq. 16.3.30 is small in a traveling wave amplifier but not negligible. From Eq. 16.3.30,

$$(\dot{\gamma} + \dot{\gamma}_1)(\dot{\gamma} - \dot{\gamma}_1)(\dot{\gamma} - j\beta_e + \beta_p e^{j\pi/4})(\dot{\gamma} - j\beta_e - \beta_p e^{j\pi/4}) = \frac{-2k\dot{Z}\omega\varepsilon_o\beta_p^2 S\dot{E}_r}{r_o\dot{V}}$$

$$(\dot{\gamma} - \dot{\gamma}_1)\left[\dot{\gamma} - (j\beta_e + \beta_p e^{j\pi/4})\right]\left[\dot{\gamma} - (j\beta_e - \beta_p e^{j\pi/4})\right] = \frac{-2k\dot{Z}\omega\varepsilon_o\beta_p^2 S\dot{E}_r}{r_o\dot{V}(\dot{\gamma} + \dot{\gamma}_1)} \qquad (16.3.47)$$

In practical traveling wave amplifiers, the right-hand side becomes a negative quantity. Let

$$-jb^3 \equiv -\frac{2k\dot{Z}\omega\varepsilon_o\beta_p^2 S\dot{E}_r}{r_o\dot{V}(\dot{\gamma} + \dot{\gamma}_1)} \qquad (16.3.48)$$

In practical traveling wave tubes, by design,

$$\beta_e \approx \beta_1 \qquad (16.3.49)$$

and

$$\beta_p \ll \beta_e \qquad (16.3.50)$$

Substituting approximations 16.3.49 and 16.3.50 and definition 16.3.48 in Eq. 16.3.47,

$$(\dot{\gamma} - \dot{\gamma}_1)^3 = -jb^3 \qquad (16.3.51)$$

$$\dot{\gamma} - \dot{\gamma}_1 = b(-j)^{1/3} \qquad (16.3.52)$$

or

$$\dot{\gamma} - \dot{\gamma}_1 = be^{-j(5/6)\pi} \qquad (16.3.53)$$

Equation 16.3.52 can be further stated as

$$\dot{\gamma} - \dot{\gamma}_1 = be^{j\pi/2} = jb \tag{16.3.54}$$

or

$$\dot{\gamma} - \dot{\gamma}_1 = be^{-j\pi/6} \tag{16.3.55}$$

In the case of Eq. 16.3.53,

$$\begin{aligned}\dot{\gamma} &= \dot{\gamma}_1 + b\cos[(5/6)\pi] - jb\sin[(5/6)\pi] \\ &= \left(\alpha_c - (\sqrt{3}/2)b\right) + j\left(\beta_c - \tfrac{1}{2}b\right)\end{aligned} \tag{16.3.56}$$

Applying Eq. 16.3.39 in Eq. 16.3.56,

$$\dot{\gamma} = \left(\alpha_c - (\sqrt{3}/2)b\right) + j\left(\beta_c - \tfrac{1}{2}b\right) \tag{16.3.57}$$

Substituting Eq. 16.3.57 in Eq. 16.3.9,

$$\dot{V} = \dot{V}_0 e^{-(\alpha_c - (\sqrt{3}/2)b)z} e^{j[\omega t - (\beta_c - \frac{1}{2}b)z]} \tag{16.3.58}$$

If $\alpha_c > (\sqrt{3}/2)b$, this wave is attenuating as it travels. On the other hand, if $\alpha_c < (\sqrt{3}/2)b$, this wave is growing. With the phase constant $\beta_c - \frac{1}{2}b$, the propagation velocity is

$$u = \frac{\omega}{\beta_c - \frac{1}{2}b} > \frac{\omega}{\beta_c} \tag{16.3.59}$$

This wave is faster than the waves in the "cold circuit." The cold circuit is a wave-slowing circuit or transmission line without the electron beam.

In the case of Eq. 16.3.54,

$$\dot{\gamma} = \dot{\gamma}_1 + jb = \alpha_c + j(\beta_c + b) \tag{16.3.60}$$

Substituting Eq. 16.3.60 in Eq. 16.3.9,

$$\dot{V} = \dot{V}_0 e^{-\alpha_c z} e^{j[\omega t - (\beta_c + b)z]} \tag{16.3.61}$$

This wave is attenuating. As it travels, the amplitude decays exponentially. This wave has a larger phase constant.

$$\beta_c + b > \beta_c$$

The propagation velocity is then

$$u = \frac{\omega}{\beta_c + b} < \frac{\omega}{\beta_c} \tag{16.3.62}$$

This is a slow forward attenuating wave. That is, this wave propagates slower than in the case of a "cold circuit." Similarly, in the case of Eq. 16.3.55,

$$\dot{\gamma} = \dot{\gamma}_1 + be^{-j\pi/6} = \left(\alpha_c + (\sqrt{3}/2)b\right) + j(\beta_c - b/2) \quad (16.3.63)$$

Substituting Eq. 16.3.63 in Eq. 16.3.9,

$$\dot{V} = \dot{V}_o e^{-(\alpha_c + (\sqrt{3}/2)b)z} e^{j[\omega t - (\beta_c - b/2)]} \quad (16.3.64)$$

This is the attenuating wave. The amplitude is attenuating exponentially as it travels. The propagation velocity of this decaying wave is

$$u = \frac{\omega}{\beta_c - b/2} > \frac{\omega}{\beta_c} \quad (16.3.65)$$

The propagation velocity is greater than the cold circuit wave velocity.

As seen from Eqs. 16.3.58, 16.3.61, and 16.3.64, in a traveling wave–electron beam interaction system, there are decaying fast waves, amplifying fast waves, and decaying slow waves. Traveling wave tubes utilize the amplifying fast waves.

The plasma oscillation frequency of the traveling wave tube in examples presented in Sections 16.1 and 16.2 is, from Eq. 16.3.24,

$$\omega_p = \sqrt{\frac{q\rho_o}{m\varepsilon_o}} = \sqrt{\frac{1.602 \times 10^{-19}\ \text{C} \times 1.18 \times 10^{-4}\ \text{C/m}^3}{9.109 \times 10^{-31}\ \text{kg} \times 8.854 \times 10^{-12}\ \text{F/m}}}$$

$$= 0.1896 \times 10^{10}\ \text{rad/s}$$

In terms of frequency,

$$f_p = \frac{\omega_p}{2\pi} = \frac{1896}{2\pi} \times 10^6\ \text{rad/s} = 302\ \text{MHz}$$

This contrasts with the operating frequency

$$f = 3000\ \text{MHz}$$

or

$$\omega = 2\pi f = 28.27 \times 10^{10}\ \text{rad/s}$$

From Eq. 16.3.46, normalized electron beam velocities are

$$\frac{u}{u_o} = \frac{\omega}{\omega \pm \omega_p/\sqrt{2}}$$

$$= \frac{28.27 \times 10^{10}\ \text{rad/s}}{28.27 \times 10^{10}\ \text{rad/s} \pm (0.1896 \times 10^{10}\ \text{rad/s})/\sqrt{2}}$$

$$= 0.9953 \text{ or } 1.0048$$

Since $u_o = 1.876 \times 10^7$ m/s,

$$u = 1.876 \times 10^7 \text{ m/s} \times 0.9953 = 1.867 \times 10^7 \text{ m/s}$$

or

$$1.876 \times 10^7 \text{ m/s} \times 1.0048 = 1.885 \times 10^7 \text{ m/s}$$

These are the velocities of the fast wave and slow wave. It follows that within the dc velocity transit time for the entire tube $\tau_o = L/u_o = 15.99 \times 10^{-9}$ s; the fast waves reach at $z_{\text{fast}} = 1.885 \times 10^7 \text{ m/s} \times 15.99 \times 10^{-9} \text{ s} = 30.14 \times 10^{-2}$ m and the slow waves at $z_{\text{slow}} = 1.867 \times 10^7 \text{ m/s} \times 15.99 \times 10^{-9} \text{ s} = 29.85 \times 10^{-2}$ m. The difference is $z_{\text{fast}} - z_{\text{slow}} = 0.29 \times 10^{-2} \text{ m} < 0.625 \times 10^{-2} \text{ m} = \lambda_l$.

Electrically speaking, both the slow wave and fast wave enter the traveling wave tube at the same time and reach the output end with about a half-period difference.

The size of b in Eq. 16.3.48 can be estimated by knowing the gain of practical traveling wave tubes. In this example, a traveling wave tube with a power gain of 20 dB is considered. Then, from Eq. 16.3.58,

$$e^{-(\alpha_c - \sqrt{3}/2b)L} = 10$$

If the wave-slowing structure is made of low loss material and $\alpha_c \approx 0$, then

$$e^{\sqrt{3}/2bL} = 10$$

$$\frac{\sqrt{3}}{2} bL = \ln 10$$

$$b = \frac{1}{L} \times \frac{2}{\sqrt{3}} \ln 10$$

$$= \frac{1}{30 \times 10^{-2} \text{ m}} \times \frac{2}{\sqrt{3}} \times 2.30 = 8.86 \text{ m}^{-1}$$

This phase constant

$$\beta_c \approx \beta_e = 1.005 \times 10^3 \text{ m}^{-1}$$

Therefore, in this approach of the b-parameter, the cold tube velocity of propagation is

$$u_c = \frac{\omega}{\beta_c} \approx \frac{2\pi f}{\beta_e} = \frac{2\pi \times 3 \times 10^9 \text{ s}^{-1}}{1.005 \times 10^3 \text{ m}^{-1}} = 1.876 \times 10^7 \text{ m/s}$$

From Eq. 16.3.59, the fast wave velocity is

$$u_{\text{fast}} = \frac{\omega}{\beta_c - \dfrac{b}{2}} = \frac{2\pi \times 3 \times 10^9\ \text{s}^{-1}}{1.005 \times 10^3\ \text{m}^{-1} - \dfrac{1}{2} \times 8.86\ \text{m}^{-1}}$$

$$= 1.884 \times 10^7\ \text{m/s}$$

From Eq. 16.3.62, the slow wave velocity is

$$u_{\text{slow}} = \frac{\omega}{\beta_c + b} = \frac{2\pi \times 3 \times 10^9\ \text{s}^{-1}}{1.005 \times 10^3\ \text{m}^{-1} + 8.86\ \text{m}^{-1}}$$

$$= 1.859 \times 10^7\ \text{m/s}$$

The growing wave is, from Eqs. 16.3.58 and 16.3.59, a fast wave of $u_{\text{fast}} = 1.884 \times 10^7$ m/s, with the parameters $\alpha_c = 0$ and $b = 8.86$ neper/m.

If the transmission loss is neglected as $\alpha_c = 0$ on the wave-slowing structure of the helix line then, according to transmission line theory [6,7],

$$\dot{Z} = j\omega L$$

where L is the inductance per unit length of the helix line and

$$\dot{Y} = j\omega C$$

where C is the capacitance per unit length of the helix line. According to transmission line theory [6,7],

$$Z_o = \sqrt{\frac{\dot{Z}}{\dot{Y}}} = \sqrt{\frac{L}{C}}$$

and

$$\dot{\gamma}_c = \sqrt{\dot{Z}\dot{Y}} = j\omega\sqrt{LC} = j\beta_c$$

Therefore $\dot{Z}_o\beta_c = \omega L$, or

$$L = \frac{\dot{Z}_o\beta_c}{\omega} = \frac{\dot{Z}_o\beta_c}{2\pi f}$$

If $\dot{Z}_o = 50\ \Omega$, then

$$L = \frac{50\ \Omega \times 1.005 \times 10^3\ \text{rad/m}}{2\pi \times 3 \times 10^9\ \text{s}^{-1}} = 2.666 \times 10^{-6}\ \text{H/m}$$

and

$$C = \frac{L}{Z_o^2} = \frac{2.666 \times 10^{-6}\ \text{H/m}}{(50\ \Omega)^2} = 1.0664 \times 10^{-9}\ \text{F/m}$$

These are the equivalent circuit parameters of the helix line of $Z_o = 50\ \Omega$ and $\beta_c = 1.005 \times 10^3$ rad/m.

16.4 Backward Wave Interactions

When microwaves and an electron beam are traveling in antiparallel directions, the microwaves can, if properly adjusted, receive energy from the oppositely directed electron beam and the microwaves will be amplified. This type of electron tube is called a backward wave tube.

An electron beam–microwave interaction scheme as illustrated in Fig. 16.3 will take place, except that the voltage wave, the current wave, and the electron wave are backward or antiparallel to the electron beam velocity $\dot{u}_o$. Microwaves are propagating in the $-z$-direction and the electron beam is moving in the z-direction. In this case, in Eqs. 16.3.9 and 16.3.10 the negative value of $\dot{\gamma}$ should be used. In Eq. 16.3.30,

$$\left(\dot{\gamma}^2 - \dot{\gamma}_1^2\right) = -\frac{2\text{k}\dot{Z}\omega\varepsilon_o\beta_p^2 S\dot{E}_r}{r_o\dot{V}\left[(\dot{\gamma} - j\beta_e)^2 - j\beta_p^2\right]}$$

or

$$(\dot{\gamma} - \dot{\gamma}_1)(\dot{\gamma} + \dot{\gamma}_1) = -\frac{2\text{k}\dot{Z}\omega\varepsilon_o\beta_p^2 S\dot{E}_r}{r_o\dot{V}\left[(\dot{\gamma} - j\beta_e^2) - j\beta_p^2\right]} \tag{16.4.1}$$

Using approximations 16.3.49 and 16.3.50,

$$\dot{\gamma} = -\dot{\gamma}_1 - \frac{2\text{k}\dot{Z}\omega\varepsilon_o\beta_p^2 S\dot{E}_r}{r_o\dot{V}(\dot{\gamma} - \dot{\gamma}_1)^3} \tag{16.4.2}$$

Let

$$\dot{a} \equiv \frac{-2\text{k}\dot{Z}\omega\varepsilon_o\beta_p^2 S\dot{E}_r}{r_o\dot{V}(\dot{\gamma} - \dot{\gamma}_1)^3} \tag{16.4.3}$$

Then

$$\dot{\gamma} = -\dot{\gamma}_1 + \dot{a} \tag{16.4.4}$$

Substituting Eq. 16.3.36 in Eq. 16.4.4,

$$\dot{\gamma} = -\alpha_c + Re\,\dot{a} + j(Im\,\dot{a} + \beta_c) \quad (16.4.5)$$

Substituting Eq. 16.4.5 in Eq. 16.3.9,

$$\dot{V} = \dot{V}_o e^{(Re\,\dot{a} - \alpha_c)z} e^{j[\omega t - (Im\,\dot{a} + \beta_c)z]} \quad (16.4.6)$$

This is a backward wave as long as

$$\beta_c < -Im\,\dot{a} \quad (16.4.7)$$

It is amplifying as it propagates backward in the $-z$-direction as long as

$$\alpha_c > Re\,\dot{a} \quad (16.4.8)$$

A favorable backward wave condition for high amplification is when both

$$Re\,\dot{a},\ Im\,\dot{a} < 0$$

Under this condition, if $\dot{\gamma} = \dot{\gamma}_1$, then $\dot{a} \to \infty$, and the system is in oscillation. In practice, backward wave interaction is used for oscillation rather than for amplification.

It is interesting to review the oscillation condition of this backward wave oscillator.
If

$$\dot{\gamma} = \dot{\gamma}_1 = \alpha_1 + j\beta_1 \quad (16.4.9)$$

then

$$\alpha = \alpha_1 \quad (16.4.10)$$

and

$$\beta = \beta_1 \quad (16.4.11)$$

If the helix line is low loss, then

$$\beta = \beta_1 = \omega\sqrt{LC} = 2\pi f\sqrt{LC} \quad (16.4.12)$$

The oscillation frequency is then

$$f = \frac{\beta}{2\pi\sqrt{LC}} = \frac{\beta_1}{2\pi\sqrt{LC}} \quad (16.4.13)$$

For the particular traveling wave tube in this example,

$$f = \frac{1.005 \times 10^3\ \text{rad/m}}{2\pi\sqrt{2.666 \times 10^{-6}\ \text{H/m} \times 0.10664 \times 10^{-8}\ \text{F/m}}}$$

$$= 3 \times 10^9\ \text{s}^{-1} = 3\ \text{GHz}$$

The oscillation frequency is 3 GHz. At this time, the acceleration voltage for the electron beam is $V_a = 1000$ V.

If the oscillation frequency of 6 GHz is desired then, from Eq. 16.4.11,

$$\beta = \frac{\omega}{u} = \beta_1 \tag{16.4.14}$$

$$u = \frac{\omega}{\beta_1} \approx \sqrt{\frac{2qV_a}{m}} \tag{16.4.15}$$

$$u = \frac{\omega}{\omega\sqrt{LC}} = \frac{1}{\sqrt{LC}} = \sqrt{\frac{2qV_a}{m}} \tag{16.4.16}$$

The acceleration voltage is therefore determined by $1/\sqrt{LC}$, which means that the operating voltage is determined by the circuit which itself is determined by an operating frequency. In this example, $1/\sqrt{LC}$ is determined for 3 GHz.

It does not operate well at 6 GHz even if V_a is adjusted. If V_a is changed, it violates the oscillation condition in Eq. 16.4.11. This difficulty is introduced because an ideal lossless line was assumed. In reality, the line has loss. If the line loss is represented by a shunt conductance per unit length of the line G (S/m) then, according to transmission line theory, the phase constant is [6]

$$\beta_1 = \omega\left[\frac{CL}{2}\left(\sqrt{1+\frac{G^2}{\omega^2C^2}}+1\right)\right]^{1/2} \tag{16.4.17}$$

Substituting Eq. 16.4.17 in Eq. 16.4.15,

$$\frac{1}{\left[\frac{CL}{2}\left(\sqrt{1+\frac{G^2}{\omega^2C^2}}+1\right)\right]^{1/2}} = \sqrt{\frac{2qV_a}{m}} \tag{16.4.18}$$

Solving for V_a,

$$V_a = \frac{\frac{m}{2q}}{\frac{CL}{2}\left(\sqrt{1+\frac{G^2}{\omega^2C^2}}+1\right)} \tag{16.4.19}$$

This equation shows that the oscillation frequency ω is controlled by the acceleration voltage V_a. In this example, $L = 2.666 \times 10^{-6}$

H/m and $C = 1.0664 \times 10^{-9}$ F/m. If a small loss term $G/\omega C = 0.1$ is therefore assumed at $f = 3$ GHz, then

$$G = 0.1\omega C = 2\pi f C \times 0.1$$
$$= 2\pi \times 3 \times 10^{9}\ \mathrm{s}^{-1} \times 10^{-9}\ \mathrm{F/m} \times 0.1$$
$$= 0.67\ \mathrm{S/m}$$

The acceleration voltage is calculated for 3 GHz oscillation from Eq. 16.4.19

$$V_a = \frac{\dfrac{9.1085 \times 10^{-31}\ \mathrm{kg}}{2 \times 1.602 \times 10^{-19}\ \mathrm{C}}}{\dfrac{1.0664 \times 10^{-9}\ \mathrm{F/m} \times 2.666 \times 10^{-6}\ \mathrm{H/m}}{2} \times \left(\sqrt{1 + (0.1)^2} + 1\right)}$$

$$= \frac{5.69 \times 10^{-12}\ (\mathrm{kg/C})}{2.84 \times 10^{-15}\ \mathrm{F\text{-}H/m^2} \times 2.005} = \frac{2.0035 \times 10^{3}\ \mathrm{V}}{2.005}$$

$$= 999.3\ \mathrm{V}$$

For 1.5 GHz oscillation,

$$V_a = \frac{2.0035 \times 10^{3}\ \mathrm{V}}{\sqrt{1 + \left(\dfrac{6.67\ \mathrm{S/m}}{2\pi \times 1.5 \times 10^{9}\ \mathrm{s}^{-1} \times 1.0664 \times 10^{-9}\ \mathrm{F/m}}\right)^2} + 1}$$

$$= \frac{2.0035 \times 10^{3}\ \mathrm{V}}{\sqrt{1 + \left(\dfrac{0.1}{0.5}\right)^2} + 1}$$

$$= 991.9\ \mathrm{V}$$

For 6 GHz oscillation,

$$V_a = \frac{2.0035 \times 10^{3}\ \mathrm{V}}{\sqrt{1 + \left(\dfrac{0.1}{2}\right)^2} + 1}$$

$$= 1001.1\ \mathrm{V}$$

By controlling the acceleration voltage, the oscillation frequency is controlled.

16.5 Gyrotrons

Ordinary traveling wave tubes are longitudinal interaction devices. Gyrotrons are transverse interaction devices. In ordinary traveling wave tubes, the longitudinal microwave electric fields interact with the electron beam, both of which are parallel to each other. To make the interaction effective, the propagation velocity of the microwave field along with the longitudinal electron beam must be slowed to approximate the speed of the electron beam using the wave-slowing structure. In gyrotrons, the interaction is transverse. The propagation velocity of the microwave field is therefore not controlled at all. Microwaves propagate down a waveguide with their natural speed. Electron beams, however, must have a transverse component which is in synchronization with the variation of the transverse microwave electric field [14].

One method of creating a transverse component in the direction of the electron beam is to create a single coil helical beam or a double coil helical beam using a longitudinally applied dc magnetic flux and a longitudinally applied dc electric field as shown in Fig. 16.4. To avoid congestion and confusion, only one representative electron orbit is sketched in these figures. Since gyrotrons are transverse interaction devices, it does not matter much if there is a standing wave or a traveling wave inside the waveguide.

For the case of a single coil helical beam gyrotron, the end view of the interaction between the microwave electric field in the waveguide and the electrons is shown in Fig. 16.5. Here again only one representative electron orbit is sketched. When electrons first enter the waveguide, microwave electric fields of the TE_{o1}^{0} mode and electrons in the orbits are not bunched and are not synchronized [6,7]. Some electrons are decelerated and other electrons are accelerated. If the electron motion and microwave field polarity change are synchronized as shown in Fig. 16.5 A, then some electrons are kept decelerated and others are kept accelerated as seen from the illustration. Thus electrons receive velocity modulation. This velocity modulation takes place when electrons are in the neighborhood of the cathode. By the time the electrons approach the output of the waveguide, the electrons bunch, as shown in Fig. 16.5 B, due to the velocity modulation given to them earlier.

If the bunched electrons and the microwave electric field are synchronized in such a way that the bunched electrons are kept decelerated as illustrated in Fig. 16.5 B, the lost kinetic energy of the bunched electrons is converted to microwave potential energy

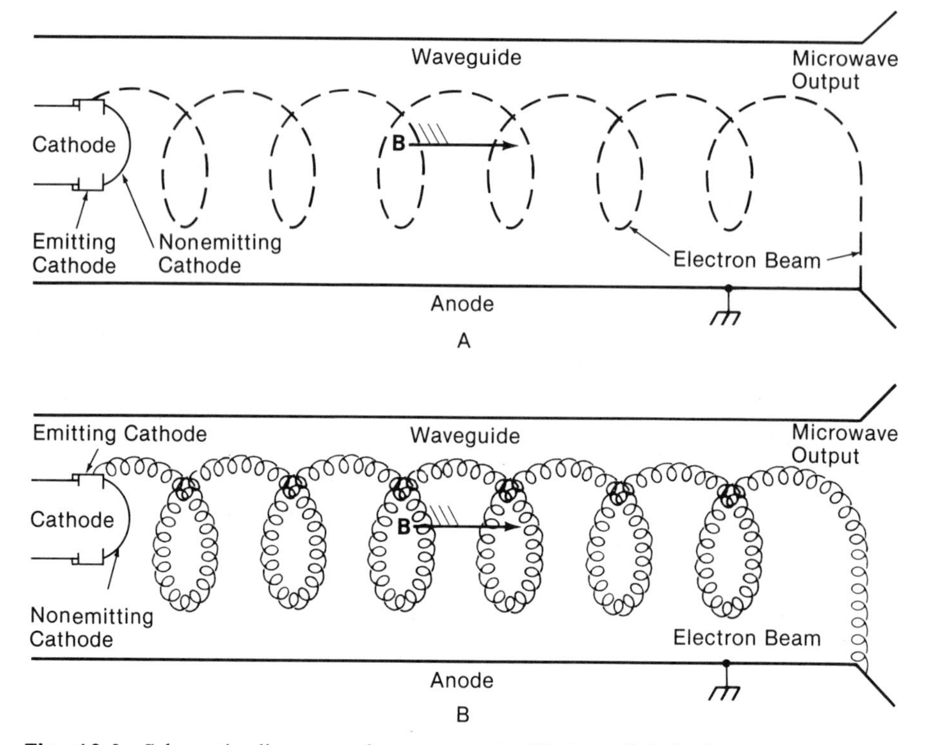

Fig. 16.4 Schematic diagrams of gyrotrons. A. Single coil helical beam gyrotron. B. Double coil helical beam gyrotron.

and the microwave field gets stronger. This is how microwave amplification or oscillation is accomplished in gyrotrons.

The feature of a gyrotron is that velocity modulation and catching are done without grids or gaps. Actually, this single coil helical electron beam gyrotron is, in principle, the same as old helical beam tubes [8,9] and the later peniotron oscillators [10].

The circular motion of a single coil helical beam is basically cyclotron oscillation. The cyclotron oscillation angular frequency ω_c (rad/s) is determined by the applied dc magnetic flux density B (T) as [11]

$$\omega_c = \frac{qB}{m} \tag{16.5.1}$$

where q is the electron charge 1.602×10^{-19} C and m is the electron mass 9.109×10^{-31} kg. The cyclotron frequency f_c is then

$$f_c = \frac{qB}{2\pi m} \tag{16.5.2}$$

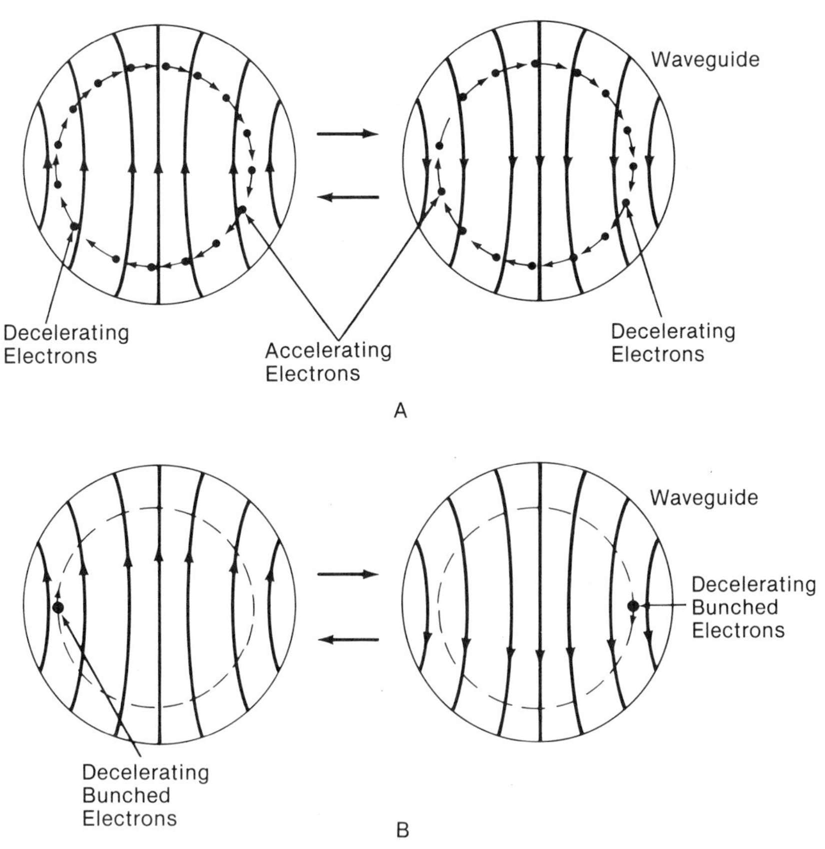

Fig. 16.5 A. Velocity modulation in a single coil helical beam gyrotron. B. Bunching and catching electrons in a single coil helical beam gyrotron.

The necessary flux density to operate the gyrotron at a desired frequency f_c is then

$$B = \frac{2\pi m f_c}{q} \tag{16.5.3}$$

For example, for $f_c = 100$ GHz operation,

$$B = \frac{2\pi \times 9.109 \times 10^{-31}\ \text{kg} \times 100 \times 10^{9}\ \text{s}^{-1}}{1.602 \times 10^{-19}\ \text{C}}$$

$$= 5.7\ \text{T}$$

This is a large amount of flux density. If the operating frequency is 10 GHz, then $B = 0.57$ T, which is more manageable. If operated at 1 GHz, then $B = 0.057$ T.

The radius of curvature r of the cyclotron motion and the cyclotron frequency ω_c are related by

$$\omega_c = \frac{qB}{m} = \frac{u_\phi}{r} \tag{16.5.4}$$

where u_ϕ is the tangential velocity of the electron [11]. Therefore,

$$r = \frac{u_\phi}{\omega_c} = \frac{u_\phi}{2\pi f_c} \tag{16.5.5}$$

In most gyrotrons, u_ϕ is about 10^7 m/s. For $f_c = 100$ GHz operation,

$$r = \frac{10^7 \text{ m/s}}{2\pi \times 100 \times 10^9 \text{ s}^{-1}} = 0.0159 \times 10^{-3} \text{ m}$$

This is considered small.

If $f_c = 10$ GHz, then $r = 0.159 \times 10^{-3}$ m. If $f_c = 1$ GHz, then $r = 1.5 \times 10^{-3}$ m.

At any rate, the radius of the cyclotron motion is small in comparison with the dimensions of the waveguide. The diameter of the waveguide is about equal to one half-wavelength of microwaves. Unless microwave electric fields are focused in the example of the peniotron structure as shown in Fig. 16.6, the interaction between the electrons and the microwave field is not effective [10]. A peniotron structure consists of a double fin structure as seen in Fig. 16.6. These double metallic fins concentrate the microwave electric field in a small, restricted region where the interaction to a single coil helical electron beam takes place.

An alternative method takes on the form of a double coil helical beam as shown in Fig. 16.4 B. When an electron orbit is

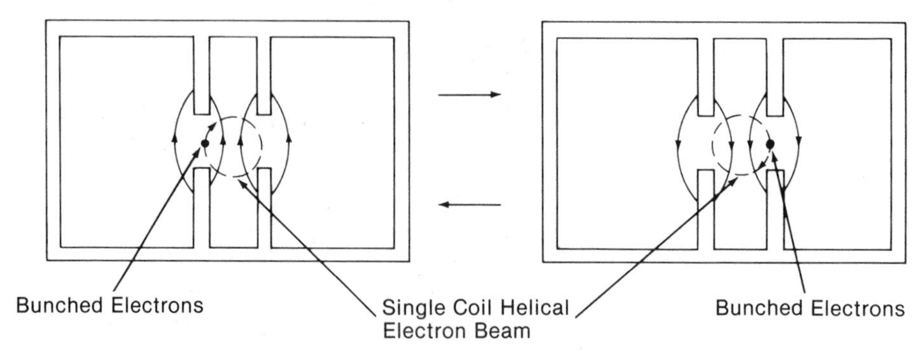

Fig. 16.6 Peniotron structure.

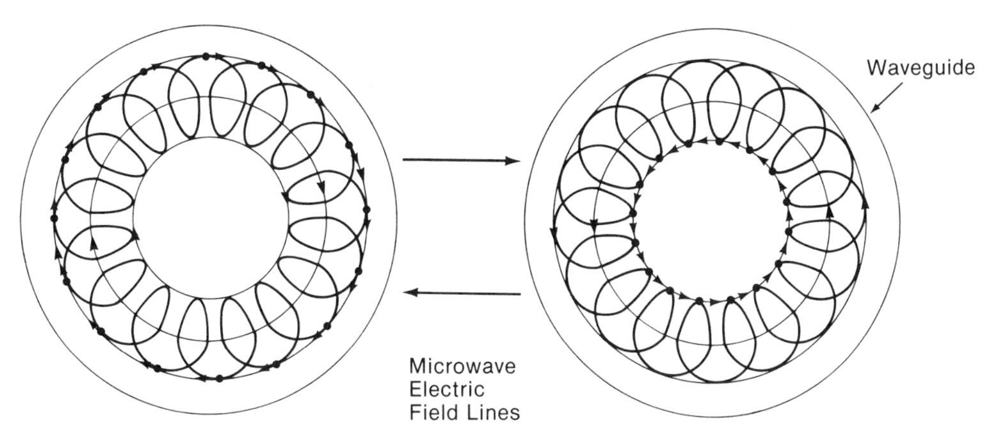

Fig. 16.7 End view of a double coil helical beam with bunched electrons.

viewed from the end of the cylindrical waveguide near the output, electrons are bunched by this time and the situation looks like that shown in Fig. 16.7. The electron cycloidal motion is adjusted so that the bunched electrons are always in phase of deceleration. Decreased kinetic energy of the electrons is transferred to microwave energy and thus amplification or oscillation takes place. If the period of the cycloidal motion of bunched electrons is τ, and if the period of the microwave frequency is T, then the synchronous condition for a double coil helical beam is

$$\tau = T \tag{16.5.6}$$

By definition,

$$\omega\tau = \omega T = 2\pi \tag{16.5.7}$$

Using Eq. 16.5.1,

$$\frac{qB}{m} T = 2\pi \tag{16.5.8}$$

The operating frequency must be

$$2\pi f = \frac{2\pi}{T} = \frac{qB}{m} \tag{16.5.9}$$

This means that, by adjusting B, the cyclotron frequency of the cycloidal orbit must match the operating frequency *and* the cycloidal motion must be synchronized with the polarity change of the microwave electric field for oscillation or amplification under an optimum operating condition.

As seen from Fig. 16.7, the double coil helical beam occupies a large volume in the waveguide. The interaction with a microwave electric field takes place in a larger area than a single coil helical

Fig. 16.8 Example of a commercial 140 GHz, 100 kW continuous wave gyrotron. Courtesy of Varian Electron Device Group, Palo Alto, California [12].

beam, though an individual cycloidal orbit is small. Most high power and high frequency gyrotrons are in the form of double coil helical beam gyrotrons.

An example of a commercial VGT–8014 gyrotron is shown in Fig. 16.8 [12]. This gyrotron produces an output of 100 kW, continuous wave at 140 GHz with a beam voltage of 80 kV, an anode current of 8 A, a heater voltage of 15 V, a heater current of 5 A, and is magnetically focused by a supercooled VYW 8014 magnet using liquid helium and liquid nitrogen.

16.6 Traveling Wave Interactions

In traveling wave electron devices, traveling microwaves interact with electron beams. Traveling microwave electric fields cause velocity modulation in the electron beam. The velocity modulated electrons form electron bunches. If properly phased, the bunched electrons are decelerated by the microwave electric field which is traveling with the bunched electrons. The decreasing kinetic energy of the bunched electrons results in increased microwave potential energy. This is how microwave oscillation or amplification takes place.

The interaction configuration between traveling microwave electric fields and the electron beam can be forward, backward, or gyrating. In a forward traveling wave tube, microwaves are slowed by the wave-slowing circuit as they travel in the same direction as the electron beam. In the backward tube, microwaves travel in the opposite direction of the electron beam. Due to the strong feedback of backward tubes, most backward tubes are backward oscillators.

In gyrotrons, electron beams interact with transverse microwave electric fields. It does not matter much if it is forward wave interaction, backward wave interaction, or standing wave interaction, as long as the synchronization between the polarity of the microwave electric field and the direction of the cycloidal motion of the bunched electrons is kept in a positively interacting phase for microwave energy production.

One feature of traveling wave devices is the wide frequency bandwidth. A wide frequency bandwidth means that there is greater channel capacity for communications.

Problems

16.1 Define traveling waves.

16.2 Estimate the acceleration voltage of an electron to the velocity of light.

16.3 Explain why $u \ll c$ is not convenient for the interaction of electrons and microwaves.

16.4 An electron beam is accelerated by 15 kV. Find the angle of the helix for a helical wave-slowing structure.

16.5 The angle of a helix is 5°. Find the electron beam acceleration voltage for the maximum interaction between the electron beam and microwaves.

16.6 Qualitatively explain the mechanism of electron bunching in a traveling wave tube.

16.7 Write down an equation for the velocity modulated electron velocity, electron density, and electron beam current.

16.8 Show that traveling waves contain second harmonics.

16.9 Define the beam coupling coefficient for a traveling wave amplifier.

16.10 Differentiate between beam coupling in a klystron and in a traveling wave amplifier.

16.11 Define the electronic wavelength of an electron beam.

16.12 Compare the interaction distance l in a klystron and in a traveling wave tube.

16.13 Show that, when the electron beam and microwaves interact in a traveling wave tube, there will be a fast wave and slow wave.

16.14 Identify all existing modes of propagation in a traveling wave tube.

16.15 By inspecting electron waves in traveling wave tubes, find methods to electronically control the gain.

16.16 State the backward wave oscillation conditions.

16.17 Explain backward wave amplification using the kinetic power theory.

16.18 Design a gyrotron that operates at 300 GHz. Estimate the possible output power.

References

1 J. R. Pierce, Traveling wave tubes. *Bell Syst. Tech. J.* **29**, 1–59 (January 1950), 189–250 (April 1950), 390–460 (July 1950), and 608–672 (October 1950).

2 J. R. Pierce, "Traveling Wave Tubes." Van Nostrand, Princeton, New Jersey, 1950.

3 R. Kompfner, The traveling wave tube as an amplifier of microwaves. *Proc. IRE* **35**, 124–217 (February 1947).

4 R. G. E. Hutter, "Beam and Wave Electronics in Microwave Tubes." Van Nostrand, Princeton, New Jersey, 1960.

5 R. F. Soohoo, "Microwave Electronics." Addison-Wesley, Reading, Massachusetts, 1971.

6 T. K. Ishii, "Microwave Engineering," 2nd ed. Technology Publ., Washington, D.C., 1989. (Originally published by Ronald Press, New York, 1966.)

7 S. Y. Liao, "Microwave Devices and Circuits," 2nd ed. Prentice-Hall, Englewood Cliffs, New Jersey, 1985.

8 R. H. Pautell, Backward-wave oscillations in an unloaded waveguide. *Proc. IRE* **47**, 1146 (June 1959).

9 D. H. Davis and K. F. Sander, *J. Electron. Control* **5**, 114–128 (August 1958).

10 G. Dohler, D. Gallapher, F. Scafuri, and R. Moats, The peniotron: A fast wave device for efficient high power millimeter wave generation. *Tech. Dig.—Int. Electron Devices Meet.* pp. 400–403 (1978).

11 S. Y. Liao, "Microwave Electron-Tube Devices." Prentice-Hall, Englewood Cliffs, New Jersey, 1988.

12 Varian Electron Device Group, "Varian... At the leading edge of fusion science. *RF Design* **10**, 18–19 (March 1988).
13 Hughes Electron Dynamics Division, A 50 KW peak power, 4 KW average power, confined-flow, PPM-focused X band TWT. *Microwave J.* **32**(4), 140–141 (1989).
14 V. A. Flyagin and G. S. Nusinovich, Gyrotron oscillators. *Proc. IEEE* **76**(6), 647–656 (1988).

17

Fundamental Principles of Microwave Electron Devices

17.1 Basic Principles Involved

In reviewing microwave electron devices in this book, the basic principles common to microwave electron devices are the transit time effect, the principle of velocity modulation, beam coupling, electron bunching, negative admittance, spin wave interaction, nonlinear admittance, quantum mechanical transitions, and power or voltage transfer principles [1–8]. Various microwave electron devices operate using one or more of these fundamental principles. How these fundamental principles take effect in microwave electron devices is reviewed in this chapter.

17.2 Electron Transit Time Effect

In an electron device, a drifting electron takes a finite time to move from one point to another point. The time necessary for an electron to drift between two specified locations is called the electron transit time. In most microwave electron devices, the electron transit time between two important locations in the device is comparable to the period of the operating microwave frequency. If the distance between the two locations in the microwave electron device is l (m) and the electron drift velocity between the two locations is u (m/s), then the electron transit time between the two locations is

$$\tau \equiv \frac{l}{u} \quad \text{(s)} \tag{17.2.1}$$

If the operating microwave frequency is f (Hz), the period of the microwave frequency is

$$T = \frac{1}{f} \quad \text{(s)} \tag{17.2.2}$$

In low radio frequency or in audio frequency electron devices

$$\tau \ll T \tag{17.2.3}$$

Thus, the electron transit time is negligible in comparison with the period of the operating frequency. In most microwave electron devices, both τ and T are on the same order of magnitude. As a result, the effect of the electron transit time produces distinct principles in microwave frequency electron devices in contrast to their low frequency counter parts. The electron transit time effect is involved in other microwave principles such as velocity modulation, beam coupling, electron bunching, and negative admittance. The electron transit time effect is therefore considered to be one of the major principles of microwave electron devices.

For example, the channel length l of a low frequency FET from the source to the drain is about 10 μm [1]. If the drift velocity of an electron in the channel is 100 km/s, then the electron transit time τ from the source to the drain is 0.1 ns. If the operating frequency is 20 kHz, the period T is 0.05 ms or 50000 ns. In this case $\tau \ll T$ and τ is therefore negligible. If the same FET is operated at 20 GHz, then the period T is 0.05 ns. Now that $\tau > T$, before the electrons complete a trip from the source to the drain, the polarity of the gate changes twice. Under this condition, transistor action is impossible. The gate does not have coherent control over the electrons in the channel. For coherent control of the gate over the electrons in the channel, τ must be negligible in comparison with the operating period T. This is the reason that the gate length l of microwave FETs is small (as small as 0.5 μm). As a result, the transit time is only 0.005 ns, which is one tenth T

for 20 GHz. The concept of the electron beam coupling coefficient developed with a klystron is applicable here. Using Eq. 14.3.12, for $l = 0.5$ μm and $f = 20$ GHz, $k_o = 0.98$. For $l = 10$ μm and $f = 20$ GHz, $k_o = -0.0005$. This simple calculation shows dramatically why the low frequency FET with $l = 10$ μm does not work at microwave frequencies. The beam coupling coefficient is a measure of the coupling of energy between the electron beam and microwaves on the electrode. A large beam coupling coefficient means better coupling. In a klystron, the gap of the buncher or catcher is on the order of 10^{-4} m. The electron velocity is on the order of 10^7 m/s. Then, by Eq. 14.3.12, at 10 GHz, $k_o = 0.98$.

To make the beam coupling coefficient larger, a shorter distance l or faster drift velocity u is required. The electron drift velocity u is limited by the relativistic velocity 3×10^8 m/s. As long as the device relies on the electron drift current—as in FETs, transferred electron devices, avalanche diodes, parametric devices, and microwave thermionic devices—the device ceases to function before the relativistic velocity sets in. In contrast, tunnel diodes are not based on drift current and are therefore not limited by a relativistic drift velocity. A narrow tunnel junction and a fast propagation velocity of the de Broglie wave function make the high frequency adaptability of a tunnel diode better.

The electron transit time effect plays vitally important roles in electron energy transfer devices such as IMPATT diodes, Gunn diodes, and TRAPATT diodes. The oscillation and amplification frequencies are closely related to the electron transit time across the drift regions of these devices as are the drift regions of klystrons and interpole drifting in magnetrons. Electrons must travel across the drift region in specified times to produce output in the output electrode at the correct phase. To do this, the electron transit time across the drift region, or the electron transit cycle, must be precisely adjusted.

The electron transit time effect is the foundation of electron bunching. In klystrons, magnetrons, and traveling wave tubes, electron bunching is created by the electron transit time effect. The electron transit time effect in the drift region produces electron bunching. The bunched electrons produce the microwave output in the output circuit.

Minimal electron transit time is desired in the interaction gap. An adequate electron transit time effect is desired in the drift region. Either in solid-state devices or in thermionic devices, microwave electron devices take advantage of electron transit time effect in the drift region to produce microwave energy at the output from the kinetic energy of drifting electrons.

17.3 Velocity Modulation

Velocity modulation produces electron bunching after drifting for a proper time period. This is an important basic principle of all microwave thermionic devices, including klystrons, magnetrons, and traveling wave tubes. In solid-state devices, the drift velocity of the electrons is modulated by the presence of microwave fields. However, the electron transit time is relatively short when compared with the thermionic counterpart. Therefore, the velocity modulation itself does not cause electron bunching as exemplified by the GaAs FET. On the other hand, the formation of the high field domain in Gunn diodes is caused by the coexistence of high mobility electrons and low mobility electrons. This is a variation of velocity modulation in which the charge accumulation region is in the bulk of the semiconductor.

17.4 Beam Coupling

Beam coupling is the coupling of energy between the microwave electric field in the circuit and the electron beam of the stream of electrons in the electron device. At the input of a microwave amplifier, beam coupling refers to the transfer of microwave energy in the circuit to the kinetic energy of electrons in the stream of electrons. At the output of a microwave amplifier or a microwave oscillator, beam coupling refers to the transfer of the kinetic energy of electrons in the device to the microwave field energy of the output circuit. Beam coupling takes place in almost all active and passive microwave electron devices except masers and parametric devices. In FETs, Gunn diodes, IMPATT diodes and all microwave thermionic devices such as magnetrons, klystrons, and traveling wave tubes, the output is produced by electrons actually striking the output electrode or by electrons passing in proximity to the output electrodes. Actually, by the principle of dynamic induction, the output microwave current starts to flow in the output circuit even before the electrons reach the output electrode via the displacement current. The displacement current, created by the bunched electrons in the device, transfers to the conduction current at the output electrodes due to the law of continuity of electric current, even before the bunched electrons reach the output electrodes. In microwave solid-state devices, the bunched electrons reach the output electrode and discharge there, creating a large pulse current. In microwave thermionic devices, some electrons actually strike the output electrodes but the majority of the bunched electrons just pass by in proximity of the output electrodes and produce microwave output current by induction.

17.5 Electron Bunching

Bunched electrons produce microwave currents at the output by induction or discharge, or by both induction and discharge. In microwave electron devices, bunched electrons are created by direct density modulation as exemplified by the GaAs MESFET, by velocity modulation as exemplified by klystrons, magnetrons, and traveling wave tubes, by periodical intermittent avalanche breakdown as exemplified by the IMPATT diode, and by periodical local quantum energy transfer as exemplified by the Gunn effect diode. Regardless of the mechanism used to create electron bunching, if the device depends upon the electron transit time effect, the bunch created in the beginning of the electron transit grows as it travels across the device. The concept of the bunching parameter in Eq. 14.2.15 can be widely applied to any electron transit time device, whether solid-state or thermionic. By suitable adaptation of the parameters in Eq. 14.2.15 for individual specific devices, the bunching parameter in Eq. 14.2.15 can be defined for transit time devices besides klystrons.

17.6 Negative Admittance

The concepts of negative admittance and negative impedance can be widely applied to active microwave electron devices. Positive admittance consumes energy and negative admittance generates energy. If the positive admittance is $Y = G + jB$ and the microwave voltage across the admittance is V, then the power consumed by the positive admittance is

$$P = \tfrac{1}{2} G V^2 \tag{17.6.1}$$

where the microwave voltage is assumed to be a sinusoidal waveform and V is the peak value. If a negative admittance is $Y = -(G + jB)$ and the microwave voltage across the admittance is V, then the power consumed by the negative admittance is

$$P = \tfrac{1}{2} G V^2 \tag{17.6.2}$$

This is the negatively consumed power by the negative impedance. Negatively consumed actually means the generation of microwave power. In all microwave active electron devices, including microwave transistors and active diodes, producing the power P with a terminal peak voltage V, the magnitude of the negative conductance is

$$G = \frac{2P}{V^2} \tag{17.6.3}$$

This type of admittance, or negative admittance, defined by the actual microwave output power and voltage is called the dynamic negative admittance. An absolutely necessary condition in obtaining negative admittance is to have negative conductance. The sign of the susceptance does not matter. For example,

$$-G \pm jB = -(G \pm jB) = -\dot{Y} \tag{17.6.4}$$

This is a negative admittance. However,

$$G - jB = -(jB - G) = -(-G + jB) \tag{17.6.5}$$

will not make a negative admittance. This is because $\frac{1}{2}V^2$ multiplied with the conductance will produce positively consumed power.

The dynamic negative admittance appears wherever the microwave radiation exists: at the drain of a GaAs MESFET; across an activated Gunn diode, IMPATT diode, tunnel diode, or parametric diode; at a klystron's catcher gap or a magnetron's gap; or at the output of a traveling wave tube. It will even appear at the open end of a radiating waveguide or across an antenna aperture. This concept of dynamic negative conductance coincides with the differential negative conductance as discussed with tunnel diodes, Gunn diodes, and IMPATT diodes. In conjunction with the positive admittance $\dot{Y}_L$ which is connected in parellel to the negative conductance, in the steady state of oscillation,

$$-\dot{Y} + \dot{Y}_L = 0 \text{ or } \mathrm{Y}^2 = \mathrm{Y}_L^2 \tag{17.6.6}$$

If

$$G < G_L \tag{17.6.7}$$

the oscillation will die down. If

$$G > G_L \tag{17.6.8}$$

the amplitude of the voltage across the negative admittance will grow until Eq. 17.6.6 is reached by nonlinear effects. As pointed out in the discussion of tunnel diodes and IMPATT diodes, 17.6.7 can be used for amplification of microwave signals if Eq. 3.4.2 is met.

At this point, a question arises about dynamic negative conductance. Does negative admittance produce microwave energy or does microwave energy create dynamic negative conductance? It is obvious in the case of the tunnel diode and the avalanche diode that dynamic negative conductance generates microwave power. It is less obvious for the Gunn diode and the IMPATT diode and is not obvious at all for klystrons, magnetrons, and traveling wave tubes. In fact, the observed current–voltage curves of the Gunn diode and the IMPATT diode generally exhibit smaller negative differential conductances than the tunnel diode, yet they produce a large oscillation output exhibiting a large dynamic negative conductance. The negative differential resistance will not appear under the normal operation of klystrons, magnetrons, and traveling wave tubes, yet it produces large amounts of microwave power, exhibiting high dynamic negative conductance. This clearly indicates that the differential conductance on the static current–voltage curve, $\Delta I/\Delta V$, is the static differential conductance. It is not equal to the dynamic negative conductance all the time. The existence of the negative static differential conductance as seen in tunnel diodes, avalanche diodes, and Gunn diodes assures the possibility of amplification and oscillation. However, the actual power balance is accounted for in the dynamic negative conductance. Except for tunnel diodes and avalanche diodes, it is safe to say that, in most active microwave electron devices, oscillation or amplification creates the *equivalent* dynamic admittance rather than that the dynamic negative admittance generates oscillation or amplification.

As seen from Eq. 17.6.6, the dynamic negative admittance can be measured by observing the load admittance under steady state oscillation. A plot of the load admittance under constant output power on a Smith chart is called a Rieke diagram [2,3]. A mirror image of the complex conjugate of the load admittance on the Rieke diagram is the negative dynamic admittance.

17.7 Spinwave Interaction

Ferrimagnetic material is used for isolators and circulators. Paramagnetic material is used as the paramagnetic crystal for crystal masers exemplified by ruby masers.

The magnetic properties of a material depend on the spin of unpaired electrons of some particular atom in the host material, exemplified by Cr atoms in Al_2O_3 crystals in the ruby maser. The Fe^{3+} ion has an unpaired electron spin in Fe_3O_4 in the case of ferrite. When the ferrimagnetic material or the paramagnetic material is magnetically biased by an externally applied dc magnetic field and a local anisotropic magnetic field, the electron spins precess around the magnetic bias field due to the magnetic dipole moment generated by the spin motion of the electron itself. Then the magnetic dipole moment of spinning and precessing electrons and the local biasing magnetic field interact with each other. In ferrimagnetic material such an interaction is classical and in paramagnetic material such a interaction is quantum mechanical. In ferrites, the magnetic dipole moment of the spin and the interaction energy of the bias magnetic fields are defined and continuous. In contrast, for paramagnetic material the spin energy is discretely described by quantum numbers.

When these electron spins are irradiated by microwaves, they change energy by absorbing the energy from the microwaves or emitting the energy in the form of microwaves. In ferrimagnetic material, the transition of the energy is continuous while in paramagnetic material, the transition of the energy is discrete. The energy of the magnetic dipole moment of the spin or magnetization is described by the Landau–Lifschitz gyromagnetic equation in the ferrimagnetic–microwave interaction. The energy of the spin angular momentum and microwaves in the paramagnetic material is described by the Schrödinger equation. The solution of the Landau–Lifschitz equation is continuous while the solution of the Schrödinger equation is discrete, defined by quantum numbers. Though one is classical and the other quantum mechanical, the principles of isolators, circulators, and masers are based on the microwave interaction of electron spin waves in the host material.

17.8 Nonlinear Impedance

If the working ranges of the input excitation are not restricted, then existing electron devices are all nonlinear. If they are not nonlinear, there would be a runaway condition. The device will either explode in a way similar to the "big bang" theory or it will implode in a way similar to the "black hole" theory. Fortunately, due to the nonlinearity of existing electron devices, neither explosion nor implostion occurs.

In GaAs FETs, the drain current is nonlinear with respect to the drain voltage. The drain impedance is nonlinear and the drain impedance is controlled by the gate voltage. When the FET oscillates or amplifies, the dynamic drain impedance is considered to be negative even though the static drain resistance is positive. The magnitude of the negative dynamic impedance of the drain is controlled by the gate voltage. The magnitude of the dynamic negative impedance determines the gain and frequency of the GaAs FET. The current–voltage curves for tunnel diodes, Gunn diodes, IMPATT diodes, TRAPATT diodes, and BARRITT diodes show strong nonlinearity. These active diodes show not only static negative resistance, but also show dynamic negative impedance. This is the reason they are used for amplification and generation of microwaves.

Schottky barrier diodes, back diodes, step recovery diodes, varactor diodes, and PIN diodes all exhibit distinct nonlinear current–voltage curves. They have static nonlinear resistance and dynamic positive nonlinear impedance. The gain is always less than unity, though dynamic positive nonlinear impedance, detection, mixing, harmonic generation, and microwave switching are accomplished. The varactor diode has positive nonlinear impedance when it is not pumped. When pumped, it has negative dynamic impedance and amplifies microwaves or oscillates at microwave frequencies. This relationship is analogous to the case of the maser. The maser material, a ruby crystal for example, is not active by itself. When pumped, it becomes active and amplifies microwaves or oscillates at a microwave frequency. Under this condition, or under the pumping condition, with proper microwave circuitry, varactor diodes and the maser material present dynamic negative nonlinear impedance. A switching PIN diode is a typical nonlinear diode. If it were not nonlinear, switching would be impossible. This is the situation with all of the detector and mixer diodes.

The anode voltage versus anode current characteristics of microwave thermionic devices such as klystrons, magnetrons, and traveling wave tubes are nonlinear. The characteristic curves saturate due to the limited availability of thermionically emitted electrons. The finite amount of available electron density creates nonlinearity in the dynamic negative impedance of the microwave thermionic device. The nonlinear electronic impedance of a thermionic device generates harmonics of its own.

In ferrimagnetic devices and masers, electron devices exhibit nonlinearity for a large input signal. Ferrites exhibit high nonlinearity for high input. Experimentally, the nonlinear impedance of a ferrite is utilized as the time varying inductance parametric amplifier rather than as a time varying capacitance parametric amplifier. Nonlinearity in masers is caused mainly by the availability of vacancies or atoms in a desirable quantum state. For a large input signal, masers exhibit nonlinearity due to the saturation of available energy states for transition.

17.9 Quantum Mechanical Transition

All solid-state devices are based on quantum mechanical transitions. Microwave solid-state electron devices such as FETs, diodes, and bulk effect devices are based on the energy band theory. The energy band theory stems from quantum mechanical transitions which can be described by the Schrödinger equation. Obviously, masers are based on quantum mechanical principles. Even microwave thermionic devices such as klystrons, magnetrons, and traveling wave tubes rely on quantum mechanical transitions at their cathodes. Thermionic emission is a quantum mechanical phenomenon. After all, the exact description of electrons in solids is impossible by classical mechanics. This is the reason that quantum mechanical concepts are used in solid-state electron devices.

In tunnel diodes, the quantum mechanical tunneling effect produces negative resistance. In Gunn effect diodes, the quantum mechanical energy transition of electrons within the conduction band generates the traveling high field domain or the charge accumulation layer.

In masers, quantum mechanical transitions of the quantum states of atoms or molecules emit microwave radiation for amplification. In most microwave electron devices except for masers, the quantum mechanical energy transfer is done between the kinetic energy of the electrons and the potential energy in some form. The current–voltage characteristics of electron devices describe such energy transfer. In masers, the quantum mechanical energy transfer is between the potential energy difference of the two energy states and the photon, or $\Delta E = hf$, where h is Planck's constant.

In a double coil helical beam gyrotron with a beam voltage on the order of hundreds of kV and a magnetic flux density of several Tesla, the electron mass and speed are relativistic. When the bunched electrons are in an outer orbit of cycloidal motion, the electrons are in the high energy state. When the bunched electrons are in an inner orbit of cycloidal motion, they are in the low energy state. These high energy states and low energy states are quantized. The energy states can be described by discrete specific quantum numbers when the bunched electrons make a downward transition from the high energy state to low energy state. If the energy difference is ΔE, then the radiation frequency of emission is $f = \Delta E/h$. When a gyrotron is operated in this mode, it is called a free electron maser. To enhance stimulation, the waveguide is a cavity resonator. The pumping, which transforms lower energy state electrons to high energy state electrons, is done by both a dc beam voltage and a dc magnetic field [4,6].

For this quantum mechanical energy transition, the electron beam does not need to be a double coil helical beam. It can be a "zig–zag" orbiting or meandering electron beam. These beams can be formed by shooting an electron beam into a spatially periodically alternating dc magnetic or electric field. Every time an electron changes its direction of motion, there is a discrete amount of energy change. If the amount of the energy change is ΔE, then the radiation frequency of emission is $f = \Delta E/h$. Again, the waveguide with the periodical static field must form a cavity resonator which resonates at the emission frequency for efficient stimulation. This type of free electron maser is called an ubitron [4,6].

Problems

17.1 List various basic principles involved with the interaction of microwaves and various electron devices.

17.2 List microwave electron devices in which the principle of the electron transit time plays a dominant role.

17.3 Analytically express the relationship among drift velocity, drift distance, electron transit time, and electron transit angle. If the drift velocity is a function of location within the drift distance, analytically express the electron transit time.

17.4 State the reason that the electron transit time effect is considered to be one of the principles of microwave electron devices.

17.5 If $\tau = 0.1T$, find the electron transit angle. If $\tau = \mathrm{P}T$ where P is a constant, find a general relationship between the electron transit angle and the constant P.

17.6 List the reasons which prevent the drift velocities of electrons in solids from reaching the relativistic value.

17.7 List the reasons for the high frequency adaptability of a tunnel diode. Point out the reasons for the existence of the upper operating frequency limit.

17.8 Define the wave function in the Schrödinger equation.

17.9 Compare a Gunn diode and a two-cavity klystron. Point out the similarities and differences in the principles of operation.

17.10 Obtain the interpole drift cycles for a magnetron following a model of a klystron.

17.11 Obtain the beam coupling coefficient for the interaction gap of a magnetron following the model of the klystron beam coupling coefficient.

17.12 Explain negative impedance from the view point of drift cycles. Formulate conditions on drift cycles to produce negative impedance.

17.13 List microwave electron devices in which electron beam coupling takes place.

17.14 Interrelate the dynamic induction principle and the electron beam coupling concept.

17.15 Analytically compare the induced current in the output circuit due to pure dynamic induction of bunched electrons passing in proximity of the output electrode and the case where the bunched electrons are actually striking the output electrode.

17.16 List various mechanisms of forming bunched electrons in various microwave electron devices.

17.17 Define a bunching parameter for a Gunn diode. Compare the concept with the tunnel diode.

17.18 Define negative admittance.

17.19 Formulate the conditions for oscillation and amplification as well as the conditions of no oscillation or amplification with respect to negative admittance.

17.20 List microwave electron devices which exhibit dynamic negative conductance.

17.21 List microwave electron devices which exhibit static negative conductance.

17.22 Analytically relate static negative conductance to dynamic negative conductance. This is only possible for certain types of microwave electron devices. Identify such devices and explain the reason theoretically.

17.23 Describe where negative conductance appears in microwave electron devices.

17.24 Describe a method to construct a Rieke diagram.

17.25 Describe a method to obtain the negative admittance of a given microwave electron device from a Rieke diagram. Theoretically justify the procedure.

17.26 List microwave electron devices in which the electron spin and microwaves interact with each other.

17.27 Differentiate the mode of interaction between electron spins and microwaves in masers and isolators.

17.28 Compare ferrimagnetism and paramagnetism.

17.29 Explain why ferrimagnetism–microwave interaction is explained by the Landau–Lifschitz gyromagnetic equation and paramagnetism–microwave interaction is explained by the Schrödinger wave equation.

17.30 Explain why all electron devices are nonlinear.

17.31 Define nonlinearity in microwave electronics.

17.32 List the uses of nonlinearity in various microwave electron devices.

17.33 Point out the cause of nonlinearity in microwave thermionic devices.

17.34 Describe the basic principles of a variable inductance parametric amplifier.

17.35 List all microwave electron devices which are based on the action of quantum mechanical transitions. Also list the way quantum mechanics contribute to particular electron devices.

17.36 Categorize the possible modes of quantum mechanical energy transfer in microwave electron devices.

References

1 A. Bar-Lev, "Semiconductors and Electronic Devices." Prentice-Hall, London, 1979.

2 P. H. Smith, Transmission line calculator. *Electronics* **12**, 29–31 (January 1939); see also *Electronics* **17**, 130–133 and 318–325 (January 1944).

3 H. J. Reich, P. F. Ordung, H. L. Krauss, and J. G. Skalnik, "Microwave Theory and Techniques." Van Nostrand, Princeton, New Jersey, 1953.

4 S. Y. Liao, "Microwave Electron-Tube Devices." Prentice-Hall, Englewood Cliffs, New Jersey, 1988.

5 B. G. Streetman, "Solid-State Electronic Devices," 2nd ed. Prentice-Hall, Englewood Cliffs, New Jersey, 1980.

6 J. T. Coleman, "Microwave Devices." Reston Publ., Reston, Virginia, 1982.

7 E. S. Yang, "Fundamentals of Semiconductor Devices." McGraw-Hill, New York, 1978.

8 T. K. Ishii, "Maser and Laser Engineering." Krieger, Huntington, New York, 1980.

Appendices

1 Microstripline Principles

The microstripline technique [1] is widely used in microwave integrated circuits and is often associated with microwave electron devices. Fundamental principles of microstripline techniques closely associated with microwave electron devices are reviewed.

A–1.1 Characteristic Impedance

In most practical striplines, a conducting strip is sandwiched by dielectric layers which are themselves sandwiched by conducting layers as shown in the cross-sectional view in Fig. A–1.1. If the line voltage is $\dot{V}$ and the line current is $\dot{I}$, assuming the microstripline is infinitely long or perfectly impedance-matched with no reflection, then the characteristic impedance of the stripline is defined as

$$Z_o = \frac{\dot{V}}{\dot{I}} \tag{A–1.1}$$

If the electric field and magnetic field between the conducting strip and the ground plate are $\dot{\mathbf{E}}$ and $\dot{\mathbf{H}}$, respectively,

$$\dot{V} = -\int_{\text{ground plate}}^{\text{conducting strip}} \dot{\mathbf{E}} \cdot d\mathbf{l} \tag{A–1.2}$$

$$\dot{I} = -\int_{c} \dot{\mathbf{H}} \cdot d\mathbf{l} \tag{A–1.3}$$

where c is an integral contour which encircles the conducting strip. Various authors publish practical calculations, computer software packages, and nomographs for characteristic impedances of striplines [1,2,11,12]. To make use of Eqs. A–1.2 and A–1.3, $\dot{\mathbf{E}}$

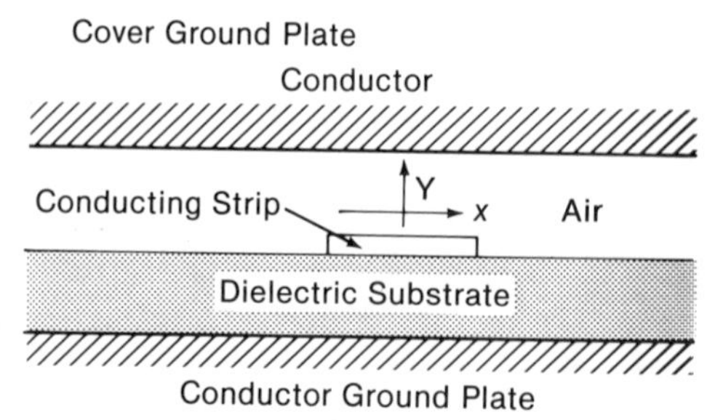

Fig. A–1.1 Cross-sectional view of a stripline.

and $\dot{\mathbf{H}}$ must be known. To obtain $\dot{\mathbf{E}}$ and $\dot{\mathbf{H}}$, Maxwell's equations must be solved under given boundary conditions.

A–1.2 Propagation Constant

Traveling voltage waves on a microstripline are expressed in the form of

$$\dot{V} = V_o e^{-\dot{\gamma}z + j\omega t} \tag{A–1.4}$$

where $\dot{\gamma}$ is a complex constant called the propagation constant:

$$\dot{\gamma} = \alpha + j\beta \tag{A–1.5}$$

The real part α is called the attenuation constant and the imaginary part β is the phase constant. The attenuation constant α is determined by the voltage loss per unit length of the line due to ohmic loss in the conducting strip, dielectric loss, and small radiation loss. In practical microwave striplines, the value of α itself is significant but the actual distance of traveling is so small that the total attenuation is usually negligible [1]. The phase constant is the phase shift per unit distance of propagation. The voltage wave shifts its phase 2π radians for a full wavelength distance of propagation. By definition of the phase constant,

$$\beta = \frac{2\pi}{\lambda} \tag{A–1.6}$$

where λ is the wavelength.

Maxwell's equation of Ohm's law at the surface of a stripline states that

$$\frac{\partial E_y}{\partial z} = -(\rho + j\omega\mu)H_x \tag{A–1.7}$$

where ρ is the resistivity of the microstrip which represents ohmic loss and μ is the permeability of the dielectric [3]. A rectangular coordinate system is set up in Fig. A–1.1, where the x-axis is parallel to the surface of the microstrip with the y-axis perpendicular to it and the z-axis coincides with the direction of the microstrip. Maxwell's equation of Faraday's law at the same location states that

$$\frac{\partial H_x}{\partial z} = -(\sigma + j\omega\varepsilon)E_y \qquad \text{(A–1.8)}$$

where σ is the equivalent conductivity of the dielectric including the representation of the dielectric loss and ε is the dielectric constant of the dielectric supporting the stripline [3].

Differentiating Eq. A–1.7,

$$\frac{\partial^2 E_y}{\partial z^2} = (\rho + j\omega\mu)\frac{\partial H_x}{\partial z} \qquad \text{(A–1.9)}$$

Substituting Eq. A–1.8 in Eq. A–1.9,

$$\frac{\partial^2 E_y}{\partial z^2} = (\sigma + j\omega\varepsilon)(\rho + j\omega\mu)E_y \qquad \text{(A–1.10)}$$

In a practical microstrip, the dielectric is thin enough so that in the middle of the strip, E_y is considered to be almost uniform. If the distance between the microstrip to the ground plate is h, then the magnitude of the voltage wave on the stripline is

$$V = hE_y \qquad \text{(A–1.11)}$$

Multiplying h to Eq. A–1.10,

$$\frac{\partial^2 hE_y}{\partial z^2} = (\sigma + j\omega\varepsilon)(\rho + j\omega\mu)hE_y = 0 \qquad \text{(A–1.12)}$$

Substituting Eq. A–1.11 in Eq. A–1.12,

$$\frac{\partial^2 V}{\partial z^2} - (\sigma + j\omega\varepsilon)(\rho + j\omega\mu)V = 0 \qquad \text{(A–1.13)}$$

Substituting Eq. A–1.4 in Eq. A–1.13,

$$\dot{\gamma}^2 V - (\sigma + j\omega\varepsilon)(\rho + j\omega\mu)V = 0 \qquad \text{(A–1.14)}$$

$$\dot{\gamma}^2 = (\sigma + j\omega\varepsilon)(\rho + j\omega\mu) \qquad \text{(A–1.15)}$$

$$\dot{\gamma} = \pm\sqrt{(\sigma + j\omega\varepsilon)(\rho + j\omega\mu)} \qquad \text{(A–1.16)}$$

Substituting Eq. A–1.5 and Eq. A–1.15,

$$\alpha^2 - \beta^2 + j2\alpha\beta = (\sigma\rho - \omega^2\varepsilon\mu) + j\omega(\varepsilon\rho - \mu\sigma)$$

or

$$\alpha^2 - \beta^2 = \sigma\rho - \omega^2\varepsilon\mu \tag{A–1.17}$$

$$2\alpha\beta = \omega(\varepsilon\rho - \mu\sigma) \tag{A–1.18}$$

Solving Eq. A–1.17 and A–1.18 simultaneously,

$$\alpha = \sqrt{\frac{\sigma\rho - \omega^2\varepsilon\mu + (\sigma\rho - \omega^2\varepsilon\mu)^2 + \omega^2(\varepsilon\rho - \mu\sigma)^2}{2}} \tag{A–1.19}$$

$$\beta = \sqrt{\frac{-\sigma\rho + \omega^2\varepsilon\mu + (\sigma\rho - \omega^2\varepsilon\mu)^2 + \omega^2(\varepsilon\rho - \mu\sigma)^2}{2}} \tag{A–1.20}$$

In the above equations, the negative sign of the plus or minus sign should be disregarded because the negative sign is physically meaningless. If the loss is negligible,

$$\sigma = 0 \text{ and } \rho = 0$$

$$\alpha = 0 \tag{A–1.21}$$

$$\beta = \omega\sqrt{\varepsilon_o\mu_o} \tag{A–1.22}$$

From Eq. A–1.6,

$$\lambda = \frac{2\pi}{\beta} = \frac{2\pi}{\omega\sqrt{\varepsilon_o\mu_o}} = \frac{2\pi}{2\pi f\sqrt{\varepsilon_o\mu_o}} = \frac{1}{f\sqrt{\varepsilon_r\varepsilon_o\mu}} = \frac{1}{f\sqrt{\varepsilon_o\mu_o}\sqrt{\varepsilon_r}}$$

$$= \frac{c}{f\sqrt{\varepsilon_r}} = \frac{\lambda_o}{\sqrt{\varepsilon_r}} \tag{A–1.23}$$

where ε_r is the relative permittivity of the dielectric, ε_o is the permittivity of free space, $c = 1/\sqrt{\varepsilon_o\mu_o}$ is the phase velocity of light in free space, and λ_o is the wavelength in free space. This free space wavelength λ_o should be distinguished from λ, the wavelength actually on the stripline.

A–1.3 Input Impedance

If the transmission loss is neglected in a microstripline, then the forward voltage wave can be represented by $\dot{V}_f(z)$ and the backward voltage wave is represented by $\dot{V}_b(z)$, where z is the coordinate along the stripline. Then the voltage wave at any location on the stripline is represented by

$$\dot{V}(z) = \dot{V}_f(z) + \dot{V}_b(z) \tag{A–1.24}$$

On the microstripline, the forward current wave is represented by $\dot{I}_f(z)$ and the backward current wave is represented by $\dot{I}_b(z)$. The resultant of the forward current wave and the backward current wave on the stripline at any location is

$$\dot{I}(z) = \dot{I}_f(z) - \dot{I}_b(z) \tag{A–1.25}$$

The transmission line impedance at any location on the stripline is then

$$\dot{Z}(z) = \frac{\dot{V}(z)}{\dot{I}(z)} = \frac{\dot{V}_f(z) + \dot{V}_b(z)}{\dot{I}_f(z) - \dot{I}_b(z)} \tag{A–1.26}$$

By definition, the characteristic impedance Z_o of the stripline is represented by

$$Z_o \equiv \frac{\dot{V}_f(z)}{\dot{I}_f(z)} = \frac{\dot{V}_b(z)}{\dot{I}_b(z)} \tag{A–1.27}$$

Then

$$\dot{I}_f(z) = \frac{\dot{V}_f(z)}{Z_o} \tag{A–1.28}$$

$$\dot{I}_b(z) = \frac{\dot{V}_b(z)}{Z_o} \tag{A–1.29}$$

Substituting Eq. A–1.28 and A–1.29 in Eq. A–1.26,

$$\dot{Z}(z) = Z_o \frac{\dot{V}_f(z) + \dot{V}_b(z)}{\dot{V}_f(z) - \dot{V}_b(z)} \tag{A–1.30}$$

If both the numerator and the denominator are divided by $\dot{V}_f(z)$,

$$\dot{Z}(z) = Z_o \frac{\left(1 + \dfrac{\dot{V}_b(z)}{\dot{V}_f(z)}\right)}{\left(1 - \dfrac{\dot{V}_b(z)}{\dot{V}_f(z)}\right)} \tag{A–1.30A}$$

By definition,

$$\dot{\rho}(z) \equiv \frac{\dot{V}_b(z)}{\dot{V}_f(z)} \quad \text{(A–1.31)}$$

is called the voltage reflection coefficient. Then the stripline impedance at any point is

$$\dot{Z}(z) = Z_o \frac{1 + \dot{\rho}(z)}{1 - \dot{\rho}(z)} \quad \text{(A–1.32)}$$

Assuming the microstripline stretches from $z = 0$ to $z = l$, the input impedance is by definition expressed by $\dot{Z}(0)$ at $z = 0$ and

$$\dot{Z}(0) = Z_o \frac{1 + \dot{\rho}(0)}{1 - \dot{\rho}(0)} \quad \text{(A–1.33)}$$

The load impedance is defined by $\dot{Z}(l)$ at $z = l$ and

$$\dot{Z}(l) = Z_o \frac{1 + \dot{\rho}(l)}{1 - \dot{\rho}(l)} \quad \text{(A–1.34)}$$

$\dot{\rho}(0)$ and $\dot{\rho}(l)$ can be related by

$$\dot{\rho}(0) = \frac{\dot{V}_b(0)}{\dot{V}_f(0)} = \frac{\dot{V}_b(l)e^{-j\beta l}}{\dot{V}_f(l)e^{j\beta l}} = \dot{\rho}(l)e^{-2j\beta l} \quad \text{(A–1.35)}$$

where β is the phase constant of the stripline. The stripline attenuation constant α is omitted this time for simplicity.

Substituting Eq. A–1.35 in Eq. A–1.33, the input impedance is

$$\dot{Z}(0) = Z_o = \frac{1 + \dot{\rho}(l)e^{-2j\beta l}}{1 - \dot{\rho}(l)e^{-2j\beta l}} \quad \text{(A–1.36)}$$

In microstriplines, the line is often open at $z = l$. If so, then $\dot{V}_b(l) = \dot{V}_f(l)$ and

$$\dot{\rho}(l) = \dot{V}_b(l)/\dot{V}_f(l) = 1 \quad \text{(A–1.37)}$$

Substituting Eq. A–1.37 in Eq. A–1.36,

$$\dot{Z}(0) = Z_o \frac{1 + e^{-2j\beta l}}{1 - e^{-2j\beta l}} \quad \text{(A–1.38)}$$

Multiplying both the numerator and the denominator by $e^{j\beta l}$

$$\dot{Z}(0) = Z_o \frac{e^{j\beta l} + e^{-j\beta l}}{e^{j\beta l} - e^{-j\beta l}} \quad \text{(A–1.39)}$$

Noting Euler's identities,

$$\left.\begin{aligned}\sin x &= \frac{-j}{2}\left(e^{jx} - e^{-jx}\right)\\ \cos x &= \frac{1}{2}\left(e^{jx} + e^{-jx}\right)\end{aligned}\right\} \tag{A–1.40}$$

Applying Eq. A–1.40 to Eq. A–1.39,

$$\dot{Z}(0) = jZ_o \frac{\cos \beta l}{\sin \beta l} = jZ_o \cot \beta l \tag{A–1.41}$$

Substituting Eq. A–1.6 in Eq. A–1.41,

$$\dot{Z}(0) = jZ_o \cot\left(\frac{2\pi l}{\lambda}\right) \tag{A–1.42}$$

For $0 < l < \lambda/4$, $\dot{Z}(0)$ is the capacitive reactance. If the equivalent capacitance presented at $z = 0$ by the open circuited lossless stripline of length l_c is C, then

$$\dot{Z}(0) \equiv \frac{1}{j\omega C} \equiv -jZ_o \cot\left(\frac{2\pi l_c}{\lambda}\right) \tag{A–1.43}$$

or the equivalent capacitance

$$C = \frac{1}{\omega Z_o} \tan\left(\frac{2\pi l_c}{\lambda}\right) \tag{A–1.44}$$

Evidently the equivalent capacitance is frequency dependent.

For $\lambda/4 < l < \lambda/2$, $\dot{Z}(0)$ is inductive reactance. If the equivalent inductance presented at $z = 0$ by the open circuited lossless stripline of length l_i is L, then

$$\dot{Z}(0) = j\omega L = -jZ_o \cot\left(\frac{2\pi l_i}{\lambda}\right)$$

or

$$L = -\frac{Z_o \cot\left(\frac{2\pi l_i}{\lambda}\right)}{\omega} \tag{A–1.45}$$

If the equivalent capacitance and inductance in Eqs. A–1.44 and A–1.45 are connected in parallel, there will be a parallel resonance circuit and the resonance frequency is given by a solution of the following transcendental equation:

$$\omega^2 = \frac{1}{LC} = -\omega^2 \cot \frac{2\pi l_c}{\lambda} \tan \frac{2\pi l_i}{\lambda} \tag{A-1.46}$$

or

$$\cot \frac{2\pi l_c}{\lambda} = -\cot \frac{2\pi l_i}{\lambda} \tag{A-1.47}$$

This means that

$$l_c + l_i = \frac{\lambda}{2} \tag{A-1.48}$$

with conditions that

$$\frac{\lambda}{4} \leqq l_c \leqq \frac{\lambda}{2}$$

and (A–1.49)

$$0 < l_i \leqq \frac{\lambda}{4}$$

In Eq. A–1.42, if

$$l = \frac{\lambda}{2} \tag{A-1.50}$$

then

$$Z(0) \to j\infty \tag{A-1.51}$$

The open circuit at $z = l$ therefore creates an open circuit at $z = 0$.

In Eq. A–1.42, if

$$l = \frac{\lambda}{4} \tag{A-1.52}$$

then

$$Z(0) = 0 \tag{A-1.53}$$

Therefore, an open circuit at $z = l$ creates a short circuit at $z = 0$.

By adjusting the length of the open circuited microstripline, any amount of inductance or capacitance and open circuit or short circuit can be created at the other end of the line.

2 Basics of Smith Chart and Rieke Diagram

Dividing Eq. A–1.32 by the characteristic impedance of the transmission line Z_o,

$$\frac{\dot{Z}(z)}{Z_o} = \frac{1 + \dot{\rho}(z)}{1 - \dot{\rho}(z)} \tag{A–2.1}$$

$\dot{Z}(z)/Z_o$ is called the normalized impedance and is represented by $\tilde{\dot{Z}}(z)$. Then

$$\tilde{\dot{Z}}(z) = \frac{1 + \dot{\rho}(z)}{1 - \dot{\rho}(z)} \tag{A–2.2}$$

Solving the above equation with respect to $\dot{\rho}(z)$,

$$\dot{\rho}(z) = \frac{\tilde{\dot{Z}}(z) - 1}{\tilde{\dot{Z}}(z) + 1} \tag{A–2.3}$$

If Eq. A–2.3 is plotted in the $\dot{\rho}(z)$ plane for all possible values of $\tilde{\dot{Z}}(z)$, from $\dot{Z}(z) = 0$, the short circuit to $\tilde{\dot{Z}}(z) = \infty$, the open

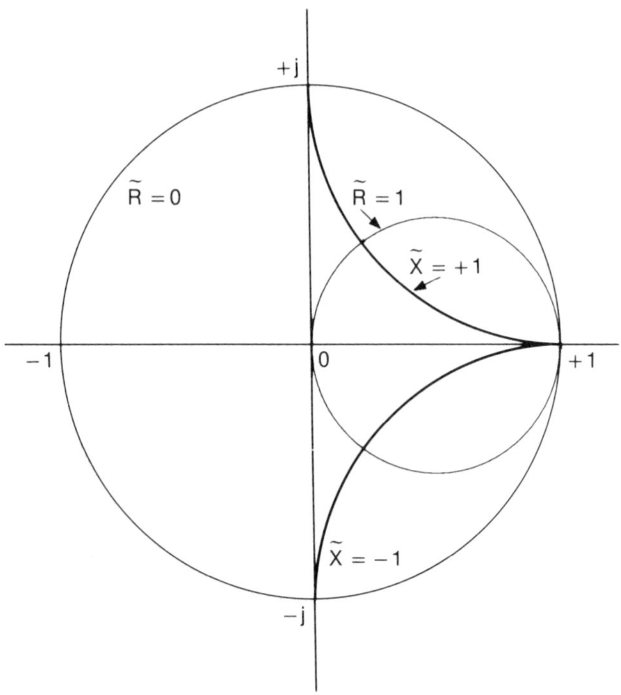

Fig. A–2.1 Plot of the voltage reflection coefficient $\rho(z)$ for various normalized impedances $\tilde{\dot{Z}}(z)$.

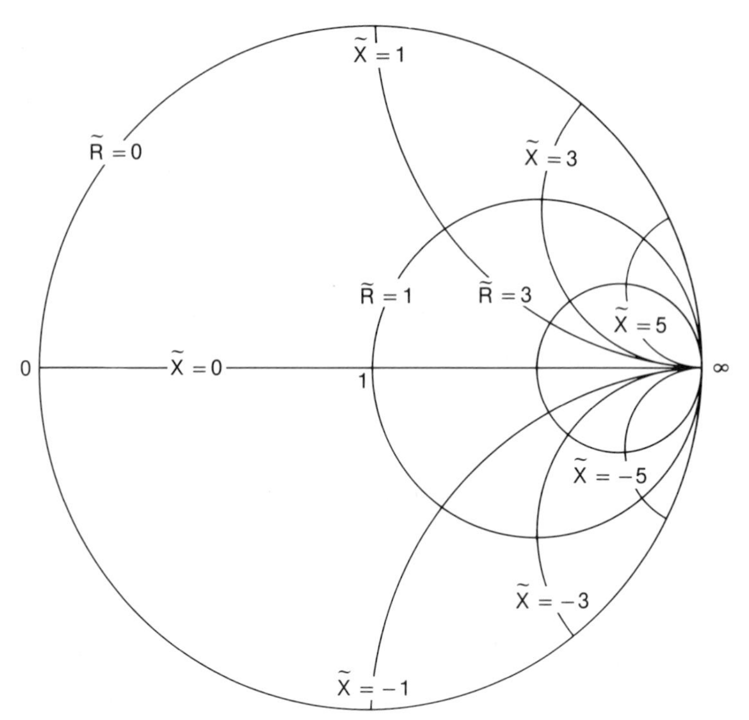

Fig. A–2.2 Simplified presentation of a Smith chart.

circuit, a plot shown in Fig. A–2.1 will be obtained. To avoid confusion, only a few lines are shown.

When the complex coordinate is removed, Fig. A–2.2 is obtained. This Smith chart [4] is a plot of voltage reflection coefficients for various normalized transmission line impedances with the complex coordinate removed. In practical Smith charts, various additional scales are attached around the outside of the outermost circle for convenience. The normalized admittance is

$$\tilde{\dot{Y}}(z) = \frac{1}{\tilde{\dot{Z}}(z)} \tag{A–2.4}$$

In a Smith chart, the value of $\tilde{\dot{Y}}(z)$ is always found at diametrically and symmetrically opposite sides of the center of the diagram for the point $\tilde{\dot{Z}}(z)$. The Smith chart can therefore be used to obtain the normalized impedance $\tilde{\dot{Z}}(z)$ and also the normalized admittance $\tilde{\dot{Y}}(z)$.

If an electronic oscillator is connected with a transmission line at a location z on the microstrip and the oscillation is in a steady state, then

$$\dot{\tilde{Y}}(z) + \dot{\tilde{Y}}_e = 0 \tag{A–2.5}$$

where $\dot{\tilde{Y}}_e$ is the electronic admittance of the electronic oscillator presented at z. Then

$$\dot{\tilde{Y}}(z) = -\dot{\tilde{Y}}_e \tag{A–2.6}$$

Therefore, knowing $\dot{\tilde{Y}}(z)$, $\dot{\tilde{Y}}_e$ is known. Various values of $\dot{\tilde{Y}}(z)$ produce the same oscillation frequency. A plot of a constant frequency contour is produced on the Smith chart. By changing $\dot{\tilde{Y}}(z)$, a family of constant power contours and a family of constant frequency contours can be produced on the Smith chart. The chart of the family of constant power contours and constant frequency contours of an electronic oscillator is called the Rieke diagram [5]. As seen from Eq. A–2.6, the Rieke diagram shows the negative electronic admittance presented at the location z.

3 Cavity Resonator Principles

Originally, microwave cavity resonators were meant to be spaces surrounded by conducting walls. When microwaves are introduced into the cavity, the stored microwave energy reaches its maximum at a certain microwave frequency and the operating frequency causing resonance is called the resonance frequency.

The original concept of the cavity resonator is now extended to almost any microwave resonator which is not necessarily made of closed conducting walls. For example, a pair of two parallel conducting plates is considered to be a cavity resonator. A stripline section with both ends open is also called a cavity resonator as well as a rectangular or circular patch on a microstrip circuit if the size is comparable to the operating wavelength.

The electromagnetic fields $\dot{\mathbf{E}}$ and $\dot{\mathbf{H}}$ in the cavity resonator are a set of solutions to Maxwell's equations

$$\nabla \times \dot{\mathbf{H}} = \sigma\dot{\mathbf{E}} + j\omega\varepsilon\dot{\mathbf{E}} \tag{A–3.1}$$

$$\nabla \times \dot{\mathbf{E}} = -j\omega\mu\dot{\mathbf{H}} \tag{A–3.2}$$

and of the wave equations

$$\nabla^2 \dot{\mathbf{H}} + \omega^2\varepsilon\mu\left(1 - j\frac{\sigma}{\omega\varepsilon}\right)\dot{\mathbf{H}} = 0 \tag{A–3.3}$$

$$\nabla^2 \dot{\mathbf{E}} + \omega^2\varepsilon\mu\left(1 - j\frac{\sigma}{\omega\varepsilon}\right)\dot{\mathbf{E}} = 0 \tag{A–3.4}$$

which can be derived from both Eqs. A–3.1 and A–3.2 by eliminating either $\dot{\mathbf{E}}$ or $\dot{\mathbf{H}}$ from them [3]. Once $\dot{\mathbf{E}}$ and $\dot{\mathbf{H}}$ are known by solving Maxwell's equations, the resonance frequency of the cavity resonator is obtained from Eq. A–3.3 or Eq. A–3.4.

$$\omega = \frac{1}{\sqrt{\varepsilon\mu}}\sqrt{\frac{-\nabla^2\dot{\mathbf{H}}}{\left(1 - j\dfrac{\sigma}{\omega\varepsilon}\right)\dot{\mathbf{H}}}} = \frac{1}{\sqrt{\varepsilon\mu}}\sqrt{\frac{-\nabla^2\dot{\mathbf{E}}}{\left(1 - j\dfrac{\sigma}{\omega\varepsilon}\right)\dot{\mathbf{E}}}} \tag{A–3.5}$$

Various microwave rectangular cavity resonators are illustrated in Fig. A–3.1. The cavities shown in Fig. A–3.1 are surrounded by conducting walls. To put microwaves into this metal box, a coupling hole or a small antenna will be constructed on the conducting wall.

Various modes of electromagnetic field distribution in the rectangular box shown in Fig. A–3.1 are conceivable. One of the simplest electric field distributions conceivable is

$$\left.\begin{aligned} \dot{E}_y &= E_o \sin\left(\frac{\pi x}{a}\right)\sin\left(\frac{\pi z}{c}\right)e^{j\omega t} \\ \dot{E}_x &= 0 \\ \dot{E}_z &= 0 \end{aligned}\right\} \tag{A–3.6}$$

The boundary conditions of the electric field at the conducting surface require that the tangential electric field on the conducting surface be zero. The above electric field distribution satisfies

$$\dot{E}_y\Big|_{\substack{x=0,\,a\\ z=0,\,c}} = 0 \tag{A–3.7}$$

$$\begin{aligned} |\nabla^2\dot{\mathbf{E}}| &= \nabla^2\dot{\mathbf{E}}_y \\ &= \frac{\partial^2\dot{E}_y}{\partial x^2} + \frac{\partial^2\dot{E}_y}{\partial y^2} + \frac{\partial^2\dot{E}_y}{\partial z^2} = -\left[\left(\frac{\pi}{a}\right)^2 + \left(\frac{\pi}{c}\right)^2\right]\dot{E}_y \end{aligned} \tag{A–3.8}$$

Substituting Eq. A–3.8 in Eq. A–3.5 with $\sigma = 0$ for the air,

$$\omega = \frac{1}{\sqrt{\varepsilon_o\mu_o}}\sqrt{\left(\frac{\pi}{a}\right)^2 + \left(\frac{\pi}{c}\right)^2} \tag{A–3.9}$$

Since $\omega = 2\pi f$,

$$f = \frac{1}{2\sqrt{\varepsilon_o\mu_o}}\sqrt{\left(\frac{1}{a}\right)^2 + \left(\frac{1}{c}\right)^2} \tag{A–3.10}$$

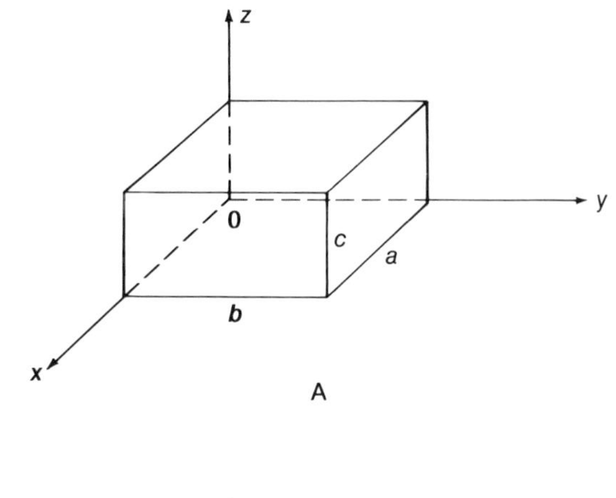

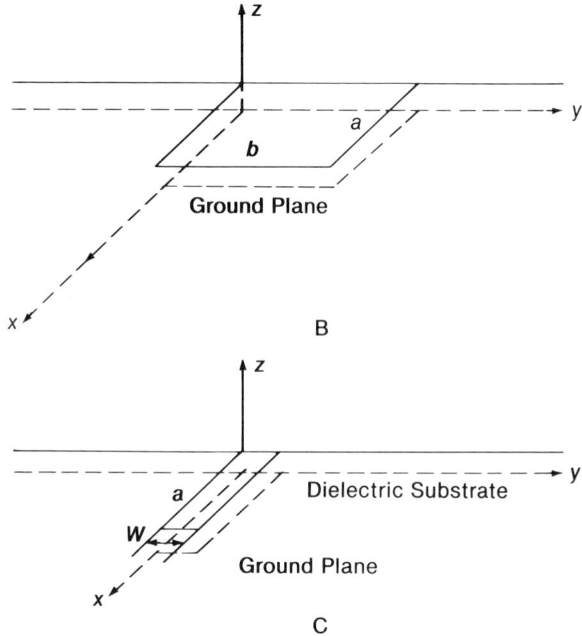

Fig. A–3.1 A. Rectangular cavity resonator. B. Rectangular microstrip cavity resonator. C. Stripline resonator.

If the cavity is filled with air, $1/\sqrt{\varepsilon_o \mu_o}$ is equal to the velocity of light c_o, then the resonance frequency is

$$f = \frac{c_o}{2}\sqrt{\left(\frac{1}{a}\right)^2 + \left(\frac{1}{c}\right)^2} \qquad \text{(A–3.11)}$$

For the rectangular microstrip cavity resonator shown in Fig. A–3.1B one of the most simple electric field distributions will be

$$\left.\begin{aligned} \dot{E}_z &= E_o \cos\left(\frac{\pi x}{a}\right) \cos\left(\frac{\pi y}{b}\right) e^{j\omega t} \\ \dot{E}_x &= 0 \\ \dot{E}_y &= 0 \end{aligned}\right\} \quad \text{(A–3.12)}$$

The distribution functions satisfy the boundary conditions for $\dot{E}_z$ stating that the field strength should be at a maximum at the open circuit. As seen from Fig. A–3.1B the edge of the cavity is open for the case of the microstrip cavity resonator. The cavity is open at $x = 0$ and a and $y = 0$ and b. Then

$$|\nabla^2 \dot{\mathbf{E}}| = \nabla^2 \dot{\mathbf{E}}_z = \frac{\partial^2 \dot{E}_z}{\partial x^2} + \frac{\partial^2 \dot{E}_z}{\partial y^2} + \frac{\partial^2 \dot{E}_z}{\partial z^2}$$

$$= -\left[\left(\frac{\pi}{a}\right)^2 + \left(\frac{\pi}{b}\right)^2\right] \dot{E}_z \quad \text{(A–3.13)}$$

Substituting Eq. A–3.13 in Eq. A–3.5, with $\sigma = 0$ for the insulating substrate,

$$\omega = \frac{1}{\sqrt{\varepsilon\mu}} \sqrt{\left(\frac{\pi}{a}\right)^2 + \left(\frac{\pi}{b}\right)^2} \quad \text{(A–3.14)}$$

The resonance frequency is then

$$f = \frac{1}{2\sqrt{\varepsilon\mu}} \sqrt{\left(\frac{1}{a}\right)^2 + \left(\frac{1}{b}\right)^2} \quad \text{(A–3.15)}$$

When the cavity is a narrow strip as shown in Fig. A–3.1C one of the simplest electric field distributions is

$$\dot{E}_z = E_o \cos\left(\frac{\pi x}{a}\right) e^{j\omega t} \quad \text{(A–3.16)}$$

$$|\nabla^2 \dot{\mathbf{E}}| = \nabla^2 \dot{\mathbf{E}}_z = \frac{\partial^2 E_z}{\partial x^2} + \frac{\partial^2 \dot{E}_z}{\partial y^2} + \frac{\partial^2 \dot{E}_z}{\partial z^2}$$

$$= -\left(\frac{\pi}{a}\right)^2 \dot{E}_z \quad \text{(A–3.17)}$$

Substituting Eq. A–3.17 in Eq. A–3.5, with $\sigma = 0$ for the dielectric substrate,

$$\omega = \frac{1}{\sqrt{\varepsilon\mu}}\frac{\pi}{a} \tag{A–3.18}$$

$$f = \frac{1}{2\sqrt{\varepsilon\mu}}\frac{1}{a} \tag{A–3.19}$$

The phase velocity of the electromagnetic waves in the dielectric substrate is

$$v = \frac{1}{\sqrt{\varepsilon\mu}} = f\lambda \tag{A–3.20}$$

where λ is the wavelength on the microstrip. Substituting Eq. A–3.19 in Eq. A–3.20,

$$\frac{1}{\sqrt{\varepsilon\mu}} = \frac{1}{2\sqrt{\varepsilon\mu}}\frac{\lambda}{a}$$

or

$$\lambda = 2a \tag{A–3.21}$$

The resonance wavelength in the dielectric substrate is twice the size of the microstrip, or the size of the microstrip is one half the resonance wavelength.

Some cylindrical cavity resonators are illustrated in Fig. A–3.2. A practical coordinate system to describe the electromagnetic fields of the cylindrical cavity resonator is the cylindrical system. The simplest electric field distribution in a closed cylindrical cavity resonator as shown in Fig. A–3.2 A is

$$\dot{E}_\phi = E_o J_1(\mathrm{k}_c r)\sin\left(\frac{\pi z}{h}\right)e^{j\omega t} \tag{A–3.22}$$

Actually, this is one particular solution of Eq. A–3.4 with $\sigma = 0$ in cylindrical coordinates with the boundary condition that

$$\dot{E}_\phi\Big|_{\substack{r=a\\ z=0,\,h}} = 0 \tag{A–3.23}$$

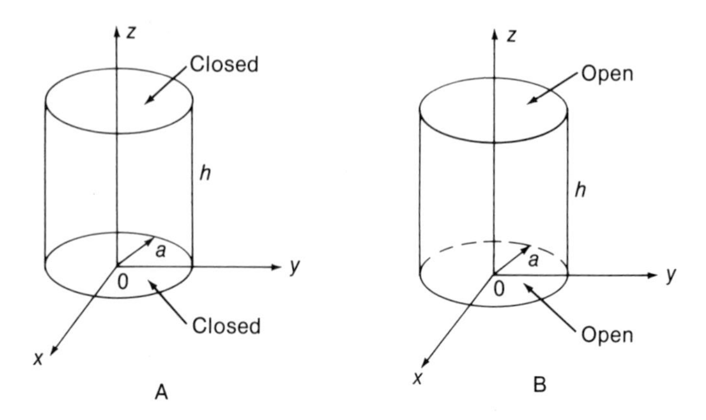

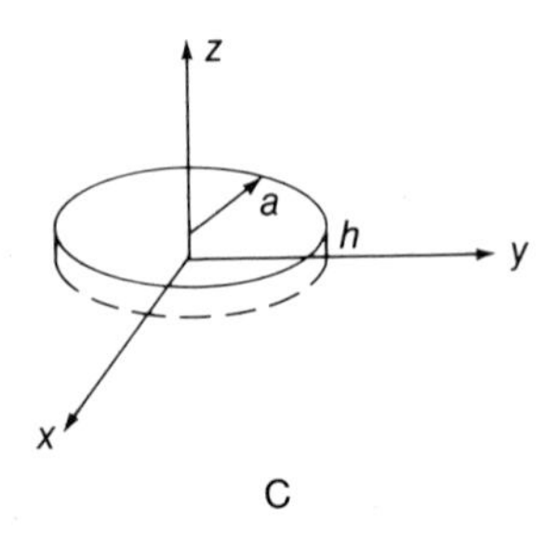

Fig. A–3.2 Cylindrical cavity resonators. A. Closed. B. Open. C. Microstrip cavity.

In Eq. A–3.22, $J_1(\mathrm{k}_c r)$ is a Bessel function of the first order with the argument $\mathrm{k}_c r$. To meet the boundary condition A–3.23,

$$J_1(\mathrm{k}_c a) = 0 \qquad \text{(A–3.24)}$$

A nontrivial first root of the above equation is, according to the table of Bessel functions,

$$\mathrm{k}_c a = 3.832 \qquad \text{(A–3.25)}$$

Then

$$\mathrm{k}_c = \frac{3.832}{a} \qquad \text{(A–3.26)}$$

Substituting Eq. A–3.22 in Eq. A–3.5 with $\sigma = 0$,

$$-|\nabla^2 \dot{\mathbf{E}}| = -\nabla^2 \dot{E}_\phi$$

$$= -\frac{1}{r}\frac{\partial}{\partial r}\left(r\frac{\partial \dot{E}_\phi}{\partial r}\right) - \frac{\partial^2 \dot{E}_\phi}{\partial z^2}$$

$$= -\frac{1}{r}\left(\frac{\partial \dot{E}_\phi}{\partial r} + r\frac{\partial^2 \dot{E}_\phi}{\partial r^2}\right) - \frac{\partial^2 \dot{E}_\phi}{\partial z^2}$$

$$= -\frac{1}{r}\left[-\frac{1}{\mathrm{k_c} r}J_1(\mathrm{k_c} r) + J_o(\mathrm{k_c} r)\right]\mathrm{k_c} E_o \sin\left(\frac{\pi z}{h}\right)\cdot e^{j\omega t}$$

$$-\left[\frac{1}{(\mathrm{k_c} r)^2}J_1(\mathrm{k_c} r) - \frac{1}{\mathrm{k_c} r}\left(-\frac{1}{\mathrm{k_c} r}J_1(\mathrm{k_c} r) + J_0(\mathrm{k_c} r)\right)\right]$$

$$\times \mathrm{k_c^2} E_o \sin\left(\frac{\pi z}{h}\right)\cdot e^{j\omega t} + \left(\frac{\pi}{h}\right)^2 \dot{E}_\phi$$

$$= \frac{1}{r^2}\dot{E}_\phi + \left(\frac{\pi}{h}\right)^2 \dot{E}_\phi = \left[\left(\frac{1}{r}\right)^2 + \left(\frac{\pi}{h}\right)^2\right]\dot{E}_\phi \qquad \text{(A–3.27)}$$

For computation of the resonance frequency, ω must be evaluated for $r = a$. Then

$$-\nabla^2 \dot{E}_\phi = \left[\left(\frac{1}{a}\right)^2 + \left(\frac{\pi}{h}\right)^2\right]\dot{E}_\phi \qquad \text{(A–3.28)}$$

From Eq. A–3.5, with $\sigma = 0$,

$$\omega = \frac{1}{\sqrt{\varepsilon\mu}}\sqrt{\left(\frac{1}{a}\right)^2 + \left(\frac{\pi}{h}\right)^2} \qquad \text{(A–3.29)}$$

or

$$f = \frac{1}{\sqrt{\varepsilon\mu}}\sqrt{\left(\frac{1}{2\pi a}\right)^2 + \left(\frac{1}{2\pi h}\right)^2} \qquad \text{(A–3.30)}$$

If the top and the bottom walls are open, as shown in Fig. A–3.2, the electric field distribution is, instead of Eq. A–3.22,

$$\dot{E}_\phi = E_o J_1(\mathrm{k}_c r)\cos\left(\frac{\pi z}{h}\right)e^{j\omega t} \qquad \text{(A–3.31)}$$

Repeating the same procedure to obtain Eq. A–3.30, the same resonance frequency as in Eq. A–3.30 is obtained for the open cavity. This type of open cavity is common for magnetrons.

For microstrip resonators, a circular cavity resonator as illustrated in Fig. A–3.2 C is common. In this type of open cavity resonator, the electric field distribution is described as

$$\dot{E}_z = E_o J_o(\mathrm{k}_c r)e^{j\omega t} \qquad \text{(A–3.32)}$$

Then

$$|\nabla^2 \dot{\mathbf{E}}| = \nabla^2 \dot{E}_z = \frac{1}{r}\frac{\partial}{\partial r}\left(r\frac{\partial \dot{E}_z}{\partial r}\right) = \frac{1}{r}\left(\frac{\partial \dot{E}_z}{\partial r} + r\frac{\partial^2 \dot{E}_z}{\partial r^2}\right)$$

$$= \frac{1}{r}\frac{\partial \dot{E}_z}{\partial r} + \frac{\partial^2 \dot{E}_z}{\partial r^2}$$

$$= -\frac{1}{\mathrm{k}_c r}J_1(\mathrm{k}_c r)E_o e^{j\omega t} + \left(\frac{1}{\mathrm{k}_c r}J_1(\mathrm{k}_c r) - J_o(\mathrm{k}_c r)\mathrm{k}_c^2 E_o e^{j\omega t}\right)$$

$$= -\mathrm{k}_c^2 \dot{E}_z \qquad \text{(A–3.33)}$$

Substituting Eq. A–3.33 in A–3.5 with $\sigma = 0$,

$$\omega = \frac{\mathrm{k}_c}{\sqrt{\varepsilon\mu}} \qquad \text{(A–3.34)}$$

For the open circular microstrip resonator of radius a, $\dot{E}_z$ is maximum at the edge at $r = a$:

$$\left.\frac{\partial \dot{E}_z}{\partial r}\right|_{r=a} = 0 \qquad \text{(A–3.35)}$$

Then, with Eq. A–3.32,

$$J_o'(\mathrm{k}_c a) = 0 \qquad \text{(A–3.36)}$$

According to the table of Bessel functions, the first root of Eq. A–3.36 is [3]

$$\mathrm{k}_c a = 3.832 \qquad \text{(A–3.37)}$$

or

$$k_c = \frac{3.832}{a} \qquad \text{(A–3.38)}$$

Substituting Eq. A–3.38 in Eq. A–3.34,

$$f = \frac{3.832}{2\pi a\sqrt{\varepsilon\mu}} \qquad \text{(A–3.39)}$$

"Cavity resonators" can be formed using sections of open or short circuited transmission lines. The transmission line can be a coaxial line, a two-wire line, a waveguide, or a microstripline. Whatever the type of the transmission line, as shown in Eq. A–1.41, the input impedance of an open circuited line is given by

$$\dot{Z}(0) = jZ_o \cot\left(\frac{2\pi l}{\lambda}\right) \qquad \text{(A–3.40)}$$

If

$$l = \frac{\lambda}{4} \qquad \text{(A–3.41)}$$

$$\dot{Z}(0) = 0 \qquad \text{(A–3.42)}$$

This is a resonance condition. The resonance wavelength is therefore, from Eq. A–3.41,

$$l = \frac{\lambda}{4} \quad \text{or} \quad \lambda = 4l \qquad \text{(A–3.43)}$$

In Eq. A–3.40, if

$$l = \frac{\lambda}{2} \qquad \text{(A–3.44)}$$

$$\dot{Z}(0) \to j\infty \qquad \text{(A–3.45)}$$

This is the antiresonance condition. The antiresonance wavelength is therefore given, from Eq. A–3.44, as

$$\lambda = 2l \qquad \text{(A–3.46)}$$

If the transmission line is a short circuited transmission line at one end, repeating a similar procedure to obtain Eq. A–1.41, it can be shown that the input impedance of the short circuited transmission line is given by [3]

$$\dot{Z}(0) = jZ_o \tan\left(\frac{2\pi l}{\lambda}\right) \tag{A–3.47}$$

If

$$l = \frac{\lambda}{2} \quad \text{or} \quad \lambda = 2l \tag{A–3.48}$$

then

$$\dot{Z}(0) = 0 \tag{A–3.49}$$

This is a resonance condition. If

$$l = \frac{\lambda}{4} \quad \text{or} \quad \lambda = 4l \tag{A–3.50}$$

then

$$\dot{Z}(0) \to j\infty \tag{A–3.51}$$

This is an antiresonance condition.

Another important factor in a cavity resonator is a quality factor. The quality factor Q of a cavity resonator is defined as a factor which is proportional to the ratio of the stored electromagnetic energy inside the cavity resonator to the power loss P in the cavity resonator. The proportionality constant is the angular frequency ω:

$$Q = \omega\frac{W}{P} \tag{A–3.52}$$

According to electromagnetic field theory,

$$W = \int_v \left(\frac{1}{4}\varepsilon E^2 + \frac{1}{4}\mu H^2\right) \mathrm{d}v \tag{A–3.53}$$

where v is the volume of the cavity resonator and

$$P = \int_s \frac{1}{2}\rho_\omega H_t^2 \,\mathrm{d}s \tag{A–3.54}$$

where ρ_ω is the resistivity of the conducting wall of the cavity resonator, s is the cavity surface, and H_t is the magnetic field tangential to the cavity wall [3]. In this simplified assumption, the loss was assumed to be only ohmic at the cavity wall. If the radiation loss, the power loss due to the coupling to the external circuit, or the dielectric loss are involved, then these losses must be added to Eq. A–3.54.

4 *S*–Parameters

In a three-port waveguide junction, as shown in Fig. A–4.1 A, if the incident electric field strength of Port-1, Port-2, and Port-3 are represented by E_1^{i}, E_2^{i} and E_3^{i}, respectively, and the electric field strength of the reflected waves for Port-1, Port-2, and Port-3 are represented by E_1^{r}, E_2^{r}, and E_3^{r}, respectively, then the reflected waves are related to the incident waves by

$$\left.\begin{aligned}\dot{E}_1^{\mathrm{r}} &= \dot{S}_{11}\dot{E}_1^{\mathrm{i}} + \dot{S}_{12}\dot{E}_2^{\mathrm{i}} + \dot{S}_{13}\dot{E}_3^{\mathrm{i}}\\ \dot{E}_2^{\mathrm{r}} &= \dot{S}_{21}\dot{E}_1^{\mathrm{i}} + \dot{S}_{22}\dot{E}_2^{\mathrm{i}} + \dot{S}_{23}\dot{E}_3^{\mathrm{i}}\\ \dot{E}_3^{\mathrm{r}} &= \dot{S}_{31}\dot{E}_1^{\mathrm{i}} + \dot{S}_{32}\dot{E}_2^{\mathrm{i}} + \dot{S}_{33}\dot{E}_3^{\mathrm{i}}\end{aligned}\right\} \qquad \text{(A–4.1 [3])}$$

Proportionality constants $\dot{S}_{11}, \dot{S}_{12}, \dot{S}_{13}, \ldots, \dot{S}_{33}$ are called scattering parameters, or *S*-parameters. The *S*-parameters are usually

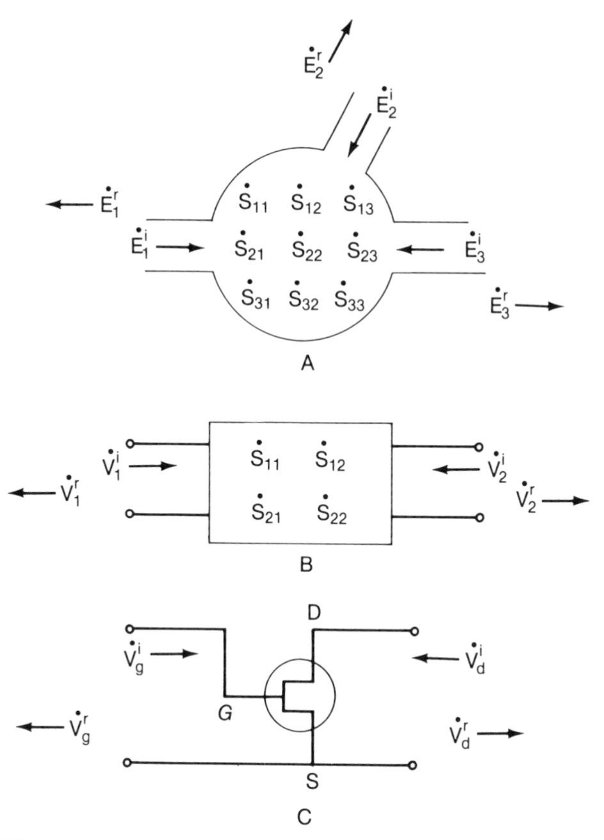

Fig. A–4.1 S-parameters. A. Waveguide network. B. Four-terminal network. C. Four-terminal FET.

expressed in the form of a matrix.

$$\begin{pmatrix} \dot{E}_1^{\mathrm{r}} \\ \dot{E}_2^{\mathrm{r}} \\ \dot{E}_3^{\mathrm{r}} \end{pmatrix} = \begin{pmatrix} \dot{S}_{11} & \dot{S}_{12} & \dot{S}_{13} \\ \dot{S}_{21} & \dot{S}_{22} & \dot{S}_{23} \\ \dot{S}_{31} & \dot{S}_{32} & \dot{S}_{33} \end{pmatrix} \begin{pmatrix} \dot{E}_1^{\mathrm{i}} \\ \dot{E}_2^{\mathrm{i}} \\ \dot{E}_3^{\mathrm{i}} \end{pmatrix} \tag{A–4.2}$$

The matrix consists of scattering parameters and is called the scattering matrix, or S-matrix. Actually, S-parameters $\dot{S}_{11}$, $\dot{S}_{22}$, and $\dot{S}_{33}$ are voltage reflection coefficients at Port-1, Port-2, and Port-3, respectively.

$$\left.\begin{aligned} \dot{\rho}_1 &\equiv \left.\frac{\dot{E}_1^{\mathrm{r}}}{\dot{E}_1^{\mathrm{i}}}\right|_{\dot{E}_2^{\mathrm{i}}=\dot{E}_3^{\mathrm{i}}=0} = \dot{S}_{11} \\ \dot{\rho}_2 &\equiv \left.\frac{\dot{E}_2^{\mathrm{r}}}{\dot{E}_2^{\mathrm{i}}}\right|_{\dot{E}_1^{\mathrm{i}}=\dot{E}_3^{\mathrm{i}}=0} = \dot{S}_{22} \\ \dot{\rho}_3 &\equiv \left.\frac{\dot{E}_3^{\mathrm{r}}}{\dot{E}_3^{\mathrm{i}}}\right|_{\dot{E}_1^{\mathrm{i}}=\dot{E}_2^{\mathrm{i}}=0} = \dot{S}_{33} \end{aligned}\right\} \tag{A–4.3}$$

Also, $\dot{S}_{12}, \dot{S}_{13}, \ldots, \dot{S}_{32}$ are actually the voltage transmission coefficients between two corresponding ports. For example,

$$\left.\begin{aligned} \dot{\tau}_{12} &\equiv \left.\frac{\dot{E}_1^{\mathrm{r}}}{\dot{E}_2^{\mathrm{i}}}\right|_{\dot{E}_1^{\mathrm{i}}=\dot{E}_3^{\mathrm{i}}=0} = \dot{S}_{12} \\ \dot{\tau}_{13} &\equiv \left.\frac{\dot{E}_1^{\mathrm{r}}}{\dot{E}_3^{\mathrm{i}}}\right|_{\dot{E}_1^{\mathrm{i}}=\dot{E}_2^{\mathrm{i}}=0} = \dot{S}_{13} \\ \dot{\tau}_{32} &\equiv \left.\frac{\dot{E}_3^{\mathrm{r}}}{\dot{E}_2^{\mathrm{i}}}\right|_{\dot{E}_1^{\mathrm{i}}=\dot{E}_3^{\mathrm{i}}=0} = \dot{S}_{32} \end{aligned}\right\} \tag{A–4.4}$$

In a four-terminal network, as shown in Fig. A–4.1 B,

$$\begin{pmatrix} \dot{V}_1^{\mathrm{r}} \\ \dot{V}_2^{\mathrm{r}} \end{pmatrix} = \begin{pmatrix} \dot{S}_{11} & \dot{S}_{12} \\ \dot{S}_{21} & \dot{S}_{22} \end{pmatrix} \begin{pmatrix} \dot{V}_1^{\mathrm{i}} \\ \dot{V}_2^{\mathrm{i}} \end{pmatrix} \tag{A–4.5}$$

$$\times \left.\begin{aligned} \dot{S}_{11} &= \left.\frac{\dot{V}_1^{\mathrm{r}}}{\dot{V}_1^{\mathrm{i}}}\right|_{\dot{V}_2^{\mathrm{i}}=0} & \dot{S}_{22} &= \left.\frac{\dot{V}_2^{\mathrm{r}}}{\dot{V}_2^{\mathrm{i}}}\right|_{\dot{V}_1^{\mathrm{i}}=0} \\ \dot{S}_{12} &= \left.\frac{\dot{V}_1^{\mathrm{r}}}{\dot{V}_2^{\mathrm{i}}}\right|_{\dot{V}_1^{\mathrm{i}}=0} & \dot{S}_{21} &= \left.\frac{\dot{V}_2^{\mathrm{r}}}{\dot{V}_1^{\mathrm{i}}}\right|_{\dot{V}_2^{\mathrm{i}}=0} \end{aligned}\right\} \tag{A–4.6}$$

When this two-port network or four-terminal network principle is applied to an FET circuit as shown in Fig. A–4.1 C

$$\begin{pmatrix} \dot{V}_g^r \\ \dot{V}_d^r \end{pmatrix} = \begin{pmatrix} \dot{S}_{11} & \dot{S}_{12} \\ \dot{S}_{21} & \dot{S}_{22} \end{pmatrix} \begin{pmatrix} \dot{V}_g^i \\ \dot{V}_d^i \end{pmatrix} \tag{A–4.7}$$

$$\dot{S}_{11} = \left.\frac{\dot{V}_g^r}{\dot{V}_g^i}\right|_{\dot{V}_d^i=0} \qquad \dot{S}_{22} = \left.\frac{\dot{V}_d^r}{\dot{V}_d^i}\right|_{\dot{V}_g^i=0} \tag{A–4.8}$$

$$\dot{S}_{21} = \left.\frac{\dot{V}_d^r}{\dot{V}_g^i}\right|_{\dot{V}_d^i=0} \qquad \dot{S}_{12} = \left.\frac{\dot{V}_g^r}{\dot{V}_d^i}\right|_{\dot{V}_g^i=0} \tag{A–4.9}$$

Therefore, if the FET circuit is perfectly impedance matched, then

$$\left.\begin{array}{ll} \dot{S}_{11} = 0, & \dot{S}_{22} = 0 \\ \dot{S}_{21} = \dot{A}, & \dot{S}_{12} = 0 \end{array}\right\} \tag{A–4.10}$$

where $\dot{A}$ is the voltage gain.

5 Fermi–Dirac Distribution Function

Energy states of electrons in solids are described by the electronic energy E and lattice energy E_l of the solid. By external excitation, the state (E, E_l) may transfer to the state (E', E_l'). If the probability of finding an electron with the energy E or E' is $f(E)$ or $f(E')$, respectively, and the probability of the lattice energy being E_l or E_l' is $g(E_l)$ or $g(E_l')$, the probability of transferring the state from (E, E_l) to (E', E_l') is proportional to $f(E)$, $g(E_l)$ and, due to Pauli's exclusion principle, $[1 - f(E_l')]$ [6–10]. If the proportionality constant is represented by a, the transition probability from E to E' is

$$P_{E \to E'} = af(E)g(E_l)[1 - f(E')] \tag{A–5.1}$$

For the reverse transition, if the same proportionality constant is assumed, then

$$P_{E' \to E} = af(E')g(E_l')[1 - f(E)] \tag{A–5.2}$$

In an equilibrium state,

$$P_{E' \to E} = P_{E \to E'} \tag{A–5.3}$$

Then

$$f(E)g(E)[1 - f(E')] = f(E')g(E_l')[1 - f(E)] \quad \text{(A–5.4)}$$

Dividing through by $f(E)$, $g(E_l)$, and $f(E')$,

$$\frac{1}{f(E')} - 1 = \frac{g(E_l')}{g(E_l)f(E)} - \frac{g(E_l')}{g(E_l)}$$

$$= \left[\frac{1}{f(E)} - 1\right]\frac{g(E_l')}{g(E_l)} \quad \text{(A–5.5)}$$

The Boltzmann relation states that

$$\frac{g(E_l')}{g(E_l)} = e^{-(E_l' - E_l/kT)} \quad \text{(A–5.6)}$$

Then

$$\frac{1}{f(E')} - 1 = \left[\frac{1}{f(E)} - 1\right]e^{-(E_l' - E_l/kT)} \quad \text{(A–5.7)}$$

In the equilibrium of electron–lattice collisions

$$E + E_l = E' + E_l' \quad \text{(A–5.8)}$$

Then

$$E_l' - E_l = E - E'$$

$$\frac{\dfrac{1}{f(E')} - 1}{\dfrac{1}{f(E)} - 1} = e^{E' - E/kT} = \frac{e^{(E'/kT)}}{e^{(E/kT)}} = \frac{be^{(E'/kT)}}{be^{(E/kT)}} \quad \text{(A–5.9)}$$

where b is an arbitrary constant. Then

$$\frac{1}{f(E)} - 1 = be^{(E/kT)} \quad \text{(A–5.10)}$$

$$\frac{1}{f(E)} = 1 + be^{(E/kT)}$$

$$f(E) = \frac{1}{1 + be^{(E/kT)}} \quad \text{(A–5.11)}$$

If a certain energy level E does exist at $E = E_f$,

$$f(E_f) = \tfrac{1}{2} \qquad \text{(A–5.12)}$$

Then

$$\tfrac{1}{2} = \frac{1}{1 + be^{(E_f/kT)}} \qquad \text{(A–5.13)}$$

$$1 + be^{(E_f/kT)} = 2$$

$$be^{(E_f/kT)} = 1$$

$$b = e^{-(E_f/kT)} \qquad \text{(A–5.14)}$$

Substituting Eq. A–5.14 in Eq. A–5.11,

$$f(E) = \frac{1}{1 + e^{(E - E_f/kT)}} \qquad \text{(A–5.15)}$$

This function is called a Fermi–Dirac distribution function and the energy level E_f is called the Fermi level, at which $f(E) = \frac{1}{2}$ [6–10].

6 Waveguide Principles

One of the most popular waveguides used with microwave electron devices is the rectangular waveguide [3], as illustrated in Fig. A–6.1. A rectangular waveguide is a conducting hollow pipe with a rectangular cross section. Microwaves are launched into the waveguide by a small antenna or a small solid-state oscillator, which can be placed directly inside the waveguide. A detector, mixer, varactor, or PIN diode is mounted inside the waveguide for various applications. The electromagnetic field of a hollow waveguide is governed by a wave equation which is derived from Maxwell's equations. The wave equation for the lossless system is

$$\nabla^2 \dot{\mathbf{H}} + \omega^2 \varepsilon \mu \dot{\mathbf{H}} = 0 \qquad \text{(A–6.1)}$$

where $\dot{\mathbf{H}}$ is a microwave magnetic field inside the waveguide, ε is the permittivity, and μ is the permeability of the medium inside the waveguide. In most cases, ε and μ are that of the air. In the coordinate system shown in Fig. A–6.1 for $\dot{H}_z$,

$$\nabla^2 \dot{H}_z + \omega^2 \varepsilon \mu \dot{H}_z = 0 \qquad \text{(A–6.2)}$$

$$\frac{\partial^2 \dot{H}_z}{\partial x^2} + \frac{\partial^2 \dot{H}_z}{\partial y^2} + \frac{\partial^2 \dot{H}_z}{\partial z^2} + \omega^2 \varepsilon \mu \dot{H}_z = 0 \qquad \text{(A–6.3)}$$

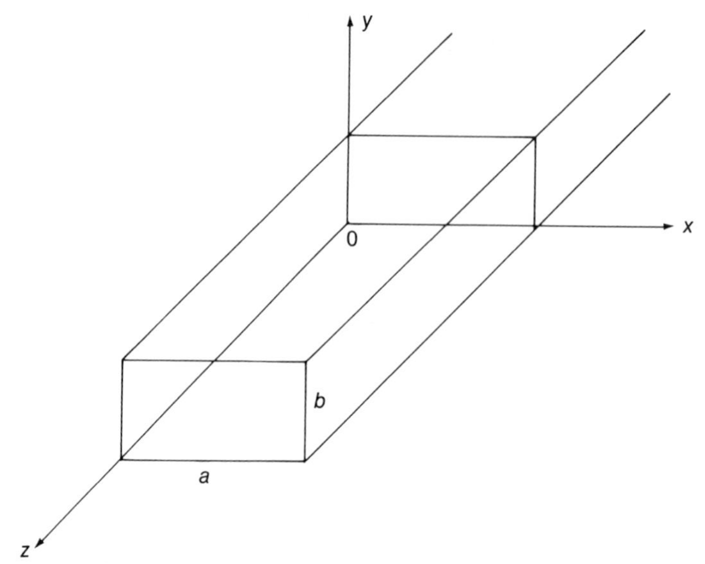

Fig. A–6.1 Rectangular waveguide.

The longitudinal magnetic field $\dot{H}_z$ is assumed to be propagating in the z-direction with the propagation constant

$$\gamma = j\beta \tag{A–6.4}$$

This neglects attenuation. β is the phase constant

$$\beta = \frac{2\pi}{\lambda_g} \tag{A–6.5}$$

where λ_g is the wavelength in the waveguide. β is therefore the phase angle shift per unit distance of propagation. Then the traveling waveform for the z-direction must be

$$\dot{H}_z = H_{oz}(x, y)e^{j(\omega t - \beta z)} \tag{A–6.6}$$

Substituting Eq. A–6.6 in Eq. A–6.3,

$$\frac{\partial^2 H_{oz}}{\partial x^2} + \frac{\partial^2 H_{oz}}{\partial y^2} - (\beta^2 - \omega^2 \varepsilon\mu)H_{oz} = 0 \tag{A–6.7}$$

On the conducting wall, the current must be maximum. Then, by Ampere's law, the magnetic field on the wall must also be maximum. The boundary conditions are

$$\left.\begin{aligned} \left.\frac{\partial H_{oz}}{\partial x}\right|_{x=0,\,a} &= 0 \\ \left.\frac{\partial H_{oz}}{\partial y}\right|_{y=0,\,b} &= 0 \end{aligned}\right\} \quad \text{(A–6.8)}$$

A function which satisfies Eq. A–6.8 is

$$H_{oz} = H_o \cos\left(\frac{\pi x}{a}\right) \quad \text{(A–6.9)}$$

Substituting Eq. A–6.9 in Eq. A–6.6,

$$\dot{H}_z = H_o \cos\left(\frac{\pi x}{a}\right) e^{j(\omega t - \beta z)} \quad \text{(A–6.10)}$$

This is the description of the longitudinal magnetic field inside of the rectangular waveguide. Substituting Eq. A–6.9 in Eq. A–6.7,

$$-\left(\frac{\pi}{a}\right)^2 - \beta^2 + \omega^2 \varepsilon\mu = 0$$

$$\beta^2 = \omega^2 \varepsilon\mu - \left(\frac{\pi}{a}\right)^2$$

$$= \left(\frac{2\pi f}{c}\right)^2 - \left(\frac{\pi}{a}\right)^2 = \left(\frac{2\pi}{\lambda}\right)^2 - \left(\frac{\pi}{a}\right)^2 \quad \text{(A–6.11)}$$

Combining Eq. A–6.5 and A–6.11,

$$\left(\frac{2\pi}{\lambda_g}\right)^2 = \left(\frac{2\pi}{\lambda}\right)^2 - \left(\frac{\pi}{a}\right)^2$$

$$\left(\frac{1}{\lambda_g}\right)^2 = \left(\frac{1}{\lambda}\right)^2 - \left(\frac{1}{2a}\right)^2 = \left(\frac{1}{\lambda}\right)^2\left[1 - \left(\frac{\lambda}{2a}\right)^2\right]$$

$$\lambda_g = \frac{\lambda}{\sqrt{1 - \left(\frac{\lambda}{2a}\right)^2}} \quad \text{(A–6.12)}$$

Here, λ is the wavelength in free space. When $\lambda > 2a$, there is no real wavelength in the waveguide, no microwave inside the waveguide, and no propagation. Therefore,

$$\lambda = 2a \quad \text{(A–6.13)}$$

is called the cutoff wavelength. In the range of propagation,

$$\lambda < 2a \tag{A–6.14}$$

If so, then

$$\lambda_g > \lambda \tag{A–6.15}$$

This means that the wavelength elongates when the microwaves are put into a conducting hollow pipe called a waveguide. If both sides of Eq. A–6.12 are multiplied by the operating frequency f,

$$f\lambda_g = \frac{f\lambda}{\sqrt{1 - \left(\frac{\lambda}{2a}\right)^2}} \tag{A–6.16}$$

$$v = \frac{c}{\sqrt{1 - \left(\frac{\lambda}{2a}\right)^2}} \tag{A–6.17}$$

This means that, in the propagation region,

$$v > c \tag{A–6.18}$$

The propagation velocity v inside the waveguide is greater than the propagation velocity in free space c.

Maxwell's equation of Ampere's law states that

$$\nabla \times \dot{\mathbf{H}} = j\omega\varepsilon\dot{\mathbf{E}} \tag{A–6.19}$$

$$\begin{aligned} j\omega\dot{E}_y &= \frac{\partial \dot{H}_x}{\partial z} - \frac{\partial \dot{H}_z}{\partial x} \\ &= -j\beta\dot{H}_x - \left(\frac{\pi}{a}\right) H_o \sin\left(\frac{\pi x}{a}\right) e^{j(\omega t - \beta z)} \\ &= j\beta\frac{\dot{E}_y}{\eta} - \left(\frac{\pi}{a}\right) H_o \sin\left(\frac{\pi x}{a}\right) e^{j(\omega t - \beta z)} \end{aligned} \tag{A–6.20}$$

where η is the wave impedance and is defined as

$$\eta = \frac{\dot{E}_y}{-\dot{H}_x} \tag{A–6.21}$$

Thus,

$$j\left(\omega\varepsilon - \frac{\beta}{\eta}\right)\dot{E}_y = -\left(\frac{\pi}{a}\right)H_o \sin\left(\frac{\pi x}{a}\right)e^{j(\omega t - \beta z)}$$

$$\dot{E}_y = j\frac{\eta(\pi/a)}{\omega\varepsilon\eta - \beta}H_o \sin\left(\frac{\pi x}{a}\right)e^{j(\omega t - \beta z)} \tag{A–6.22}$$

From Eq. A–6.19,

$$j\omega\varepsilon\dot{E}_x = \frac{\partial \dot{H}_z}{\partial y} - \frac{\partial \dot{H}_y}{\partial z}$$

$$= 0 + j\beta\dot{H}_y = j\beta\frac{\dot{E}_x}{\eta} \tag{A–6.23}$$

$$\eta = \frac{\beta}{\omega\varepsilon} \tag{A–6.24}$$

This is an impossible situation as seen from Eq. A–6.22. If Eq. A–6.24 is true, by Eq. A–6.22 $\dot{E}_y \to \infty$. $\dot{E}_y$ is in fact finite. Therefore Eq. A–6.24 is impossible. The only way to satisfy Eq. A–6.22 is if

$$\dot{E}_x = 0 \tag{A–6.25}$$

This is indeed the case in the waveguide. Then, from Eq. A–6.19,

$$j\omega\varepsilon\dot{E}_z = \frac{\partial \dot{H}_y}{\partial x} - \frac{\partial \dot{H}_x}{\partial y} = \frac{\partial}{\partial x}\left(\frac{\dot{E}_x}{\eta}\right) + \frac{\partial}{\partial y}\left(\frac{\dot{E}_y}{\eta}\right) \tag{A–6.26}$$

Substituting Eqs. A–6.25 and A–6.22 in Eq. A–6.26,

$$j\omega\varepsilon\dot{E}_z = 0 \tag{A–6.27}$$

So

$$\dot{E}_z = 0 \tag{A–6.28}$$

Maxwell's equation of Faraday's law states that

$$\nabla \times \dot{\mathbf{E}} = -j\omega\mu\dot{\mathbf{H}} \tag{A–6.29}$$

Thus,

$$-j\omega\mu\dot{H}_x = \frac{\partial \dot{E}_z}{\partial y} - \frac{\partial \dot{E}_y}{\partial z}$$

$$= 0 + \frac{\eta\beta(\pi/a)}{\omega\varepsilon\eta - \beta}H_o \sin\left(\frac{\pi x}{a}\right)e^{j(\omega t - \beta z)}$$

or

$$\dot{H}_x = j\frac{\eta\beta(\pi/a)}{\omega\mu(\omega\varepsilon\eta - \beta)}H_o \sin\left(\frac{\pi x}{a}\right)e^{j(\omega t - \beta z)} \quad \text{(A–6.30)}$$

From Eq. A–6.29,

$$-j\omega\mu\dot{H}_y = \frac{\partial\dot{E}_x}{\partial z} - \frac{\partial\dot{E}_z}{\partial x} \quad \text{(A–6.31)}$$

Substituting Eq. A–6.25 and Eq. A–6.28 in Eq. A–6.31,

$$-j\omega\mu\dot{H}_y = 0 \quad \text{or} \quad \dot{H}_y = 0 \quad \text{(A–6.32)}$$

Now the electromagnetic fields inside the rectangular waveguide are completely described. In this study, to satisfy boundary condition A–6.8, a particular solution (A–6.9) was chosen. Mathematically, instead of A–6.9,

$$H_{oz} = H_o \cos\left(\frac{m\pi x}{a}\right)\cos\left(\frac{n\pi y}{b}\right) \quad \text{(A–6.33)}$$

can be chosen. If so, the electromagnetic field distribution of this mode is called the H_{mn} mode or TE_{mn} mode, where m and n are zero or integers. Equation A–6.9 is for the case of $m = 1$ and $n = 0$ or H_{10} mode or TE_{10} mode. In most applications, the TE_{10} mode is used. TE stands for transverse electric which means $\dot{E}_z = 0$. H mode means $H_z \neq 0$.

The entire analysis can be repeated using

$$E_{oz} = E_o \cos\left(\frac{m\pi x}{a}\right)\cos\left(\frac{n\pi y}{b}\right) \quad \text{(A–6.34)}$$

This mode is the E_{mn} mode or TM_{mn} mode. TM mode, or transverse magnetic mode, is not used for practical waveguide transmission due to a higher attenuation than the TE mode. Due to mode instability, a waveguide with a circular cross section is not used often. Waveguides with elliptic cross sections are used at times and are treated approximately like rectangular waveguides.

7 Cyclotron Frequency and Plasma Oscillation Frequency

When an electron with velocity u enters a magnetic field of flux density B which is perpendicular to the electron velocity u, the amount of force exerted on the electron by the magnetic field is

$$f = Bqu \quad \text{(A–7.1)}$$

where q is the electric charge of the electron. The direction of the magnetic force is perpendicular to both B and u or, in the vector form,

$$\mathbf{f} = -q\mathbf{B} \times \mathbf{u} \tag{A–7.2}$$

As a result, the orbit of the electron is bent and the electron goes into circular motion. This circular motion is called cyclotron motion. According to mechanics, r, the radius of curvature of this circular motion, the angular frequency ω, and the centripetal force f are related to each other by

$$f = \frac{mu^2}{r} = \frac{m(r\omega)^2}{r} = mr\omega^2 \tag{A–7.3}$$

where m is the mass of the electron. Using the relation

$$u = r\omega \tag{A–7.4}$$

in Eq. A–7.1,

$$f = Bqr\omega \tag{A–7.5}$$

Equating Eqs. A–7.3 and A–7.5,

$$mr\omega^2 = Bqr\omega$$

$$\omega = \frac{Bq}{m} \tag{A–7.6}$$

This angular frequency of circular motion, or cyclotron motion, is commonly called the cyclotron frequency. The cyclotron frequency is actually the angular velocity of the cyclotron motion. The actual cyclotron frequency is obtained by dividing Eq. A–7.6 by 2π.

Weakly ionized gaseous media is called plasma. In plasma, electrons which do not recombine with positive ions oscillate around a neutral position. This oscillatory motion of electrons around a neutral location is called plasma oscillation. The plasma oscillation is generally three dimensional. If only an x-component is observed, the equation of motion of the electron is, omitting friction,

$$m\frac{\mathrm{d}^2x}{\mathrm{d}t^2} = -\mathrm{k}x \tag{A–7.7}$$

where k is a constant. Poisson's equation applied to the electron-filled space is

$$\frac{d^2V}{dx^2} = -\frac{\rho}{\varepsilon} \tag{A–7.8}$$

where ρ is the electron charge density and ε is the permittivity of the space [7]. Upon integration,

$$\frac{dV}{dx} = -\frac{\rho}{\varepsilon}x \tag{A–7.9}$$

where the integrating constant is assumed to be zero because $x = 0$ is a neutral point at which $dV/dx = 0$. Then the force exerted on an oscillating electron due to the field dV/dx by mutual repulsion among the electrons is

$$q\frac{dV}{dx} = -q\frac{\rho}{\varepsilon}x \tag{A–7.10}$$

Equating Eq. A–7.7 and Eq. A–7.10,

$$m\frac{d^2x}{dt^2} = -q\frac{\rho}{\varepsilon}x \tag{A–7.11}$$

$$\frac{d^2x}{dt^2} + \frac{q\rho}{m\varepsilon}x = 0 \tag{A–7.12}$$

The solution of this linear homogeneous differential equation is known to be of the form

$$x = Ae^{j\omega t} \tag{A–7.13}$$

Substituting Eq. A–7.13 in Eq. A–7.12,

$$-\omega^2 x + \frac{q\rho}{m\varepsilon}x = 0 \tag{A–7.14}$$

or

$$\omega = \sqrt{\frac{q\rho}{m\varepsilon}} \tag{A–7.15}$$

This is called the plasma oscillation frequency.

8 Crystallographic Axis

A crystallographic axis is represented by the Miller indices. Miller indices are formed by establishing a rectangular coordinate system in the crystal. If a crystallographic plane is given, the intersection of the plane and the three coordinate axes—namely x-axis, y-axis, and z-axis—are expressed in terms of the lattice constants. If these intersections are designated as lattice constants n_x, n_y, and n_z, the Miller indices of this plane are $(1/n_x, 1/n_y, 1/n_z) \times n_{x \text{ or } y \text{ or } z}$ where $n_{x \text{ or } y \text{ or } z}$ is the smallest number of the lattice constant values n_x, n_y, and n_z. If these sets of numbers are represented by m_x, m_y, and m_z, respectively, then

$$m_x = \left(\frac{1}{n_x}\right) n_{x \text{ or } y \text{ or } z} \tag{A-8.1}$$

$$m_y = \left(\frac{1}{n_y}\right) n_{x \text{ or } y \text{ or } z} \tag{A-8.2}$$

$$m_z = \left(\frac{1}{n_z}\right) n_{x \text{ or } y \text{ or } z} \tag{A-8.3}$$

The Miller indices of the crystallographic plane are now given by (m_x, m_y, m_z). An axis perpendicular to the crystallographic plane (m_x, m_y, m_z) is represented by the symbol $[m_x, m_y, m_z]$. If the crystal has a cubic lattice, the rectangular coordinate system is commonly set up along the cubic lattice. Sometimes the rectangular coordinate system may be set up arbitrarily, regardless of the natural crystal lattice structure. In such cases, the Miller indices of the crystallographic plane are represented by $\langle m_x, m_y, m_z \rangle$. Miller indices are integers by nature of the crystal.

For example, crystallographic axis [110] means that this axis is perpendicular to the crystallographic plane (110) and a set of the reciprocal numbers is (11∞). This plane intersects the x-axis at a distance of one lattice constant. The same plane intersects the y-axis at a distance of one lattice constant. The plane does not intersect the z-axis. So, the [110] axis direction is perpendicular to the plane (110), which means the [110] axis is 45 degrees from both the x- and y-axes and perpendicular to the z-axis.

9 Mathematical and Physical Formulas and Identities

9.1 Euler's Identity

$$e^{j\theta} = \cos\theta + j\sin\theta$$

9.2 Series

1. $$\sin x = x - \frac{x^3}{3!} + \frac{x^5}{5!} - \frac{x^7}{7!} + \cdots$$
$$= \frac{-j}{2}(e^{jx} - e^{-jx})$$
$$= -j\sinh(jx)$$

2. $$\cos x = 1 - \frac{x^2}{2!} + \frac{x^4}{4!} - \frac{x^6}{6!} + \cdots$$
$$= \frac{1}{2}(e^{jx} + e^{-jx})$$
$$= \cosh(jx)$$

3. $$\tan x = x + \frac{x^3}{3} + \frac{2x^5}{15} + \frac{17x^7}{315} + \cdots$$
$$= -j\frac{e^{jx} - e^{-jx}}{e^{jx} + e^{-jx}} = j\tanh(jx)$$

4. $$e^x = 1 + x + \frac{x^2}{2!} + \frac{x^3}{3!} + \frac{x^4}{4!} + \cdots + \frac{x^n}{n!} + \cdots$$

5. $$\sinh x = x + \frac{x^3}{3!} + \frac{x^5}{5!} + \frac{x^7}{7!} + \cdots = \frac{1}{2}(e^x - e^{-x})$$
$$\cosh x = 1 + \frac{x^2}{2!} + \frac{x^4}{4!} + \frac{x^6}{6!} + \cdots = \frac{1}{2}(e^x + e^{-x})$$

9.3 Binomial Expansion

$$(x + y)^n = x^n + nx^{n-1}y + \frac{n(n-1)}{2!}x^{n-2}y^2 + \cdots + y^n$$

9.4 Approximations ($\delta \ll 1$)

1. $(1 \pm \delta)^n \approx 1 \pm n\delta$
2. $e^{\delta} \approx 1 + \delta$
3. $\ln(1 + \delta) \approx \delta$
4. $\tan\delta \approx \sin\delta \approx \delta$
5. $\cos\delta \approx 1$

9.5 Fourier Series

$$f(x) = \frac{a_o}{2} + \sum_{n=1}^{\infty} (a_n \cos nx + b_n \sin nx)$$

$$a_n = \frac{1}{\pi}\int_{-\pi}^{\pi} f(x)\cos(nx)\,dx$$

$$b_n = \frac{1}{\pi}\int_{-\pi}^{\pi} f(x)\sin(nx)\,dx \qquad n = 0, 1, 2, 3..$$

9.6 Vector Identities

1. $\nabla \times (\nabla \times \mathbf{H}) = \nabla(\nabla \cdot \mathbf{H}) - \nabla^2 \mathbf{H}$
2. Divergence Theorem

$$\int_v \nabla \cdot \mathbf{H}\,dv = \int_s \mathbf{H} \cdot d\mathbf{S}$$

3. Stokes' Theorem

$$\int_s \nabla \times \mathbf{H} \cdot d\mathbf{S} = \int_c \mathbf{H} \cdot d\mathbf{l}$$

9.7 Vector Differential Operations

1. $\nabla f = \mathbf{a}_x \frac{\partial f}{\partial x} + \mathbf{a}_y \frac{\partial f}{\partial y} + \mathbf{a}_z \frac{\partial f}{\partial z}$
2. $\nabla \mathbf{F} = \frac{\partial F_x}{\partial x} + \frac{\partial F_y}{\partial y} + \frac{\partial F_z}{\partial z}$
3. $\nabla \times \mathbf{F} = \begin{vmatrix} \mathbf{a}_x & \mathbf{a}_y & \mathbf{a}_z \\ \frac{\partial}{\partial x} & \frac{\partial}{\partial y} & \frac{\partial}{\partial z} \\ F_x & F_y & F_z \end{vmatrix}$

4. $\nabla^2 f = \frac{\partial^2 f}{\partial x^2} + \frac{\partial^2 f}{\partial y^2} + \frac{\partial^2 f}{\partial z^2}$

5. $\nabla^2 \mathbf{F} = \frac{\partial^2 \mathbf{F}}{\partial x^2} + \frac{\partial^2 \mathbf{F}}{\partial y^2} + \frac{\partial^2 \mathbf{F}}{\partial z^2}$

6. $\nabla f = \mathbf{a}_r \frac{\partial f}{\partial r} + \mathbf{a}_\phi \frac{1}{r}\frac{\partial f}{\partial \phi} + \mathbf{a}_z \frac{\partial f}{\partial z}$

7. $\nabla \cdot \mathbf{F} = \frac{1}{r}\frac{\partial}{\partial r}(rF_r) + \frac{1}{r}\frac{\partial F_\phi}{\partial \phi} + \frac{\partial F_z}{\partial z}$

8. $\nabla \times \mathbf{F} = \begin{vmatrix} \frac{1}{r}\mathbf{a}_r & \mathbf{a}_\phi & \frac{1}{r}\mathbf{a}_z \\ \frac{\partial}{\partial r} & \frac{\partial}{\partial \phi} & \frac{\partial}{\partial z} \\ F_r & rF_\phi & F_z \end{vmatrix}$

9. $\nabla^2 f = \frac{1}{r}\frac{\partial}{\partial r}\left(r\frac{\partial f}{\partial r}\right) + \frac{1}{r^2}\frac{\partial^2 f}{\partial \phi^2} + \frac{\partial^2 f}{\partial z^2}$

10. $\nabla^2 \mathbf{F} = \frac{1}{r}\frac{\partial}{\partial r}\left(r\frac{\partial \mathbf{F}}{\partial r}\right) + \frac{1}{r^2}\frac{\partial^2 \mathbf{F}}{\partial \phi^2} + \frac{\partial^2 \mathbf{F}}{\partial z^2}$

9.8 Hyperbolic Functions

1. $\cosh x = \cos(jx)$
2. $\sinh x = -j\sin(jx)$
3. $\tanh(x \pm jy) = \frac{\sinh(2x) \pm j\sin(2y)}{\cosh(2x) + \cos(2y)}$
4. $\coth(x \pm jy) = \frac{\sinh 2x \mp j\sin(2y)}{\cosh(2x) - \cos(2y)}$

9.9 Bessel Functions

1. Bessel's equation

$$x^2\frac{d^2y}{dx^2} + x\frac{dy}{dx} + (x^2 - \nu^2)y = 0$$

Solution

$$y = AJ_\nu(x) + BN_\nu(x)$$

2. Approximations

$$J_\nu(x) \approx \frac{x^\nu}{2^\nu x!} \quad \text{for small } x$$

$$J_\nu(x) \approx \sqrt{\frac{2}{\pi x}} \cos\left(x - \frac{\nu\pi}{2} - \frac{\pi}{4}\right) \quad \text{for large } x$$

$$N_\nu(x) \approx \frac{2^\nu(\nu - 1)!}{\pi x^\nu} \quad \text{for small } x$$

$$N_\nu(x) \approx \sqrt{\frac{2}{\pi x}} \sin\left(x - \frac{\nu\pi}{2} - \frac{\pi}{4}\right) \quad \text{for large } x$$

3. Differential and Integrals

$$\begin{aligned}\frac{\mathrm{d}J_\nu(x)}{\mathrm{d}x} &= \frac{\nu}{x}J_\nu(x) - J_{\nu+1}(x) \\ &= -\frac{\nu}{x}J_\nu(x) + J_{\nu-1}(x) \\ &= \frac{1}{2}\left[J_{\nu-1}(x) - J_{\nu+1}(x)\right]\end{aligned}$$

$$J_\nu(x) = \frac{x}{2\nu}\left[J_{\nu+1}(x) + J_{\nu-1}(x)\right]$$

$$\frac{\mathrm{d}J_0(x)}{\mathrm{d}x} = J_1(x)$$

$$\int J_1(x)\,\mathrm{d}x = -J_0(x)$$

$$\int x^\nu J_{\nu-1}(x)\,\mathrm{d}x = x^\nu J_\nu(x)$$

$$\int x^{-\nu} J_{\nu+1}(x)\,\mathrm{d}x = -x^{-\nu} J_\nu(x)$$

4. Zeros

$$\begin{array}{llll} J_0(x) = 0, & x = 2.405, & 5.520, & 8.654, \ldots \\ J_1(x) = 0, & x = 0, & 3.872, & 7.016, \ldots \\ J_2(x) = 0, & x = 0, & 5.135, & 8.417, \ldots \\ N_0(x) = 0, & x = 0.894 & 3.958, & 7.086, \ldots \\ N_1(x) = 0, & x = 2.20, & 5.43\ , & 8.60, \ldots \end{array}$$

9.10 Electromagnetic Field Theory

1. Maxwell's Equations

$$\nabla \times \mathbf{H} = \sigma \mathbf{E} + \varepsilon \frac{\partial \mathbf{E}}{\partial t}$$

$$\nabla \times \mathbf{H} = -\mu \frac{\partial \mathbf{H}}{\partial t}$$

$$\nabla \cdot \mathbf{D} = \rho$$

$$\nabla \cdot \mathbf{B} = 0$$

$$\oint_c \mathbf{H} \cdot \mathrm{d}\mathbf{l} = \int_s \left(\mathbf{J}_c + \frac{\partial \mathbf{D}}{\partial t} \right) \cdot \mathrm{d}\mathbf{S}$$

$$\oint_c \mathbf{E} \cdot \mathrm{d}\mathbf{l} = -\int_s \frac{\partial \mathbf{B}}{\partial t} \cdot \mathrm{d}\mathbf{S}$$

$$\int_s \mathbf{D} \cdot \mathrm{d}\mathbf{S} = \int_v \rho \, \mathrm{d}v$$

$$\oint_s \mathbf{B} \cdot \mathrm{d}\mathbf{S} = 0$$

2. Electric Fields and Electric Potential

$$\mathbf{E} = -\nabla V - \frac{\partial \mathbf{A}}{\partial t}$$

$$V = \int_v \frac{\rho \, \mathrm{d}v}{4\pi \varepsilon r}$$

$$\mathbf{A} = \int_v \frac{\mu \mathbf{J} \, \mathrm{d}v}{4\pi r}$$

3. Magnetic Fields and Vector Potential

$$\mathbf{H} = \frac{1}{\mu} \nabla \times \mathbf{A}$$

4. Continuity of Electric Current

$$\nabla \cdot \mathbf{D} = -\frac{\partial \rho}{\partial t}$$

5. Wave Equations

$$\nabla^2 V - \mu\varepsilon \frac{\partial^2 V}{\partial t^2} = -\frac{\rho}{\varepsilon}$$

$$\nabla^2 \mathbf{A} - \mu\varepsilon \frac{\partial^2 \mathbf{A}}{\partial t^2} = -\mu \mathbf{J}$$

10 Physical Constants

Permittivity of Free Space	$\varepsilon_o = 8.854 \times 10^{-12}$ F/m
Permeability of Free Space	$\mu_o = 1.257 \times 10^{-6}$ H/m
Speed of Light in Vacuum	$c = 3 \times 10^8$ m/s
Free Space Wave Impedance	$\sqrt{\mu_o/\varepsilon_o} = 376.7\ \Omega$
Boltzmann Constant	$\mathrm{k} = 1.38054 \times 10^{-23}$ J/°K
Electron Charge	$e = 1.6021 \times 10^{-19}$ C
Electron Mass	$m = 9.1091 \times 10^{-31}$ kg
Specific Charge of Electron	$e/m = 1.7588 \times 10^{11}$ C/kg
Planck's Constant	$h = 6.6256 \times 10^{-34}$ J-s
Gyromagnetic Ratio	$\gamma = 2.8$ MHz/Oe = 28 GHz/T
Proton Rest Mass	$m_p = 1.617243 \times 10^{-27}$ kg
Bohr Magnetron	$\mu_B = 9.2732 \times 10^{-21}$ erg/Gauss
	$= 9.2732 \times 10^{-28}$ J/Gauss
	$= 9.2732 \times 10^{-24}$ J/T

Relative Permittivity [6, 8–10]

Ge	16
Si	11.8
GaAs	10.9
SiO_2	3.9
H_2O	55
Polyethylene	2.25
Polystyrene	2.54
Teflon	2.1

Forbidden Band Gap [6, 8–10]

Ge	0.68 eV
Si	1.12
GaAs	1.43
SiO_2	8

Thermal Conductivity [6, 8–10]

Ge	0.6 W/cm-°C
Si	1.5
GaAs	0.8
SiO_2	0.01

Intrinsic Mobility [6, 8–10]

Ge	$\mu_{eL} = 0.39\ m^2/V\text{-}s$
	$\mu_{eU} = 0.19$
Si	$\mu_{eL} = 0.135$
	$\mu_{eU} = 0.048$
GaAs	$\mu_{eL} = 0.86$
	$\mu_{eU} = 0.025$

where μ_{eL} is the mobility in the lower valley and μ_{eU} is the mobility in the upper valley.

References

1 H. Howe, Jr., "Stripline Circuit Design." Artech House, Dedham, Massachusetts, 1974.

2 S. Cohn, Problems in strip transmission lines. *IEEE Trans. Microwave Theory Techniques* **MTT-3**(2), 119–126 (1955).

3 T. K. Ishii, "Microwave Engineering," 2nd ed. Technology Publ., Washington, D.C., 1989. (Originally published by Ronald Press, New York, 1966.)

4 P. H. Smith, Transmission line calculator. *Electronics* **12**, 29–31 (January 1939); see also *Electronics* **17**, 130–133 and 318–325 (January 1944).

5 H. J. Reich, P. F. Ordung, H. L. Krauss, and J. G. Skalnik, "Microwave Theory and Techniques." Van Nostrand, Princeton, New Jersey, 1953.

6 A. van der Ziel, "Solid-State Physical Electronics," 2nd ed. Prentice-Hall, Englewood Cliffs, New Jersey, 1976.

7 W. H. Hayt, Jr., "Engineering Electromagnetics." McGraw-Hill, New York, 1974.

8 E. S. Yang, "Fundamentals of Semiconductor Devices." McGraw-Hill, New York, 1978.

9 S. M. Sze, "Physics of Semiconductor Devices." Wiley, New York, 1969.

10 A. S. Grove, "Physics and Technology of Semiconductor Devices." Wiley, New York, 1967.

11 E. Hammerstad and O. Jensen, Accurate models for microstrip computer-aided design. *IEEE MTT-S Int. Microwave Symp. Dig.* pp. 407–409 (1980).

12 T. Uwano, Accurate characterization of microstrip resonator or open-end with new current expression in spectral-domain approach. *IEEE Trans. Microwave Theory Tech.* **MTT-37**(3), 630–633 (1989).

Index